D1037395

Guidelines for
Design of Low-Rise Buildings Subjected to Lateral Forces

Edited by

Ajaya Kumar Gupta
North Carolina State University
Raleigh, North Carolina

Peter James Moss
University of Canterbury
Christchurch, New Zealand

National Science Foundation Grant BCS 8414476

Council on Low-Rise Buildings
North Carolina State University
Raleigh, North Carolina

CRC Press
Boca Raton Ann Arbor London Tokyo

Library of Congress Cataloging-in-Publication Data

Guidelines for design of low-rise buildings subjected to lateral forces / editors, Ajaya Kumar Gupta, Peter James Moss.
p. cm.
On t.p.: "Council on Low-Rise Buildings, North Carolina State University, Raleigh, North Carolina."
"National Science Foundation grant BCS 8414476."
"March 1992."
Includes bibliographical references.
ISBN 0-8493-8969-0
1. Structural design. 2. Wind resistant design. 3. Earthquake resistant design. I. Gupta, Ajaya K. II. Moss, Peter James.
TA658.4.G85 1993
624.1'76--dc20

93-4733
CIP

This book represents information obtained from authentic and highly regarded sources. Reprinted material is quoted with permission, and sources are indicated. A wide variety of references are listed. Every reasonable effort has been made to give reliable data and information, but the author and the publisher cannot assume responsibility for the validity of all materials or for the consequences of their use.

In no event shall North Carolina State University (NCSU) or any of its employees who have been involved in the creation, production, or delivery of this publication be liable for any damages whatsoever (including without limitation, damages for lost profits, business interruption or other pecuniary loss) arising out of the use of or inability to use the materials included in this publication, even if NCSU has been advised of the possibility of such damages.

This material is based upon work supported by the National Science Foundation under Grant No. BCS 8414476. The Government has certain rights in this material.

Any opinions, findings, and conclusions or recommendations are those of the author(s) and do not necessarily reflect the views of the National Science Foundation.

Direct all inquiries to CRC Press, Inc., 2000 Corporate Blvd., N. W., Boca Raton, Florida 33431.

International Standard Book Number 0-8493-8969-0

Library of Congress Card Number 93-4733

Printed in the United States 1 2 3 4 5 6 7 8 9 0

Printed on acid-free paper

FOREWORD

Low-rise buildings include some of the most complex structures that are built, yet they have received less attention in terms of analysis and design than have high-rise buildings. The vast majority of construction in the United States falls into the low-rise category, loosely defined as being buildings up to four or five stories in height. Improvement in the design of these buildings would have the combined benefit of using building materials more efficiently and reducing the extent of future wind and earthquake damage.

Some of the specific problems associated with the performance of low-rise buildings have long been recognized by the research and design community and led to the holding of a special workshop on the "Seismic Performance of Low-rise Buildings" in Chicago, Illinois on May 13-14, 1980 sponsored by the National Science Foundation. The Workshop participants identified the performance issues, concerns and research needs associated with low-rise buildings. From the Workshop conclusions it was clear that there was need for guidelines for the design of low-rise buildings. Accordingly, some of the participants at the Workshop formed the Council on Low-rise Buildings with the express purpose of preparing guidelines for the design of low-rise buildings subjected to wind forces and earthquake motions.

These guidelines are the result of the efforts of the Council. They have been written as a resource document for structural engineers to supplement the current codes or other design recommendations and provide a design basis where there are no current requirements. They are not intended to supersede or conflict with existing requirements but it is hoped that they will supplement and extend the present body of information relating to the effect of wind forces and earthquake loadings on low-rise buildings.

Council on Low-Rise Buildings

Steering Committee

Authors

Chapter 1

Robert D. Ewing
Ewing and Associates
Rancho Palos Verdes, California

Chapter 2

Coordinator
Yi-Kwei Wen
University of Illinois
Urbana, Illinois

Ross B. Corotis
Johns Hopkins University
Baltimore, Maryland

Carl J. Turkstra
Polytechnic Institute of New York
Brooklyn, New York

John W. Reed
Jack R. Benjamin & Associates
Mountain View, California

Jamshid Mohammadi
Illinois Institute of Technology
Chicago, Illinois

Chapter 3

Coordinator
James R. McDonald
Texas Tech University
Lubbock, Texas

Kishor C. Mehta
Texas Tech University
Lubbock, Texas

Marvin E. Criswell
Colorado State University
Fort Collins, Colorado

Chapter 4

Coordinator
Ajaya K. Gupta
North Carolina State University
Raleigh, North Carolina

John Kariotis
Kariotis and Associates
Structural Engineers
Alexandria, Virginia

Robert D. Ewing
Ewing and Associates
Rancho Palos Verdes, California

Peter J. Moss
University of Canterbury
Christchurch, New Zealand

Chapter 5*

Coordinator
Edwin G. Zacher
H.J. Brunnier Associates
Consulting Engineers
San Francisco, California

Coley Altman, Jr.
Enwright Associates, Inc.
Greenville, South Carolina

Douglas A. Foutch
University of Illinois
Urbana, Illinois

Louis F. Geschwinder
Pennsylvania State University
University Park, Pennsylvania

Melvyn Green
Melvyn Green & Associates
Manhattan Beach, California

Mark Schaefer
Carr and Associates
St. Louis, Missouri

Mete A. Sozen
University of Illinois
Urbana, Illinois

John R. Tissel
American Plywood Association
Tacoma, Washington

* listed in alphabetical order

Chapter 6

Coordinator
David R. Bonneville
H.J. Degenkolb Associates
San Francisco, California

Keith Hjelmstad
University of Illinois
Urbana, Illinois

David W. Cocke
H.J. Degenkolb Associates
San Francisco, California

Guy C. Morrow
H.J. Degenkolb Associates
San Francisco, California

James J. Mogannam
H.J. Degenkolb Associates
San Francisco, California

Richard N. White
Cornell University
Ithaca, New York

Chapter 7

Coordinator
Satwant S. Rihal
California Polytechnic State University
San Louis Obispo, California

Robert K. Reitherman
Scientific Service, Inc.
Redwood City, CA

Barry J. Goodno
Georgia Institute of Technology
Atlanta, Georgia

Hiroshi Ito

TABLE OF CONTENTS

1 Introduction, scope and performance criteria

1.1 INTRODUCTION

Background

Low-rise buildings can be defined in a number of different ways depending on the circumstances, and no one definition will necessarily satisfy every situation. While Section 1.2 gives a detailed description of low-rise buildings for the purposes of these Guidelines, a looser definition would be to describe them as those buildings less than four or five stories in height. Such buildings comprise about 85 to 90 percent of the structures that are at risk from natural hazards, such as windstorms and earthquakes. Moreover, these buildings house the majority of the population of the United States, and furnish a substantial amount of the workplaces for the manufacturing, sales, and service industries. Damage to these structures by extreme natural hazards not only poses a potential threat of substantial damage repair costs, but also poses a threat to the life and limb of the occupants and the general public. However, the disruption of the industries and critical services that utilize and occupy these buildings, the concurrent loss of employment, and the loss of housing potentially constitute a much larger economic loss than the damage repair costs.

Reduction of life safety risks and potential damage to structures has been a goal of national hazard reduction programs for more than a decade. For example, the National Earthquake Hazards Reduction Program (NEHRP) was released on June 22, 1978 following the passage by Congress of the Earthquake Hazards Reduction Act of 1977 (P.L. 95-124).

Surveys of urban areas struck by natural hazards have always indicated that the degree of significant structural damage can generally be attributed to the quality of the structural details, and how well the buildings and their parapets and ornamentations are tied together. Moreover, recent research in the field of wind and earthquake loadings of structures has indicated that current design techniques do not stress the special design problems that are posed by low-rise buildings.

The specific problems associated with the performance of low-rise buildings have long been recognized by the research and design community, and a special workshop on the "Seismic Performance of Low-rise Buildings" [Gupta, 1981], sponsored by the National Science Foundation (NSF), was held in Chicago, Illinois on May 13 and 14, 1980. The workshop participants identified the performance issues, concerns, and research needs associated with low-rise buildings. It became clear from the workshop conclusions that there was a great need for guidelines for the design of low-rise buildings. Accordingly, following the workshop some of the participants formed the

Council On Low-rise Buildings with the expressed purpose of preparing guidelines for the design of low-rise buildings subjected to wind forces and earthquake motions.

Guidelines

These guidelines are a result of the efforts of the council. They are intended to stand on their own merit as a resource document for design professionals, knowledgeable in structural engineering, who can use the guidelines as a design basis where there are no current requirements (i.e., ordinances or codes). They are also intended to be a supplement to existing design recommendations, codes, and ordinances, as well as to extend the body of available information. They are not intended to supersede or conflict with existing requirements, but rather are intended to assist the user by pin-pointing design topics that deserve special consideration, by providing simple and straightforward design techniques that encompass the uncertainties of wind and earthquake loadings, and by restating recommendations for lateral load analysis in a single document. In many cases, the format of these guidelines will vary from current design recommendations.

Users of The Guideline

The intended users of these guidelines are the design and analysis professional who prepares the construction documents, the building official that conducts the structural reviews, the local agency that assesses its risk exposure and establishes design requirements, the building industry which constructs these low-rise structures in natural hazard zones, and the universities where building structural design is taught and researched.

Identification of The Problem

Surveys of areas subjected to windstorms or earthquake motions indicate that low-rise buildings have the majority of damage loss. This is due both to the number of low-rise buildings and to the complexity and lack of homogeneity of their construction. Mathematical modeling of wind flow around these low structures is exceedingly difficult due to surface effects and adjacent structures. Response of the structural elements to earthquake shaking cannot be easily modeled. The difficulty of modeling for earthquake response is complicated by the number of possible combinations of materials and the dependency of the response on the intensity of ground shaking. These analytical problems have limited the development of design recommendations that address the special design problems of low-rise buildings and the special problems that are related to the broad range of hazard zones. These guidelines will address these special problems within the framework of current design regulations.

1.2 DEFINITION OF LOW-RISE BUILDINGS

Introduction

A precise and formal definition of low-rise buildings is difficult to state in a few words, as a low-rise building can not simply be defined by the number of stories or by its over-all height. This is partly due to the fact that low-rise buildings are not generally well understood, and are perceived differently within the engineering community. An attempt to quantify a definition as a go or no go decision for the application of these guidelines is not necessary or desirable. However, it is important and desirable to describe those characteristics that are specific to low-rise buildings and to describe how these characteristics affect the building response and performance.

Low-Rise Buildings

Low-rise buildings have low aspect ratios (i.e., ratios of their over-all height to their plan dimensions), shallow foundations, generally flexible horizontal diaphragms, and are frequently constructed with several different materials that have dissimilar stiffness, strength, and mass properties. Low-rise buildings typically utilize a broad range of building materials in their structural frame and architectural cladding. Wood, masonry, concrete, structural steel, light gauge steel, and manufactured materials often are combined into a framework that acts as a composite of these materials. Also, low-rise buildings are likely to have irregular plans, elevations, and structural systems. In contrast with the medium to high-rise buildings, these buildings can contain heavy overhead crane systems that alter their response.

This type of construction leads to response modes that are not properly accounted for in our present building codes. The current codes are based on a response model that is representative of a multi-story, homogenous, regular, medium to high-rise building that has a fixed base and rigid diaphragms. The key point here is *low-rise buildings do not fit the response model that the current building codes use as the basis for determining design loads and ductility demands*. Accordingly, the design and analysis procedures developed in these guidelines are based on response models that are representative of typical low-rise buildings.

Low-rise buildings typically are not fixed at their bases and can rock on their foundations, making foundation-soil interaction an important issue. Buildings that have a wood or metal roof diaphragm supported on and attached to heavy masonry walls will have an additional response mode, where the diaphragm couples the mass of the heavy side walls with its in-plane response and behaves as a nonrigid member. This response will be superimposed on the global response, and will increase the demand on connections and increase the interstory drift in the center span of the diaphragm, as well as placing different ductility demands on the individual components. Moreover, low-rise buildings that have a composite construction will likely have their components responding at different frequencies with different amplification factors (C_p). Due to their low aspect ratios, these buildings can be impacted by windborne missiles. Local wind effects resulting from surface irregularities and adjacent structures are difficult to predict. Finally, the base shear approach for earthquake loading will, in general, not be applicable to low-rise buildings.

1.3 SPECIAL PROBLEMS IN LOW-RISE BUILDINGS

All buildings are required to be designed for gravity loads and lateral forces. Low-rise building design has in the past been primarily concerned with gravity load effects, and the effects of lateral forces produced by windstorms and earthquakes have often been given only cursory acknowledgment.

The nature of the sources of these lateral forces is so divergent that the philosophy and methodology of design that have evolved for the effects of each are quite different (Chapters 3 and 4). This divergence is often overlooked in the design of simple buildings and an appreciation of the similarities and differences in the design approach for wind and for earthquake will lead to improved design for all buildings.

One essential difference between windstorms and earthquakes is the frequency of occurrence of the events. Nearly all buildings are subject to structurally significant wind forces every year. For typical locations in the United States, the 100-year mean recurrence interval (MRI) wind speed is only 30 to 60 percent higher than the average yearly maximum wind speed. The 500-year MRI wind speed is only 10 to 20 percent above the 100-year MRI wind speed or no more than 70 percent higher than the average yearly maximum wind speed. It is meaningless to attempt to define an average yearly maximum earthquake but the acceleration associated with a 500-year event can easily be 100 to 250 percent of that associated with a 100-year event. Any given building will almost certainly be subjected to several occurrences of significant wind forces during its lifetime, but it will be subjected to few, if any, occurrences of earthquake forces. The difference in recurrence intervals of significant forces coupled with the fact that large earthquakes produce extremely severe forces on buildings sets the stage for the philosophical divergence in the design approach.

The performance objective for seismic design, as stated in recommended earthquake criteria, is primarily life safety for the extremely rare event. Functionality and damage limitation are secondary considerations. A building assigned to a "Standard Occupancy" in the Uniform Building Code is considered to have performed acceptably if it resists the forces from a large earthquake without collapse, even though the damage may result in the building being a total loss. Although the performance objective for wind forces is also life safety, a building must resist the "design" wind (i.e. a 50 or 100-year MRI wind) with essentially no damage. The codes do not require that continued functionality be provided during a windstorm; however, it is often considered to be a criterion for design. For low-rise buildings this criterion usually translates into deflection and story drift limits. For high rise buildings the criterion may also include limits on motion to control occupant discomfort.

Earthquakes and windstorms produce both horizontal and vertical forces on buildings. The vertical forces produced by earthquakes, although significant, are rarely a factor in design since they are only a fraction of the mass acting on the structure, and structures normally possess extra strength in the vertical direction due to normal margins of safety. Certain elements of framing which are frequently found in low-rise structures, such as pre-stressed beams, are susceptible to the effects of vertical motions and are covered by design criteria in the latest codes. Vertical forces generated by windstorms are significant in the design of lightweight roof systems. There are special hold-down anchors available for wood frame construction and special bracing is often required to prevent steel roof joist member buckling caused by wind uplift.

The lateral forces generated in a building by earthquake ground motions are due to inertia resulting from the dynamic response of the structure to these motions. Very large structures could have forces generated by differential motions of the ground in different portions of the structure but this is not typically considered in design. The lateral forces due to windstorms result from the pressures, positive and negative, acting usually on the exterior surfaces of the structure. Tall structures do respond dynamically and therefore generate some inertial forces, but this is of no concern in low-rise structures. The fact that wind forces are treated for design as constant does not mean that they actually are constant. The pressures on any surface of a building vary both in time and space. It is convenient to consider wind forces as having a constant and a variable component. Buildings with structural vibration frequencies greater than 2 to 4 hertz (periods less than one half second) respond essentially statically according to studies made of the variable component. This indicates that the inertial forces are negligible and the total force is due to the sum of the pressures from the constant and variable components.

The characteristics of a structure that affect the magnitude of the wind or earthquake forces acting on the structure are distinctive. The mass and stiffness of the structure are the key items affecting the response of a structure to earthquake ground motions. For low-rise structures even stiffness is of secondary importance, since the code design force coefficients applied to the mass are constant for structures with periods below approximately one-half second. The surface area of the exterior walls and roofs (for buildings with sloped roofs) are the key items of the structure affecting the wind forces. Size and mass are related but there is not a constant direct relationship. Squat buildings constructed with high density materials will probably have seismic design forces larger than wind design forces even in areas of low seismicity and high winds. Conversely large buildings constructed with low density materials will probably have wind design forces larger than seismic design forces, even in areas of high seismicity and low winds.

Perhaps the most important consideration for design is the difference in the limit states for structural material behavior upon which the design criteria are based. Design for wind forces is based upon conventional linear behavior; yielding is not permitted. Design for earthquake forces at code force levels is based on the same concept but the real behavior is expected to be in the inelastic range of material behavior. The two reasons for this difference are (1) forces due to earthquake are so extreme that a considerable economic penalty would be incurred to maintain linear behavior and (2) since earthquake ground motions are dynamic, structures responding to the motions are storing or dissipating the energy generated. Materials going beyond yield into the inelastic range provide for temporary storage of energy, and the response of structures utilizing this property of the material is reduced to as little as one eighth of the response that would occur if the structure behaved elastically.

This difference in expected performance means that the energy absorption or dissipation capacity of the structural system (the system, not just the material) is of extreme importance for earthquake resistant design. The seismic resistant structural system must provide stable resistance over repeated cycles of inelastic strain while there is no such requirement for the wind resistant structural system. Redundancy is more critical for seismic resistance than for wind resistance, since failure of some part of the seismic resistant system is more likely where the design response is so close to ultimate capacity.

Not all structural systems possess equal capacity in the inelastic region of their materials. Design provisions in the codes recognize this by assigning different coefficients to adjust the design forces. The Uniform Building Code uses the factor R_w for this purpose, with values ranging from 4 to 12. It should also be recognized that some relatively brittle systems cannot achieve the anticipated energy absorption--dissipation level indicated by their R_w value and are prohibited in high seismic risk areas.

The reduction of the seismic design force from that which would result for a structure behaving elastically is diametrically opposed to the procedure for wind where the design pressures correspond to the expected level of pressures for a 50 year MRI event. Making a comparison of earthquake and wind design forces is like comparing apples and oranges; one is based on having a system behaving in the inelastic range in a very rare event, while the other is based on a system behaving in the elastic range in an event recurring at least once in fifty years.

The present common practice of ignoring the special seismic detailing requirements where design wind forces exceed earthquake forces is, unfortunately, incorrect and potentially unsafe. The correct approach is to compare the wind design forces with the earthquake design forces multiplied by a factor representing the reduction for inelastic action. The 1991 UBC uses $3R_w/8$ to approximate this factor. The 1991 UBC requires that the seismic detailing requirements be followed even when wind forces govern the design.

The basic tenet of good lateral force design is, as Newtonian principles tell us, that for every force there must be an equal and opposite reaction. We believe these principles apply to all design. To resist the forces induced in the structure there must be a complete and continuous load path provided from the point of origin of the force to the point of final resistance. Lateral force design, especially design for earthquake forces, has often been construed as increasing the cost of design and construction. Low-rise buildings which are well designed and detailed for gravity loads can achieve the expected level of performance for lateral forces for little or no extra cost when the design adheres to the principles stated above and quality control in construction is exercised.

1.4 NEW CONCEPTS

These guidelines incorporate recent research into the determination of lateral loadings by the use of probabilistic methods, the variability and magnitude of wind loading on different parts of the building, the response of structural systems to the full intensity range of windstorms and earthquake shaking in the United States. In addition, these guidelines incorporate several new concepts that are significant departures from existing seismic and wind design recommendations and provisions. These new concepts are introduced into the guidelines to correct and improve current practice for low-rise buildings.

The new concepts embodied in these guidelines are listed below:

- The guidelines recognize that a typical low-rise building does not fit the analytical response model used in current design codes for earthquake loading.
- The guidelines consider the effects of the structure rocking on the soil in lieu of using fixed-base assumptions.
- The guidelines are directed toward the limitation of property damage when buildings are subjected to lateral loads. Life safety risks are considered to be reduced by property damage limitation design procedures. However, special design consideration for building elements that have a higher than average probability for posing a life safety threat are given in the guidelines.
- The guidelines recognize that the base shear approach for design of all parts of the structural system for earthquake forces will, in general, provide an adequate level of strength for the main lateral force resisting system, but will not be entirely adequate for certain components of low-rise buildings.

- The guidelines recognize that all structures must be designed to be ductile (i.e., have energy dissipation capacity). This is a two-step process, where structural systems with their load paths and structural detailing of connections and intersections are treated in separate steps.

These guidelines provide for the design of low-rise buildings that leads to determining a required strength for the building's lateral load resisting system. The guidelines also recognize that the actual peak strength will in general be different from the required strength, and methods are provided for determining peak strength, as well as determining the expected post-peak strength characteristics of the structural system.

1.5 PERFORMANCE CRITERIA

General

Several issues must be taken into account when developing performance criteria for buildings, namely;

- Life safety
- Personal injury
- Loss of function
- Serviceability
- Property damage
- Cost effectiveness of the design

The threat to life safety, personal injury, and loss of function is due to the inability of the structure to sustain its lateral or vertical load-carrying capability and/or the separation of the parts of the structure from one another. The loads are induced in or delivered to the structure in a different manner for earthquakes than for windstorms, and the sequence of events leading to general collapse is different for the two natural hazards. In earthquakes, lack of adequate anchorage between the horizontal diaphragms and the walls causes separation of the walls from the diaphragms and collapse of the vertical load carrying walls. Failure induced by windstorms typically originate with loss of cladding and breach of the building envelope. Vertical load carrying walls may fail in the out-of-plane direction due to wind pressure. Collapse of lateral windforce resisting structural systems rarely occurs, but can happen if connections and anchorages are inadequate. The loss of parapets and ornamentations due to inadequate anchorage are a threat to the life safety and personal injury of people on streets and sidewalks in both earthquakes and windstorms.

The reduction of property damage (i.e., damage control) will in most cases result in a reduction in the risk to life safety and personal injury. The use of anchors between horizontal diaphragms and vertical wall elements will reduce property damage, as well as result in a reduction in the risk to life safety and personal injury. The details of anchoring and tying a structure together are different for an extreme wind than for an earthquake. For example, in an extreme wind, roof sheathing has a greater tendency to begin separation at the eaves, ridges, and gable ends; while separation of parts in an earthquake occurs between the horizontal diaphragm and the vertical wall elements and at gable ends. Accordingly, attention to different details are required when dealing with the two natural hazards.

Serviceability requirements are based on acceptable performance under normal day-to-day loadings (e.g., gravity loads, temperature, ambient wind, etc.), and involve issues of appearance, protection of the building interior from the elements, and, in some instances, human comfort.

There is a point of diminishing returns when it comes to reducing probable property damage. For a given cost, a certain level of probable property damage reduction is attained; it then becomes a problem in economics to weigh the present value of the investment costs against the costs of future property losses and the loss of use of the property. Moreover, the economics problem must be placed in a probabilistic framework, since the losses are associated with a natural hazard that has a probability of occurrence. In addition, the probabilities for the occurrence of extreme wind and earthquakes are quite different, so the point of diminishing returns is, in general, different for the two natural hazards.

Current design practice for wind loads is to use elastic analyses. In contrast, the current design practice for earthquake loadings utilizes inelastic models and procedures which recognize that inelastic response will occur to determine the base shear and, in some cases, the lateral force distribution, and relies on member ductilities (i.e., system energy dissipation capacities) to reduce the response amplification factors. This guideline will continue with these current design philosophies. Accordingly, separate performance criteria will be defined for wind loadings and earthquake loadings.

Performance Criteria for Earthquakes

The loads applied or induced in structures due to earthquakes are quite different from those due to the normal service loadings. Obviously, a large intensity earthquake will produce stresses and deformations in the structure that greatly exceed those from the normal service loads, but at the same time the probability of occurrence of such an earthquake during the service life of the structure is very low. Clearly then, special performance criteria need to be established for these types of extreme environments, and the paramount consideration in developing these criteria is to protect life safety in the event of a severe earthquake. Structural design that requires the structure to remain elastic would result in an extremely uneconomical structure for an earthquake that has such a low probability of occurrence.

The generally accepted performance criteria for the design of structures subjected to lateral forces induced by earthquakes are described as:

- the structure will remain elastic, or nearly so, when subjected to small or moderate intensity earthquakes that have a high or moderate probability of occurrence, and

- the structure will respond in the inelastic range by local yielding, but will retain its ability to sustain vertical load and be safe from collapse when subjected to the most severe probable earthquake.

The first of this two-part performance criteria is interpreted to mean the structure will show no visible signs of damage, and will not sustain any damage that would degrade its lateral or vertical load-carrying capability when subjected to a small intensity earthquake that has a high probability of occurrence. Further, the structure will sustain only repairable damage, and will not sustain any damage that would degrade its lateral or vertical load-carrying capability when subjected to more than one moderate intensity earthquake that has a moderate probability of occurrence. The last of this two-part performance criterion is interpreted to mean the structure will not sustain severe structural damage, will have toughness, and will not collapse when subjected to the most severe probable earthquake expected during its exposure period. This also implies that the structure will have serious permanent deformations requiring, as a minimum, very major repairs, but will not collapse when subjected to an earthquake somewhat larger than the most severe probable event.

This performance criterion is based on several issues including life safety, damage control, and economics, the most important of which is life safety.

Performance Criteria for Windstorms

The generally accepted performance criteria for the design of structures subjected to lateral loads induced by windstorms are described as:

- the structure should suffer little or no damage when subjected to winds associated with a 50-year mean recurrence interval.

- damage from winds associated with a 100-year mean recurrence interval should not pose a threat to a large number of people assembled in the building, nor should the damage prevent the intended function of the building from taking place.

- the structure should not collapse and there should be identifiable areas within the building for occupant protection if the building is subjected to wind speeds substantially greater than design values (e.g., an intense tornado).

The first of this three-part criterion implies that a building should sustain little or no damage when subjected to wind forces associated with a 50-year mean recurrence interval. The second criterion applies to facilities where more than 300 people are assembled in one location, such as a theater or auditorium, or, to a hospital, fire station or emergency operations center, where the function of the facility must continue after being exposed to 100-year mean recurrence interval winds. The third criterion suggests that if a building is struck by a low-probability extreme event, such as an intense tornado, some identifiable area of the building should remain structurally intact to the extent that occupants can be protected from death or serious injury. These criteria imply that the structure can be essentially destroyed, but people seeking safety within the building should be protected from collapsing walls or roofs and from flying debris. The importance of this last criterion cannot be overemphasized, because when extreme weather threatens, the public is urged to seek shelter in an "engineered" building.

1.6 LIMITATIONS ON USE OF THE GUIDELINES

These guidelines were developed for the mitigation of natural hazards of wind and earthquake for a broad range of buildings that have been classified as low-rise buildings. They were developed for new construction and are not intended to be used for existing buildings. As stated earlier in the definition of low-rise buildings, there are many different types of buildings that are classified as low-rise structures, and these guidelines are recommended for use only for those buildings that meet the definitions given herein.

2 Definition of Risk

2.1 PROBABILISTIC DEFINITION OF NATURAL HAZARDS

2.1.1 Inherent Variability

Every year, natural hazards lead to billions of dollars of economic loss for the constructed environment [NRC, 1987; Rubin, et al., 1985]. Yet, due to their high degree of inherent variability and natural triggering mechanisms, it is not feasible to design structures to resist all possible hazards. For this reason, engineers must balance their designs -- considering the probabilities of load occurrence, intensity, and structural resistance.

Structural design to resist natural hazards must start with a clear definition of the initiating event, such as a hurricane or an earthquake. Consideration must then be given to the magnitude of the event, the frequency of occurrence, and the resistance of the structure to that event. Finally, the consequences of the resulting structural performance must be assessed.

The hazard is defined here as the source of the disturbing forces, such as wind or earthquake. In a general sense, the resulting forces themselves can also be considered as the hazard. The term *risk* is used to describe the likelihood of some level of undesirable structural performance, such as unserviceability, damage or collapse. Further, this risk is defined in terms of time periods of exposure, such as the annual risk or the risk over the exposure period (e.g., 50 years). Also in common use, is the concept of risk as the consequence of performance weighted by its likelihood of occurrence.

When considering structural response to natural hazards, it is useful to distinguish two basic sources of uncertainty. One part of the uncertainty is due to the inherent variability in the natural phenomenon and is sometimes simply referred to as the variability. The variability may be better assessed, but not systematically reduced, by additional data collection or more accurate modeling. The other type of uncertainty is modeling uncertainty, which includes the form of the probabilistic and phenomenological model as well as the uncertainty in their parametric values. This uncertainty may be reduced with improved data and models.

Natural hazards are characterized by a high degree of variability, e.g., the time and duration of occurrence, the intensity during an event, its variation with time during one occurrence, as well as the spatial extent and variation. All these variations have somewhat different ramifications for wind and for seismic hazards.

The wind hazard exhibits variability in the occurrence of the storms (including hurricanes and tornadoes) and, given one of these events, a range of wind speed. Even if a particular event has been classified by its maximum wind speed there exists time and spatial variation, including gusts and turbulence, as well as other longer term fluctuations. From a design standpoint, the wind hazard must be described in terms of summary statistics, the calibration of which will be site-dependent throughout the United States.

The seismic hazard also exhibits characteristic variabilities. These includes the occurrence of the earthquake, source location, source magnitude, frequency content, amplitude rise and fall, and duration. The motion undergoes an attenuation from the source to the site and an amplification that is dependent on local soil characteristics. For design, one must consider the geographic variation of the summary statistics throughout the country.

2.1.2 Selection of Design Loads

When engineering structures to resist natural hazards it must be kept in mind that, due to the limited resources of society and the vagaries of nature, structures must be designed with some probability of failure. Therefore, design loads should be selected with full consideration of consequences. It may be desirable to have different acceptable risks for different parts of a structure or different types of structures. For example, a storage building may have a different optimal design level than an office structure which, in turn, may be different from a hospital or assembly hall. The desired safety level may depend on the consequences of failure, such as: loss of intended function; physical damage (including partial damage or total collapse); economic loss (related to both of the previous consequences); and human injury or death (dependent on the number of people at risk).

The actual safety, or risk, inherent in a structure depends on the selected design level of the hazard, the design strength, and the degree of accuracy and conservatism of the structural model. Although practical considerations require as complete a separation as possible between the load and resistances, there is in reality an interdependence. Structural characteristics such as the ability to absorb energy (e.g., through damping or nonlinear hysteresis), natural period, ductility and load redistribution all affect the manner in which the forces from the hazard are accommodated by the structure. This directly influences the risk of structural failure. Well-known examples of this include the way in which local wind pressures are averaged by diaphragm action, high local wind pressures on parts and portions in corners, the dependence of earthquake-induced forces on structural mass, and design controlled by deflection and vibration. The design of connections and other details are also important in how a structure accommodates the forces [Corotis, 1982].

The selection of actual design loads is more logically based on a level of "comparative risk" for various types of structures. Either explicitly or implicitly this risk level must consider the cost of altering this risk, the sensitivity of the risk to changes in design, the consequences of unsatisfactory performance, and society's acceptance of risk for various activities. On the basis of an individual building one would tend to expect a lower level of risk for a large, high-rise structure than for a low-rise building with lesser consequences of failure. However, on a societal scale, there may be many more low-rise buildings and hence a greater amount of total resources related to the low-rise buildings. This might mean a greater loss from a wind storm or earthquake through damage in low-rise structures. Table 2.1.1 gives a general description of the importance of different factors in selecting an allowable risk level. In cases where a particular building is affected by two or more of these factors, the combined result should be considered. For example, a family housing unit may suggest a larger accepted failure probability in the sense that it involves a low number of occupants and low potential property damage. On the other hand, the same unit may be regarded as part of a housing development that as a group involves a large number of occupants and a relatively high property damage. The selection of an appropriate risk level in such a case may require a detailed investigation.

Table 2.1.1. Relative Effectiveness of Factors Influencing the Accepted Risk

Factor	Larger Accepted Failure Probability	Smaller Accepted Failure Probability
Building Occupancy	single, isolated family unit	building complexes
Primary use	warehouses	essential facilities (hospital, fire department)
Replacement cost	low cost	high cost
Useful life	shorter life	longer life

A review of factors given in Table 2.1.1 may provide the designer some guidelines for selecting an acceptable level of risk for a given structure. Gupta, et al. [1981] give a comprehensive discussion on the risk assessment in low-rise buildings. Studies similar to this may be consulted when considering an allowable risk level for a given structure.

There will be a trade off in design between the simplicity of the code and the consistency of the structural reliability. A code that is less responsive to variations such as different structural configurations, different types of elements within a structure, changes in geographical location, and local ground conditions and topographical features will result in a wider range of actual reliability levels.

With respect to the wind and seismic hazards in particular, there are a number of different quantities that might be considered as input variables. Maximum wind speed is clearly a dominant quantity for design to resist wind. In the United States this is often taken as fastest-mile speed, although one-minute averages and hourly averages are used also [NBCC Supplement, 1985]. Rather than the wind speed itself, the engineer actually needs the forces on cladding and roofing, secondary systems and components, and main wind bracing system. The gustiness of the wind also affects the dynamic response of the structure. Local terrain conditions modify the input wind forces as compared to values obtained from a general map of design wind speeds. Section 2.4 discusses these aspects in more detail. For seismic consideration, peak ground acceleration is usually considered the basic quantity of interest for design. However, design may actually be controlled by building response that is related to frequency content, duration of shaking, maximum ground displacement, etc. Utility connections may be sensitive to displacement, foundations to differential fault displacement and settlement. Structural integrity to dynamic displacement response or slope stability (and liquefaction) -- building contents respond to floor accelerations.

Due to practical considerations, the fundamental wind and seismic hazards will be basically the same for low-rise buildings as for structures in general. Once these particular hazard characteristics are selected, it is necessary to select the design level. This might be done by a method that only implicitly reflects probabilistic ideas. Such methods would include historical values and experience, collective judgment, and public or political realities. More consistently, however, probability is a tool by which observed historical data and uncertainty may be incorporated into a mathematical model to quantify the structural risk. Due to the long time period necessary to collect reliable data for extreme events (which are precisely those of interest for structural design), it is unlikely that one will have sufficient data for direct statistical use. Rather, probabilistic models will need to be assumed, with as much physical logic and justification as possible, and then these models calibrated from available observed data.

Commonly used terms for design reference are T_0, the exposure time, P, the annual probability of exceeding a given design level of a hazard; P_0, the probability of exceeding that value in the exposure time, and T, the return period for that value. These will be discussed in Section 2.2. Other measures of design level hazard could be the number of expected exceedances during a design lifetime, the hazard rate (time rate of relative change of the reliability), and the statistics of the design lifetime maximum hazard (mean, standard deviation, fractile, etc.). For a design to resist a single hazard, characteristics which are discussed in this section are needed. For consideration of the combination of hazards (due to simultaneous occurrence), additional statistics of the hazard are likely to be needed. These are related to the time and spatial stochastic characteristics of the hazard and will be addressed in Section 2.5.

2.1.3 Comparative Risk Statistics

Once the design level hazard is selected, it is used in a structural design code. Combining this with the degree of conservativeness introduced in the design process (through the choice of analytical models, assumed strength values and quality control) results in a structure with a certain reliability or risk.

Studies on societal risks associated with various activities have established hazards in terms of fatalities per person per year [Ingles, 1979]. These vary from higher values for *voluntary* activities, such as $3x10^{-4}$ for motor vehicles and 10^{-4} for work or home accidents, to lower values or *involuntary* hazards, such as 10^{-6} for tornadoes and earthquakes and 10^{-7} for lightning. These may be compared to building failures on the order of 10^{-6}. Recognizing varying consequences of failure, values such as 10^{-4} for local structural collapse and 10^{-7} for total collapse have been suggested. Historically, structures with a low consequence of failure apparently have been designed for an annual probability around 10^{-3} for significant structural damage due to wind or earthquake. Hospitals, fire and police stations, and emergency centers seem to have an annual probability of loss-of-function around 10^{-3} to 10^{-4}.

A number of survey studies have also been conducted to determine the loss associated with the performance of structures during natural hazard events. Projections to the year 2000 estimate total annual building losses due to natural hazards of around $30 billion in current dollars. This includes about $13 billion due to hurricanes and tornadoes, and $3 billion due to earthquakes. Other sources include river flood ($7 billion) and expansive soil ($7 billion). It is estimated that somewhere around 75% of this loss would be associated with low-rise buildings.

2.2 PROBABILISTIC METHODS AND RISK IMPLICATIONS

As indicated in Section 2.1 most loadings, especially lateral forces such as those due to wind and earthquake, fluctuate in time and space and are random in nature. Their effects on low-rise buildings also have similar characteristics. When considering particular limit states being exceeded for low-rise buildings under such loadings, structural properties such as stiffness, strength, energy dissipation capacity, etc., are also important factors but are generally not perfectly known. There may be significant deviation of the actual values from the nominal ones because of the inherent variabilities in construction material properties as well as engineers' imperfect modeling of the structure in evaluating these structural properties. In view of the uncertainties that are associated with the loading environments and the building systems, when defining loads for building design and in evaluating building performances probabilistic methods and concepts are generally required. In this section the probabilistic methods and the risk implications relevant to definitions of hazards for design of low-rise buildings are summarized.

2.2.1 Prediction of Probability of Individual and Combined Loading

In design of low-rise buildings against lateral loads, the objective is to ensure that the buildings perform satisfactorily in terms of providing required services and functions on a daily basis and safety against limit states being exceeded over their lifetime. The loadings of concern for design against these limit states, (i.e., serviceability, collapse) are correspondingly those of everyday occurrence such as arbitrary-point-in-time value or daily maximum value and lifetime maximum values. These quantities are random variables. Their statistics and probability distributions can be estimated from past records if such records are available and of sufficient quantity. Known methods of statistical inference can be used for this purpose [Ang and Tang, 1984]. When the data are limited -- e.g., for the lifetime maximum values -- records have been kept, at most, for a few tens of years. The information inferred from the data, then, may not be as reliable, especially in the extreme value range. In such case, a confidence level may be attached to the estimated statistics to reflect the possible error due to small sample.

The selection of a design load is customarily based on a given probability of the load level being exceeded in a prescribed period, such as a day, a year, 30 or 50 years, etc. [NBCC, 1980; Supplement, 1985]. Such design loads can be determined once the statistics and distribution of the maximum values of the loads over the period are known. For example, a commonly used criterion is to select the design load based on an annual exceedance probability P. The mean recurrence or return period, T, of such design level load being exceeded is equal to 1/P. P can be determined from the distribution of the annual maximum values. Note that implied in the notion of return period are the assumptions that the statistical characteristics of the loads do not change and the load variation is independent from year to year which are only approximately true for most loads. Also note that the probability of the design load being exceeded in a time period equal to the return period is not equal to either zero or unity. The probability is equal to $1 - (1 - 1/T)^T \approx 1 - e^{-1} = 0.63$. Another frequently used criterion is based on a probability of exceedance over a period of n years [ATC, 1978]. If this probability is P_o, it is related to the annual exceedance probability P by the equation

$$P_o = 1 - (1 - P)^n \tag{2.2.1}$$

For example, the design earthquake intensity in [Ellingwood et al, 1980] has an exceedance probability of $P_o = 0.10$ in 50 years. The annual exceedance probability (P) of such an earthquake is therefore 0.0021, which corresponds to a return period (T) of 475 years.

When the occurrence of a load is so rare that local records do not exist at all, purely data-based statistical methods obviously cannot be used. Tornadic wind load is one example. The average waiting time for a tornado striking a given site in the most tornado prone area is of the order of several hundred years. Under these circumstances, the probability information can be estimated only with the aid of analytical modeling. In such procedures, models based on probability theory and physical characteristics and the regional statistics of the hazards are constructed and used to estimate the required probability information. These, of course, have been subjects of investigation in most seismic and severe wind risk studies. The results have been widely used as a basis for safety study and design of important structures, such as nuclear power plants, against such hazards.

Although winds and earthquakes are the two major hazards to a low-rise building, the combination of these hazards with other loadings such as dead and live load also needs to be considered. The most severe effect on the building may result from a combination with a number of such loads. The data situation here is similar to that in the case of rare load; it is just not normal practice to keep simultaneous record of different loads. Again, analytical models can be utilized to gain some insight into the problem and yield the required probability information. Good progress has been made in this area both in model developments and applications. The major considerations here are individual load occurrence frequency, duration and intensity variations and chance of coincidence of different loads when load effects are added on. As most loads are transient, at the design load level, the frequency of occurrence is generally low and the duration is generally brief. The coincidence of two or more design level loads therefore has a small but not entirely negligible probability. In most codes and standards, a reduction of the combined design load is allowed to account for this small probability. However, the amount of reduction is largely judgment-based. In general, however, the problem of combination of time varying loads is extremely complex; simple combination rules cannot be expected to yield good results for all loading situations. The analytical methods allow an appraisal of the accuracy of such combination (reduction) rules. In spite of their shortcomings, such rules may be necessary in code format for reason of simplicity. More detailed discussion will be given in Section 2.5.

2.2.2 Prediction of Probability of Building and Structure Failure

The design philosophy in most current codes and standards is to require the structure to have a strength considerably higher than the design load. Normally this is achieved by the use of allowable (working) stress or load and resistance factors. Therefore, the exceedance of the design load does not necessarily mean failure of the structure. The risk (probability) of failure is in general much lower.

To predict the failure or limit state probability, the structural response statistics and structural resistance uncertainties need also to be taken into consideration. Depending on the loading and structural characteristics, the response may be primarily static or may involve significant dynamic amplifications. Therefore, a static analysis, a probabilistic dynamic analysis or repeated time history analyses (e.g., simulation, response spectrum method) may be necessary to obtain the response statistics. While the loading uncertainties are usually large and have received most attention, recent studies show that those associated with the structural resistance can be also significant, sometimes even dominant, e.g., uncertainties in stiffness, strength, damping and especially those caused by human errors. Some of these uncertainties may be reduced or eliminated with the advancement of analysis and design methods, and by implementing better inspection and quality control procedures, others may not.

In considering the building performance under lateral forces, because of the uncertainties involved, both the capacity (e.g., strength at elastic limit or at a prescribed ductility ratio) and the demand (e.g., maximum lateral force, maximum displacement due to the lateral forces or in combination with other loads) cannot be described in deterministic terms. The successful or unsuccessful performance of a structure can be stated realistically only in terms of probability. The reliability analysis allows one to consider and treat the uncertainties rationally and determine the required probability information. In its simplest form the reliability of a structure under hazard S may be given by (see Figure 2.2.1)

$$L = 1 - P_f = 1 - \int_0^\infty P(F|s)\, f_s(s)\, ds \qquad (2.2.2)$$

where:

L = reliability, probability of no failure

P_f= probability of failure (a given limit state being reached)

$f_S(s)$ = probability density of S, describing the uncertainties in the loading environment, determined form hazard risk analysis. The discrete probability mass at S=0 corresponds to that of no occurrence of the hazard over the time period.

$P(F|s)$= probability that the structural capacity is exceeded given S=s, primarily a function of the uncertainties in the structural resistance, also known as structural "fragility".

In problems where there is a lack of data to yield enough information on the type of distributions of the random variables under consideration, the reliability may be formulated in terms of only the first two moments (mean and standard deviation) of the random variables (loads as well as resistance) and expressed in terms of a safety index β. The information on distribution can be also included in the safety index through a First Order Reliability Method (FORM) [Ang and Tang, 1984]. The relationships between β and P_f is given in Table 2.2.1.

2.2.3 Risk-Based Design and Code Calibration

It is clear from previous sections that over a given time period the design load will have a finite probability of being exceeded. The same is true of various limit states of the building being reached (failure). The risks (probability) associates with these events are the implied risks in the building design. The objective of a sound design is to ensure that such risks are kept at a level which is in balance with other risks an occupant of the building has to face in a modern society, i.e., a level too high means unsafe design whereas a level too low means over-design and waste. The risk level determined from the foregoing consideration is the implied optimal risk level from which the design load and design structural resistance should be determined. While most codes and standards give explicit risk information about the design load such as in the form of a return period, this is generally not the case with the risks of limit states although it is the latter which have direct bearing on the design decision. That such information is generally unavailable and difficult to obtain is in part due

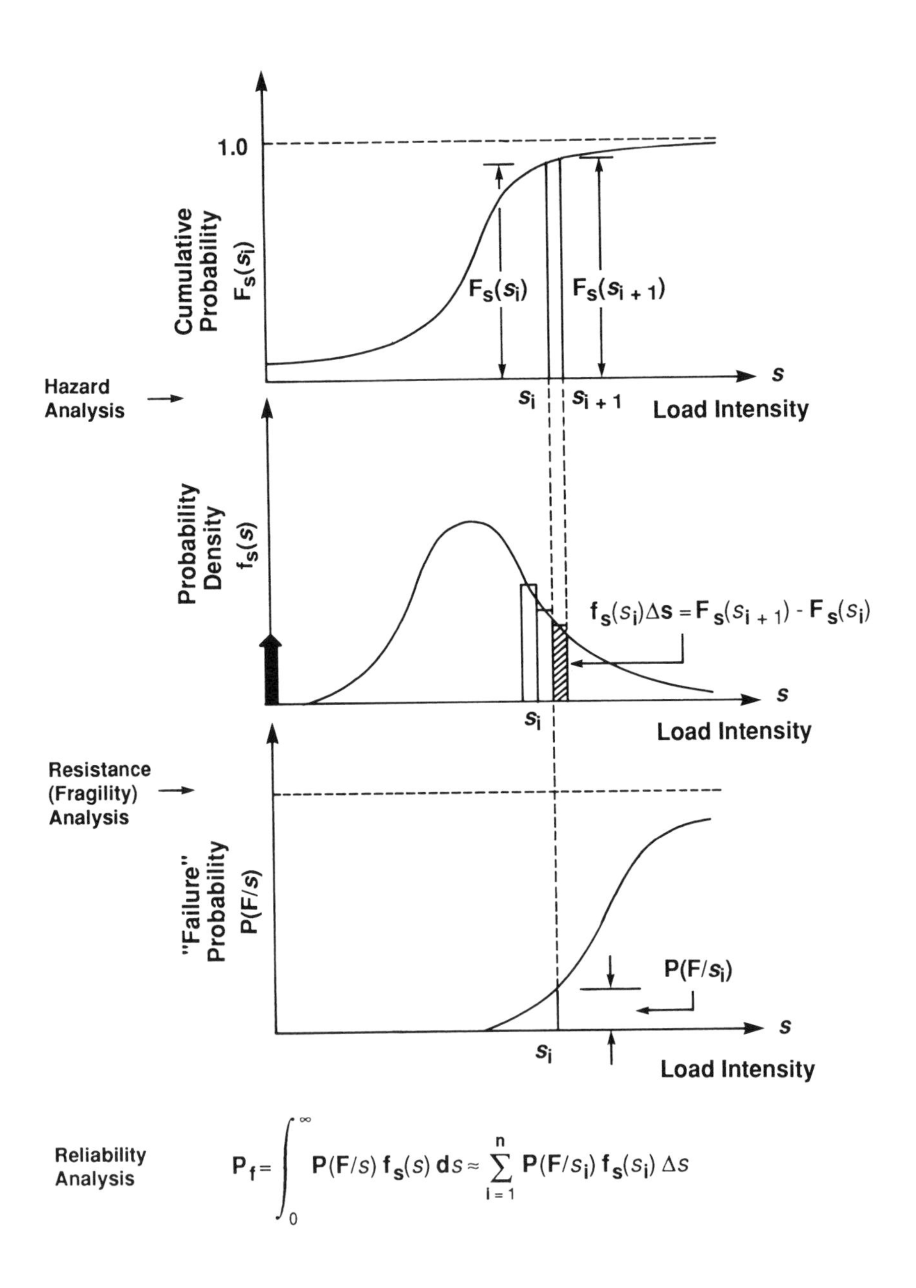

$$P_f = \int_0^\infty P(F/s)\, f_s(s)\, ds \approx \sum_{i=1}^{n} P(F/s_i)\, f_s(s_i)\, \Delta s$$

Figure 2.2.1 Elements of Prediction of Probability of Building and Structural Failure (limit state being reached).

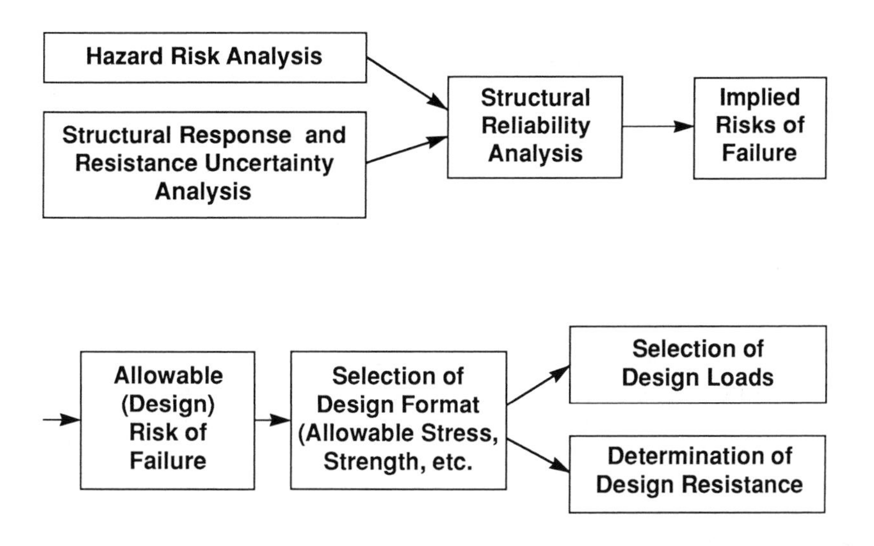

Figure 2.2.2 Code Calibration Procedures
(Top) Risk Implication of Current Practice. (Bottom) Calibration of Future Code

to the additional complexity in the traditional design procedure, the uncertainties in the load structural interaction and structural resistance which are not fully understood and the human factors which are difficult to evaluate and quantify. Recent studies [Ellingwood et al, 1980] have strived to resolve some of the difficulties and to find the implied risks of failure in current practice according to the provisions in

Table 2.2.1 Elements of Prediction of Probability of Building and Structural Failure (Limit State Being Reached)

Probability of Failure P_f	Safety Index β
1.58×10^{-1}	1
6.68×10^{-2}	1.5
2.27×10^{-2}	2
6.21×10^{-3}	2.5
1.35×10^{-3}	3
2.32×10^{-4}	3.5
3.16×10^{-5}	4
3.39×10^{-6}	4.5
2.86×10^{-7}	5

*This relationship is based on the assumption that the distributions of the random variables are normal (i.e., Gaussian); therefore, it is only approximately true for random variables that are not normal.

current codes. The results have been used to suggest improvements in the design code format such that some consistency can be achieved in the implied risk of failure. The procedure is also known as code calibration, it is shown schematically in Figure 2.2.2.

Such a procedure represents a rational way of synthesizing knowledge of recent theoretical reliability research, information on structural load and resistance uncertainties and level of safety deemed desirable by the profession and incorporating into a code format.

2.3 CONSIDERATION OF SEISMIC FORCES

2.3.1 Input Characteristics Critical to Low-Rise Building

Low-rise buildings exhibit special characteristics under seismic excitation. First of all, a large number of low-rise structures are used as residential units, shopping centers, light industries and recreational facilities. Such buildings are often irregular in shape. In most cases, they have at least one wall with large openings. Therefore, low-rise buildings behave differently in the event of an earthquake. Furthermore, the construction material used in low-rise buildings also adds to the variability of the structural behavior. For example, low-rise metal structures have been reported to perform relatively well [Gupta, 1981] in past earthquakes. On the other hand, many unreinforced masonry buildings performed rather poorly [Thiel and Zsutty, 1987]. Also, the experience with wood-framed residential buildings shows problems of inadequate connections, effect of load duration and high-strain rates (e.g., in wood) on resistance and light-weight panels [Gupta, 1981].

As a result of these special features, low-rise buildings have a wide range of period of vibrations (0.06–0.25 sec. in buildings with heights of 10–40 ft., [Gupta, 1981]). The same is also true of damping in the structure. With such a wide range of structural dynamic properties, the response may be drastically different for excitations of different frequency content and duration. Thus, these ground motion parameters are of special importance in the analysis and design of low-rise buildings.

2.3.2 Selection of Risk Level for Design Earthquake Forces

Generally speaking, to assure a desired safety level for a structure, appropriate design intensity (e.g., in terms of ground acceleration and velocity) should be selected. There is a correspondence between the design intensity and the associated risk level of the limit state being reached. To evaluate the risk, however, requires a detailed seismic risk and structural reliability analysis, specifically.

- Identification of potential sources in the area.
- Calculation of probability of occurrence of future earthquakes based on the past records of seismic activities (frequency, intensity, etc.)
- Consideration .of the "attenuation" and its variability, i.e., the local site intensity, as a function of the earthquake magnitude and the distance from the source to the site.

- Structural reliability analysis to evaluate the risk of the limit state due to future earthquakes.

The seismic risk can be described by a risk curve for the general seismic region where the structure is planned. An example is given in Fig. 2.3.1 of the intensity return periods (or inversely, the annual probability of exceedance) versus different intensity levels. The family of curves reflect the uncertainty associated with the risk curve. Methods for developing curves such as these are given, for example, in Cornell (1986).

The capacity of the structure against damage and collapse is also subject to variability and uncertainty. This is mainly due to variability associated with the material resistance, uncertainty in geometry and cross-sectional dimensions of members (especially in reinforced concrete, timber and masonry structures), inaccuracy in structural modeling, and in human errors occurring during the design, fabrication and/or construction process.

By convolution of the risk curve and the fragility curve, one obtains the risk of the limit state being reached (see also Figure 2.2.1). Since the uncertainty of the excitation generally dominates, as a crude approximation, one can regard the resistance as a deterministic quantity given by the 50-percentile value from the mean fragility curve. The acceleration level corresponding to the 50% probability is then entered into the mean hazard curve to obtain the approximate level of the risk of the limit state (see Figure 2.3.2). The selection of allowable risk level may depend on the following considerations:

- The type of structure in terms of the material used;
- Classification of the structure in terms of its primary use and the number of occupants;
- The general quality of construction which depends on the geographical area and the type of material used;
- Single, isolated buildings versus buildings as a group or complex;
- Cost of replacement of the structure;
- Expected useful life of the structure; and
- The record of past experience of similar structures in a seismic load environment.

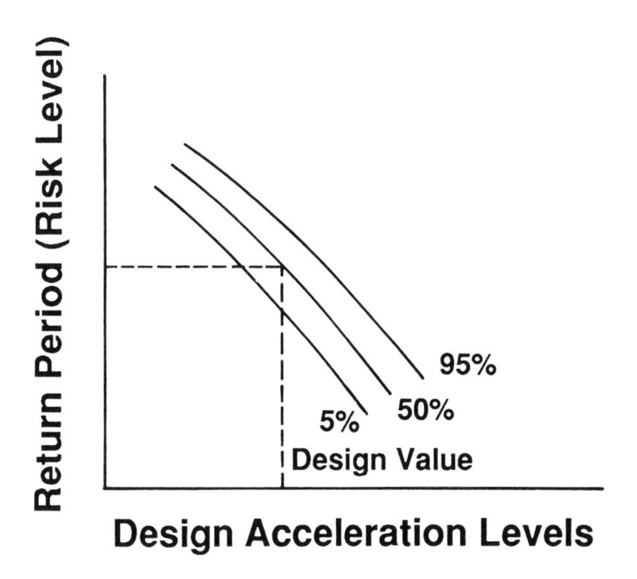

Figure 2.3.1 Design Acceleration-Risk Level Family of Curves

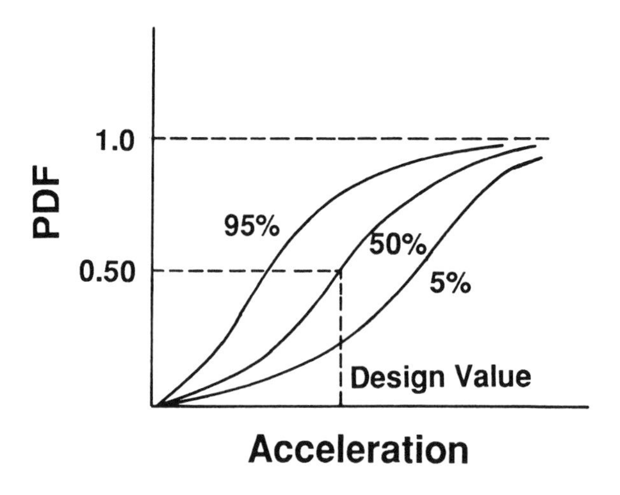

Figure 2.3.2 Effect of Uncertainty in Fragility Curves

2.3.3 Risk Implied in Current Standards

The accepted risk levels are generally implemented in recommended design values by seismic design codes. The design values, therefore, assure the life safety of the occupants and the overall safety of the structure to a certain degree as described below.

The level of safety provided by the Applied Technology Council (ATC3-06) provisions is related to the probabilities that effective peak acceleration (EPA) and effective peak velocity-related acceleration (EPV) will be exceeded in a given year [ATC3-06, 1978]. For buildings designed according to the ATC3-06 provisions, collapse should not be expected if the design ground motion were to occur. This safety level is that implied in the ATC design requirements and excludes risk associated with construction errors, inadequate inspections, etc. For a ground motion with a intensity twice as strong as the design value, a chance of collapse of 0.01 to 0.02 is expected. This value increases to 0.05 to 0.10 if the intensity becomes three times as strong as the design value. ATC3-06 utilized these results for the prediction of the risk of a single building during a 50-year period, i.e., risk associated with structural collapse as well as with the life threatening damage. In cases where there are a number of similar buildings at a potential seismic area, the probability that at least one of the buildings fails is obviously larger. Calculations are then made assuming 100 buildings in, respectively, one and five cities. The results indicates that the chance of failure could increase from 0.0001 to 0.01 for a single building to 0.05 for a group of buildings in the same city and further up to 0.15 for one or more groups of buildings in five cities.

The American National Standards in A58.1-1982 [ANSI, 1982] uses the Effective Peak Velocity-Related Acceleration Coefficient (EPV) recommend by ATC3-06. Thus, the risk associated with A58.1 provisions are the same as those in ATC. The contour map in ATC used for EPV was, however, converted into a zoning map which is utilized in seismic design calculations according to A58.1.

In a recent study [Thiel and Zsutty, 1987], an investigation of seismic performance of a variety of low-rise buildings to recent California earthquakes was carried out. Potential damage to a structures and failure probability are considered These results compared favorably with observed data. Table 2.3.1 shows the failure probabilities estimated from the study and observed data for different types of low-rise buildings. Similar tables for the relative levels of damage (observed vs. estimated) for these buildings are also given in Thiel and Zsutty (1987).

The seismic design provisions in the National Building Code of Canada (1985) use a seismic zoning map based on a statistical analysis of seismic activities in Canada since 1899. In the 1985 version, there is a 10% exceedance probability of the design acceleration values in a 50-year period (0.0021 per annum), i.e., same as that in ATC3-06. The corresponding risk value for the 1980 version of the code is 0.01 per annum. According to this code, the change made in the risk level has not produced any major change in the calculation of minimum lateral seismic forces.

The selection of a particular code and zoning map is generally governed by the local regulations and jurisdictions. Different codes, however, may not agree on recommended design values for the same geographical area.

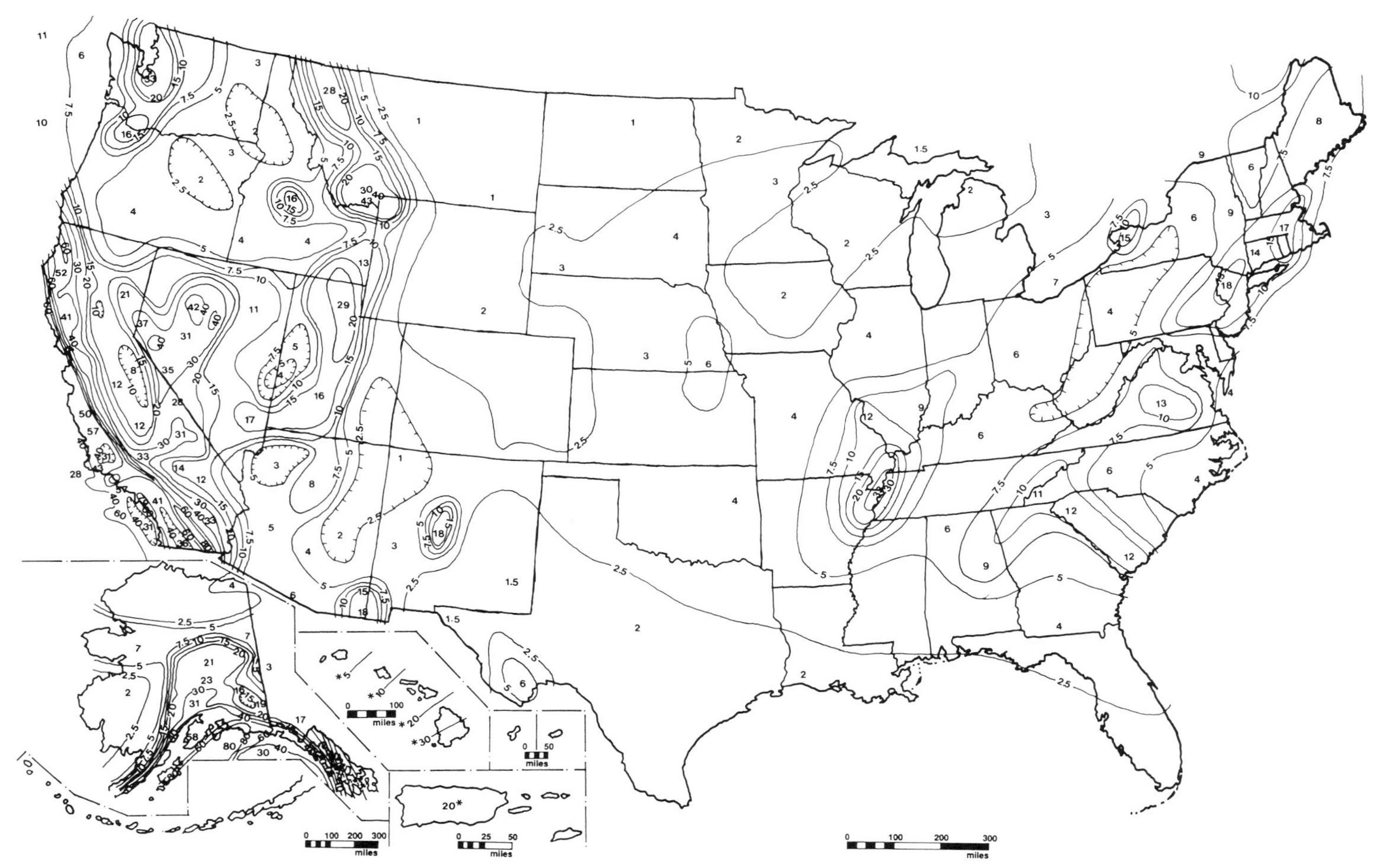

Figure 2.3.3 Map of Horizontal Acceleration, A (expressed as a percent of gravity) in Rock with 90 percent probability of not being exceeded in 50 years.

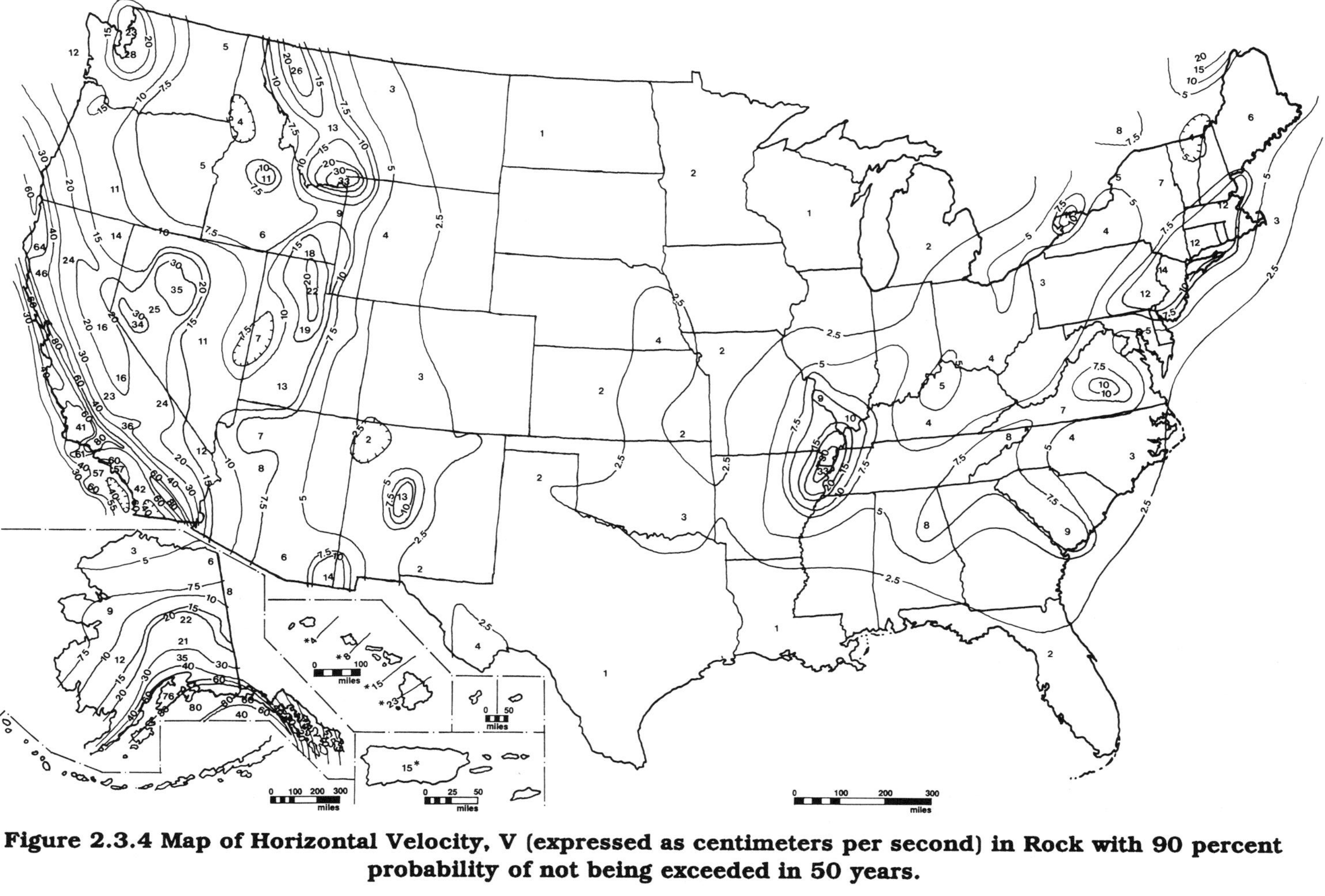

Figure 2.3.4 Map of Horizontal Velocity, V (expressed as centimeters per second) in Rock with 90 percent probability of not being exceeded in 50 years.

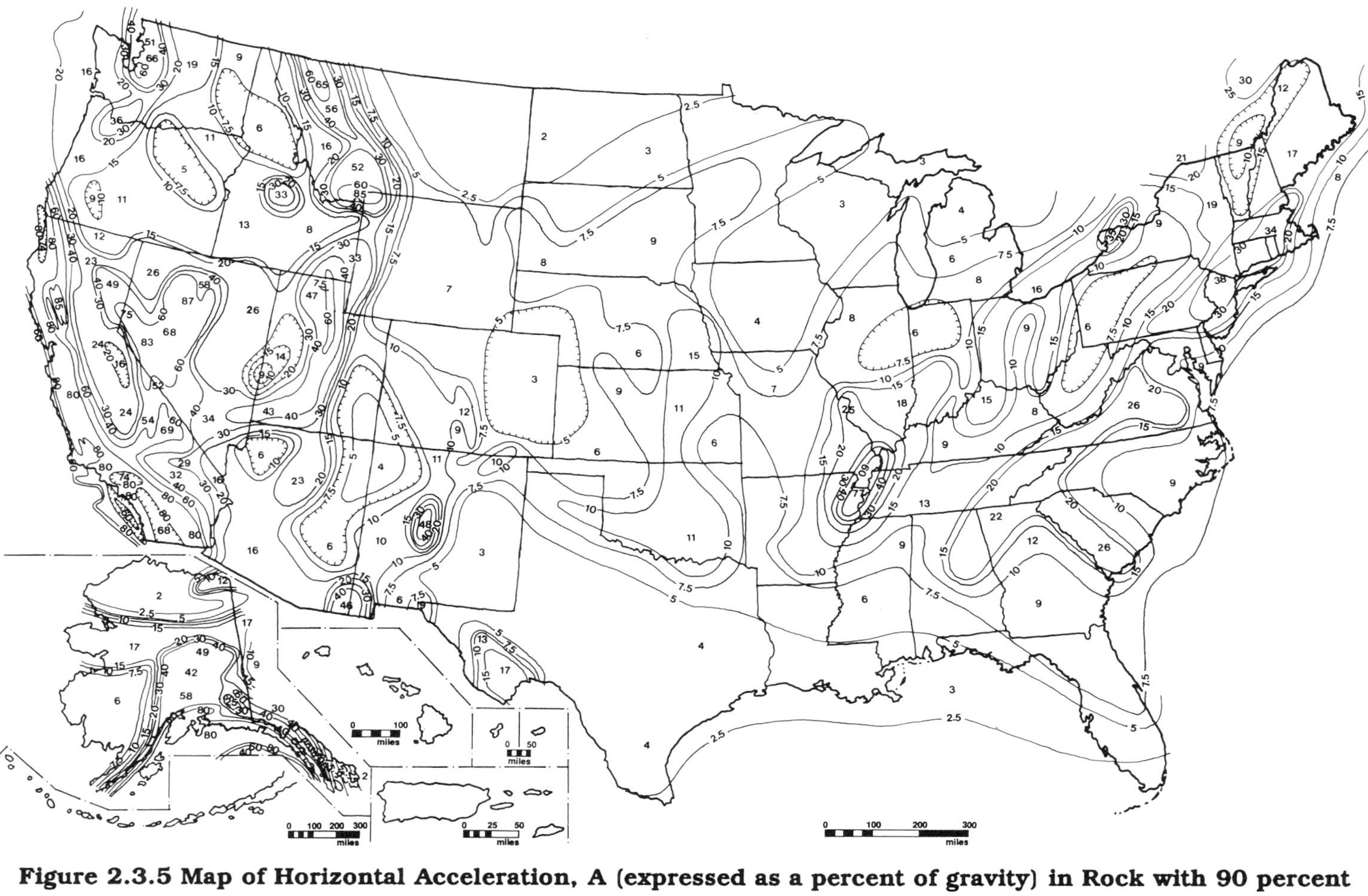

Figure 2.3.5 Map of Horizontal Acceleration, A (expressed as a percent of gravity) in Rock with 90 percent probability of not being exceeded in 250 years.

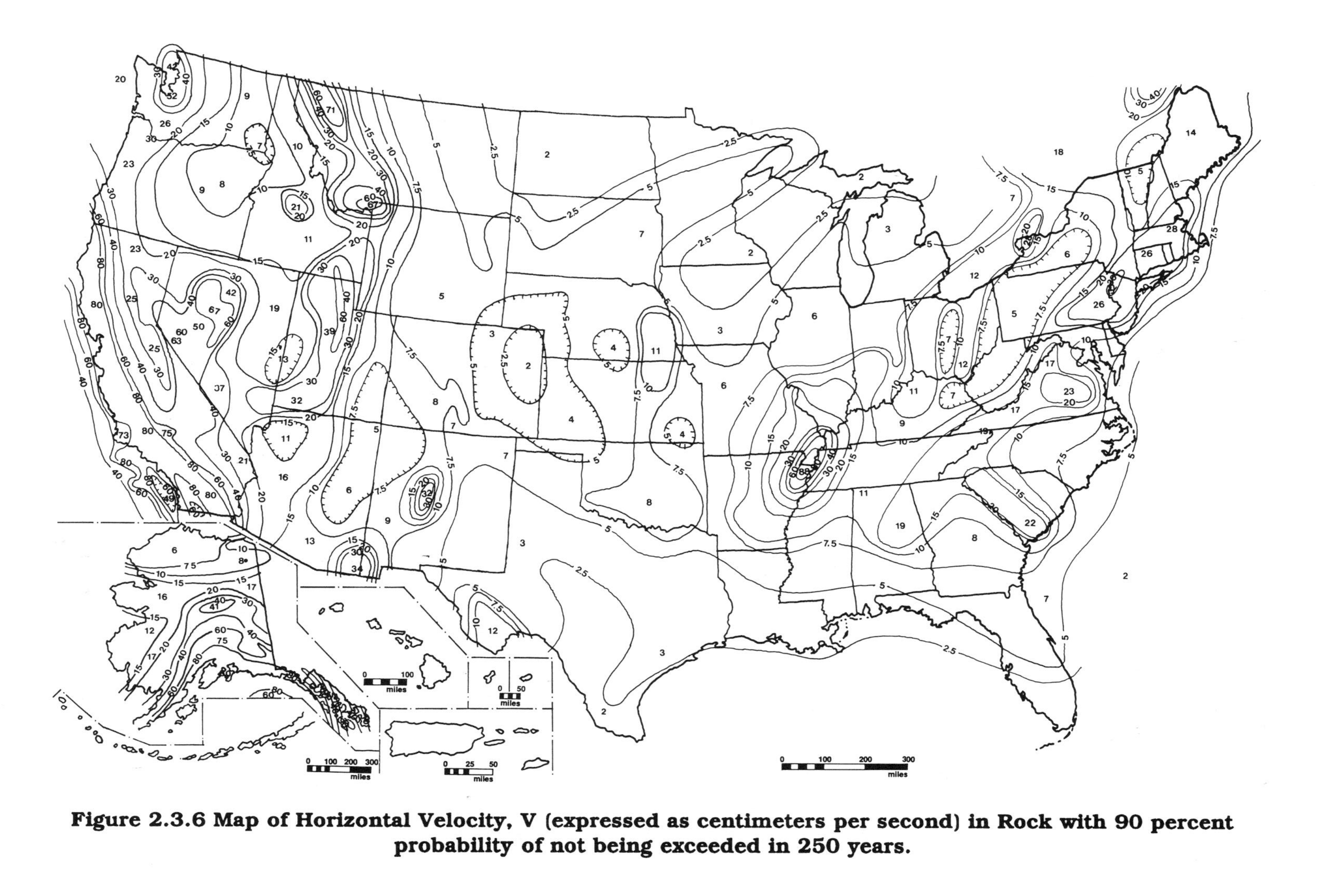

Figure 2.3.6 Map of Horizontal Velocity, V (expressed as centimeters per second) in Rock with 90 percent probability of not being exceeded in 250 years.

Table 2.3.1 Failure Probability According to Thiel and Zsutty (1987)

Data Set (Earthquake and Building Type	Estimated p from observed data	Estimated p from model
Long Beach, pre-1930 URM[1]	0.386	0.394
Long Beach, post-1930 URM	0.185	0.201
Long Beach, Compton URM	0.766	0.678
Santa Barbara URM	0.519	0.550
Kern County URM	0.610	0.674
San Fernando URM	0.375	0.421
San Francisco, Stanford Phase 1 URM	0.276	0.295
San Francisco, Stanford Phase 2 URM	0.553	0.578
San Francisco, downtown URM	0.299	0.351
Coalinga revised URM	0.633	0.671
Santa Barbara, dwellings	0.220	0.216
Santa Barbara, adobe	0.667	0.659
Santa Barbara, 4B	0.363	0.337
Long Beach, Compton URM residences	0.137	0.060
San Fernando, R/C[2] tilt-up	0.261	0.270
San Fernando, tilt-up less soil	0.200	0.215
San Fernando, wood frame, all	0.073	0.088
San Fernando, wood frame, 1&2 story	0.221	0.270
San Fernando, wood frame, pre-1940	0.104	0.110
San Fernando, wood frame, 2 story	0.165	0.185

[1] Unreinforced Masonry
[2] Reinforced Masonry

The most recent seismic ground motion risk maps in terms of peak acceleration and peak velocity [Algermissen, et al., 1987] are produced in Figs. 2.3.3 to 2.3.6.. They are based on most comprehensive earthquake records and geological information. Contour maps are given of ground motion levels corresponding to a 90 percent nonexceedance probability over a given period of time. For example, the acceleration level given in Fig. 2.3.3 corresponds to a nonexceedance probability in 50 years, or according to the conversion equation in Section 2.2.1, an annual probability of exceedance of 0.0021, i.e., a return period of 475 years. Similarly, one can show that Fig. 2.3.5 gives 2,373-year accelerations. Use of these maps in design will be given in later chapters.

2.4 CONSIDERATION OF WIND FORCES

2.4.1 Input Characteristics Critical to Low-Rise Buildings

Different parts of the US have different wind hazards because of local weather conditions. For example, the maximum wind speed which is likely to occur in California due to winter storms is less than the maximum wind speed due to hurricanes on the tip of Florida. The basic wind speed which varies from location to location is established first when designing a low-rise building for wind. To provide equal safety against wind damage at different locations, basic wind speeds should be established for equal return periods.

Fastest-mile wind speed is the standard unit of wind speed measure in the United States. Fastest-mile wind speeds are routinely measured at National Weather

Service Offices throughout the country. The set of annual extreme fastest-mile wind speed records collected over a period of years at a weather station is used to obtain the distribution of wind speeds associated with a specified return period.

Thom (1960) used a Fisher-Tippet Type II extreme value probability distribution to represent the annual extreme fastest-mile wind series. Using this distribution and wind data accumulated over a 21-year period at 138 stations across the United States, he calculated wind speeds associated with 25-, 50-, and 100-year return periods. Plotting these values at the 138 station locations, he developed wind speed contour maps for the three return periods. The maps were updated in 1968 and were used in the 1972 edition of the ANSI A58.1 wind load provisions.

Simiu, et al. (1979) reexamined the fastest-mile wind speed records. He established a criterion that required a minimum of 10 years of wind speed records, a history of the anemometer height and anemometer located in flat open country such as an airport location. Only 129 stations through the country met the established criteria.

In his analysis, Simiu found through goodness-of-fit tests that the Fisher-Tippet Type I extreme value distribution was more appropriate for the vast majority of accumulated data than the Type II distribution used by Thom. The differences between the Type I and II distributions are small for return periods less than 200 years. They give significantly different results for return periods greater than 1,000 years. The Type II distribution predicts significantly higher wind speeds than the Type I at large return periods. The Type II wind speeds are unreasonably large for return periods greater than 10,000 years.

Because of the difficulty in drawing consistent contour maps of wind speed, only one map corresponding to the 50-year return period was prepared by Simiu. Wind speeds associated with other return periods are obtained by applying a factor (called importance factor in ANSI A58.1-1982) to the 50-year value.

Simiu's analysis indicated that the hurricane-prone regions are not well represented in the extreme wind speed analysis described above. The length of data records is not sufficient to provide a representative sample of hurricane winds. To circumvent this lack of data, Batts, et al. (1980) numerically generated a set of hurricane data based on 100 years of Gulf of Mexico and Atlantic coast hurricane records. The analysis provides a larger numerically generated data base for hurricane winds at equally spaced mile posts along the coastline. Results of this analysis provide wind speeds for a 50-year return period along the hurricane-prone coastline. These were incorporated into the 1982 ANSI A58.1 windload standard contour map. The probability distribution of hurricane wind speeds has a longer tail than that for the inland stations. In order to provide the same probabilities of overloading in hurricane-prone regions as in inland regions, an importance factor (greater than or equal to one) is applied at the hurricane-prone ocean line. The hurricane-wind effects are assumed to be negligible at distances of more than 100 miles inland from the ocean line; the importance factor can be linearly interpolated between the ocean line and 100 miles inland.

Figure 2.4.1 is the ASCE 7-88 (and ANSI A58.1-1982) 50-year return period basic wind speed map for the United States. The map gives the unadjusted basic wind speed used in the design of low-rise buildings. It provides basic wind speeds as a function of geographic location for most standards and model building codes used for the design of low-rise buildings. Corrections to the basic wind speed values are discussed briefly in the following paragraphs and in more detail in Chapter 3.

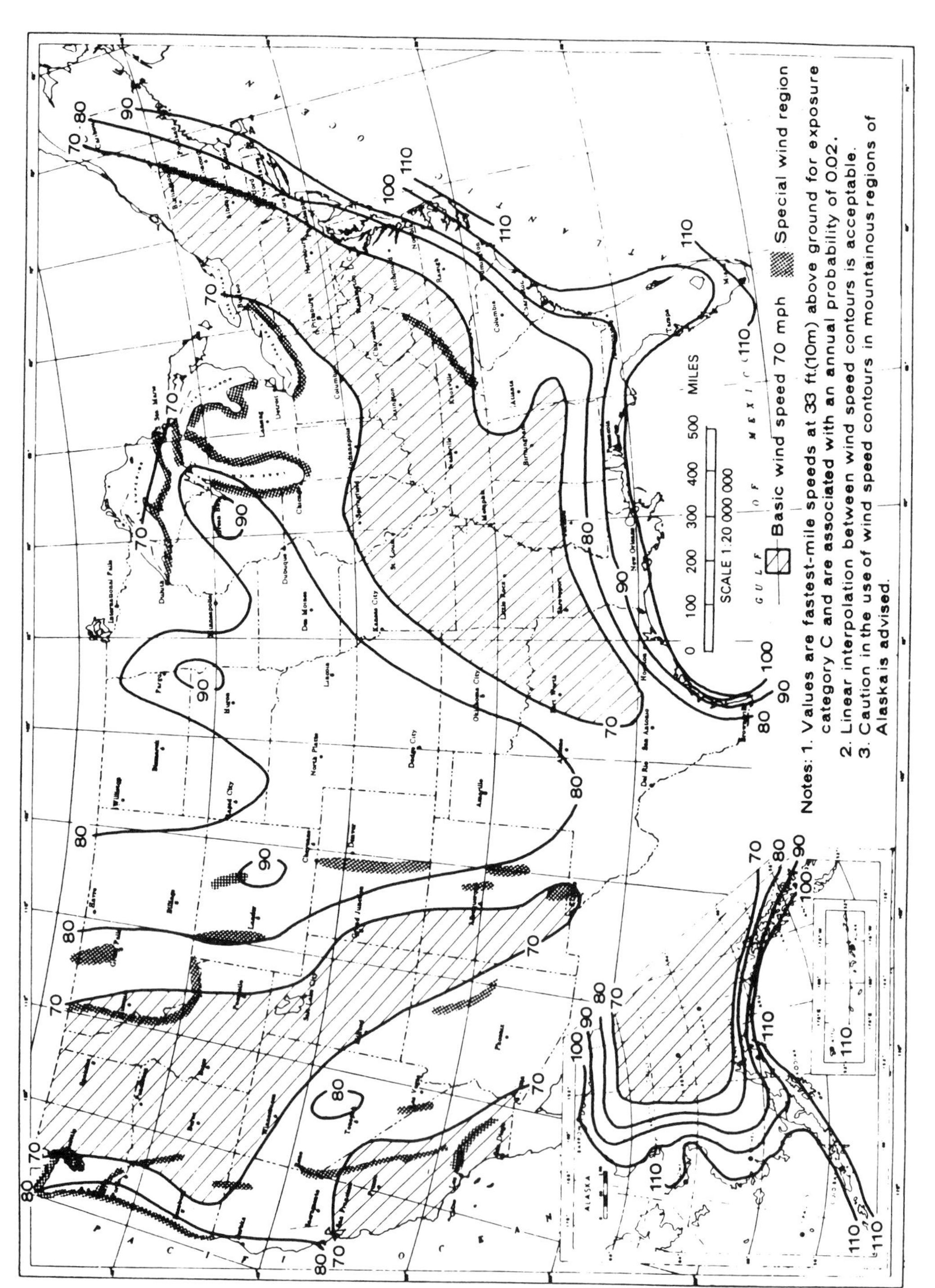

Figure 2.4.1 Basic Wind Speed (mph) [ASCE, 1988]

In addition to geographical location, the wind speed is also a function of height above ground at which it is measured, the roughness of the terrain and the time over which the wind speed is measured. These aspects are important for loading both on buildings as a whole and their cladding and other elements. As the height above the ground rises, the wind speed also increases up to a gradient elevation of approximately 700 to 1,500 feet above the ground, where the mean speed remains constant at even higher elevations. For a fixed gradient wind speed (i.e., located at a high elevation where the wind speed is unaffected by ground conditions) the corresponding mean speed near the ground (e.g., at 33 feet elevation) decreases as ground roughness increases. Thus, the ground wind speed is greater in flat areas such as farm land, as compared to "rougher" areas such as large cities. This subject is discussed in more detail in Chapter 3.

The ground roughness causes the fluctuating wind component due to gusts, about the mean speed, to increase. This is particularly important for small elements of buildings where the total wind speed (i.e., mean plus fluctuating component) can be high. In general, the fluctuating component is greater in city areas than in country areas.

These force characteristics of wind are important to the designer of low-rise buildings. In selecting the wind speed for designing the capacity of the building as a whole, or for a component such as the connection of the roof to the wall, each of the above physical aspects pertinent to wind hazard is considered. In addition, the importance of the building governs the selection of the ultimate force. A warehouse generally would have a lower design level than a hospital, which is an essential facility and needed in the event of a natural catastrophe such as a hurricane. In this manner the designer can lower or raise the over-all risk of failure of different classes of buildings according to their importance.

From the viewpoint of the physical characteristics of buildings, the properties of wind may present special problems. For example, connections of building parts are vulnerable areas for failure. Although wind forces are often thought of as the result of positive pressures on a building being buffeted by a wind storm, the idealized design pressures on a building are shown schematically in Figure 2.4.2. Note that the front end exposed to the direct impact of the wind is under positive pressure but the sides, back and roof are under negative pressure. In addition, the negative pressures at corners and eaves are even higher than the average pressure over the side or roof of the building. This creates potential points of initial failure where the roofing or even the roof or wall could separate from the main part of the building (e.g., see Figure 2.4.3). These locations must be designed for higher pressures than the "average" pressures used to design the main shear and overturning strength of the building.

High winds tend to pick up debris, which becomes missiles. This is important for tornado type winds which are discussed in more detail in Section 2.4.3. However, even in hurricanes or other fierce storms 2 x 4-size pieces of wood and pea gravel off roofs become missiles which can easily break windows, and in some cases pierce the sides and roofs of a building. Although it is difficult to design against wind-born missiles, window covers and barriers can be put up when warnings of imminent severe storms are given.

In the case of earthquake forces, a building oscillates back and forth during the ground motion. Ductility inherent in the building can absorb much of the energy of an earthquake and provide capacity above the yield level of the structural components. In contrast, wind forces are constant in direction over significant periods of time which negates the benefit of ductility. Although an occasional gust can be resisted by yielding of the structural member under stress, ultimately the member will fail if too many excursions above yield occur.

2.4.2 Selection of Risk Level for Design Wind Forces

The American Society of Civil Engineers publication ASCE 7-88 *Minimum Design Loads for Buildings and Other Structures* [ASCE, 1988], an updated version of the American National Standard ANSI A58.1-1982 *Minimum Design Loads for Buildings and Other Structures* [ANSI, 1982], is the principle national standard for specification of minimum wind loads. The Uniform Building Code is another model code which defines wind loads used in the design of low-rise buildings and is now similar to the ANSI standard. Many locals codes developed for cities or regions of the country are generally patterned after either of these two documents.

The selection of risk level is partially implied in the selection of the design wind speed. In the ASCE (ANSI) standard, a basic wind speed is selected corresponding to the geographic location of the building for an annual probability of occurrence of 0.02 (this corresponds to a return period of 50 years). Figure 2.4.1 is the ASCE (ANSI) basic wind speed map used to design buildings.

This is only the beginning of a series of steps leading to the forces used in a building design. Next, an importance factor is applied to the basic wind speed to reflect the nature of building occupancy and to add additional conservatism for buildings locate din hurricane-prone regions. For example, the design wind speed for a hospital located on the Eastern US coast would be 11 percent higher than the design wind speed for a small office building in San Francisco.

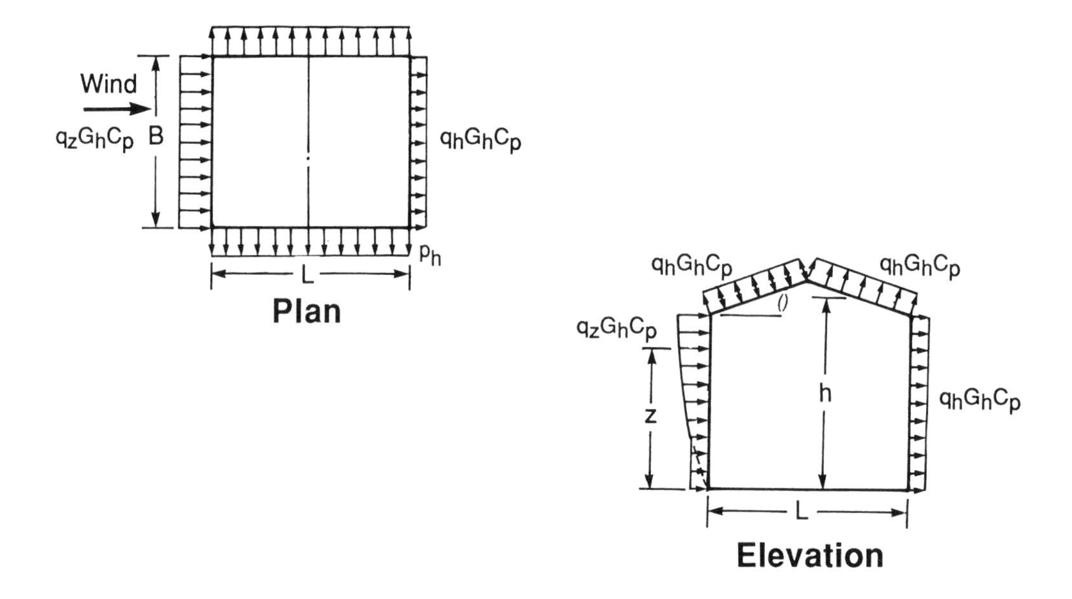

Figure 2.4.2 Wind Pressure on a Building (ASCE, 1988)

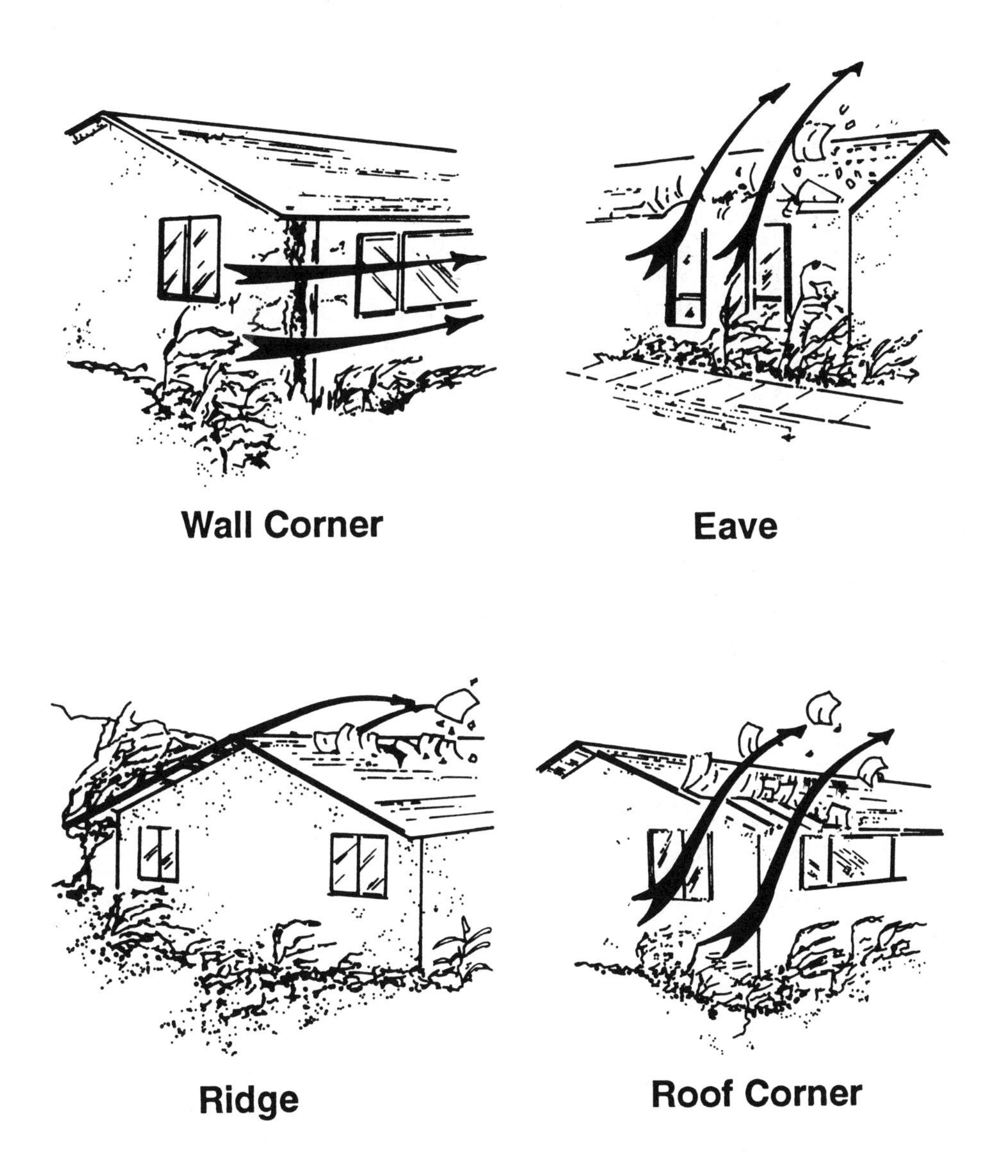

Figure 2.4.3 Local Effects of Wind [McDonald, 1985]

When the modified wind speed is converted to a pressure, it is scaled by another factor to account for the elevation of the interest above the ground and also is a function of the ground surface roughness. As discussed above, the terrain conditions affect the wind profile; thus, it is necessary to adjust the wind pressure depending on the height of the building. A second factor is applied to the wind pressure to reflect the fluctuations of pressure caused by wind gusts. This insures that peak pressures are properly accounted for in the building design. Note that this gust factor is a function of the ground roughness, height above ground, and to some extent the dimensions of the structure and the tributary area.

Finally, the wind pressure is adjusted for the building element being designed. This modification reflects the effect of the shape and configuration of the structure itself on the wind pressure. Figure 2.4.2 shows schematically the positive and negative idealized design pressures on a building. Figure 2.4.3 shows locations such as building eaves and corners where the negative pressures are higher than the average pressures. The external pressure coefficients account for these effects.

The above factors are combined in the specification of the wind pressure applied to a building as a whole or to specific locations such as at the corner of a roof. These factors ultimately lead to an implied risk which is generally different from the 0.02 annual probability level use to select the basic wind speed. Factors such as the importance factor are designed to decrease the risk for critical structures such as hospitals and other buildings which house community services needed in the event of a disaster. For example, the ASCE standard gives importance factor values of 1.07 and 0.95 which imply equivalent wind speed return periods of 100 and 25 years, respectively. This is in comparison to the importance factor of 1.00, which corresponds to the basic wind speed return period of 50 years.

Factors such as the gust factor and external pressure coefficient are designed to be risk-neutral, since they attempt to reflect physical characteristics of wind flow near the ground and around buildings. However, the approximations necessary in developing a practical, workable design code require simplification of a complex physical process and thus lead to inconsistencies which affect the underlying risk level. Because of this the risk may be larger or smaller than intended by the code provisions; however, the code is usually on the conservative side.

2.4.3 Special Considerations for Tornadoes

Although tornado occurrence is rare, a tornado strike on man-made structures can be devastating. Tornados have three principal effects. Similar to straight winds and hurricanes, their pressure effects can rip roofs off houses or flatten them to the ground. In addition, the drop in pressure as a tornado passes over a structure tends to cause the structure to blow apart, if it is not vented. Finally, tornado missiles can perforate a building, causing damage to its exterior shell and to the contents within. In contrast to the spectacular missiles which are sometimes displayed in the news media such as steel beams sticking in the ground, most missiles are small objects like a 2x4 board or pea gravel off roofs.

It is extremely difficult and expensive to tornado-proof a building. Nuclear power plants are design for the effects of the worst tornados ever experienced. Often walls are 2-ft. thick concrete barriers, and missile protection is provided at each door. This type of protection is obviously not suitable for typical low-rise buildings. It is estimated that the tornado protection provided for nuclear power plants insures that their annual probability of failure is on the order of $1x10^{-7}$ or even smaller. At best, if a building must maintain some level of protection in the event of a tornado strike it may be more economical to provide a protected island within the building which is tornado proof rather than designing the entire building to resist a tornado strike. An area in the basement would be best location, or above ground a place near the center of the building could be dedicated to this requirement.

2.4.4 Risk Implied in Current Standards

As discussed above the selection of the basic wind speed and the conversion to a design pressure leads partially to the risk of failure associated with design using a building code. The other major contribution comes from the capacity which is determined primarily by the selection of building materials and the governing allowable stresses. Together, the wind loading (i.e., wind speed selection and pressure determination), other loads (e.g., dead and live loads) and the building capacity determine the risk of failure. The influence of sustained loads which are normally present during a severe storm are included in the discussion in this section for wind. A general discussion of combinations of loads is given in more detail in Section 2.5.

Recent studies have examined the risk implied by current design practice for the basic building materials and for a variety of different loads [Ellingwood et al, 1980, 1982; Galambos et al, 1982]. For wind loads combined with live and dead loads the over-all safety index for limit state of first yield in structural member (see Section 2.2 for a discussion of safety index) is 2 to 3 with an average value of about 2.5. However, Ellingwood et al. (1982) say that historical evidence suggests that the true safety factor may in reality be even higher.

The Ellingwood et al. (1982) study was based on a 50-year time period which requires a conversion from safety factor to an annual probability of occurrence. As discussed in Section 2.2, safety index values of 2 and 3 correspond to probabilities of failure of 2.2×10^{-2} and 1.3×10^{-3}, respectively. Similarly, a safety index β of 2.5 corresponds to a probability of 6.2×10^{-3} which is in between the above two values. The conversion from a 50-year probability to an annual probability of failure is given in Table 2.4.1 using the equations presented in Section 2.2:

Table 2.4.1 Failure Probabilities

Safety Index β	50-year Probability of Failure	Annual Probability of Failure
2	2.2×10^{-2}	4.6×10^{-4}
2.5	6.2×10^{-3}	1.2×10^{-4}
3	1.3×10^{-3}	2.7×10^{-5}

This says that there is roughly a 1 in 10,000 chance of a building failing (i.e., for a safety index of 2.5) in a single year due to wind when designed according to a nationally recognized building code. Note that the basic wind speed is selected to have an annual probability of occurrence of 0.02; hence, the conservatism in the capacity selection for the structural members and connection details corresponds to a probability of failure much less than unity. It has been estimated that catastrophic failure of the building given that the design wind occurs is roughly 1 in 1,000 to 1 in 100. This example is somewhat approximate, since failure can occur at other wind speeds besides the design wind speed; thus, it is more rational to consider the possibility of failure at all wind speeds (see Section 2.2.2) in order to determine the relative contribution of the building capacity to the overall probability of failure. However, this example captures the essence of the situation.

In reality, buildings do not always fail due to extreme wind storms. Partial damage usually occurs and in most cases buildings are repaired. A study given in Hart (1976) presents estimated damage levels corresponding to various wind speeds. This data is based on a survey of selected wind experts who were asked to estimate the expected damage for different wind speeds. Figure 2.4.4 shows an example set of wind speed fragility curves for one to three story commercial or industrial buildings. As the wind speed increases, the mean percent damage also increases, until the wind speed is so great that there is 100 percent damage or total failure. As an example, consider the basic wind speed from the ASCE standard [ASCE, 1988] which ranges roughly from 70 to 110 mph in the US (e.g., see Figure 2.4.1). From Figure 2.4.4, it is seen that the expected percent damage is less than 10 percent. This is consistent since there is additional conservatism in the design of buildings to account for winds above the basic design speed as discussed above.

Tornado winds are very severe and are not addressed by the national standards since they are relatively rare. However, the public is aware of this phenomenon since tornado damage is well publicized in the news media. When a tornado strikes, man-made objects are often completely or nearly totally destroyed. As a first approximation it is instructive to assume that the probability of occurrence of a tornado striking a building is also a measure of the probability of failure of that building (i.e., it is implicitly assumed that failure is certain if a tornado strikes, which from Figure 2.4.4 is conservative). Rheinhold and Ellingwood (1987) present the results of a tornado hazard study of the U.S. Regional occurrence rates are given for four US regions which span from high-tornado areas in the midwest to low areas in the far West. The tornado occurrence rate per square mile per year ranges from 3.5×10^{-5} to 7.0×10^{-4} (a difference in hazard of a factor of 20).

Individual buildings occupy an area less than a square mile; however, the average size of tornados is generally greater than two square miles. So in an approximate sense the occurrence probabilities per square mile are equal to the probability of hitting a building. If the assumption that a tornado strike means certain failure is relaxed and the conditional probability of failure of a building given a tornado hit is assumed to be 1 in 10, then the rate of failure would be on the order of 1×10^{-5} to 1×10^{-4} per year, which is up to a factor of 10 less than the expected failure due to straight winds and hurricanes which are designed by building codes.

A general study of natural hazard risk assessment and public policy used the data from Hart (1976) to perform a loss assessment for the entire US due to wind and other hazards [Petrak and Atkinson, 1982]. The annual damage rates for the different types of winds are given in Table 2.4.2, where damage rate is expressed in terms of the fraction of damage relative to the building value.

Table 2.4.2 Annual Damage Rate

Wind Type	Building Damage Rate (per year)
Tornado	0.5×10^{-3}
Hurricane	1.0×10^{-3}
Severe Wind	0.006×10^{-3}

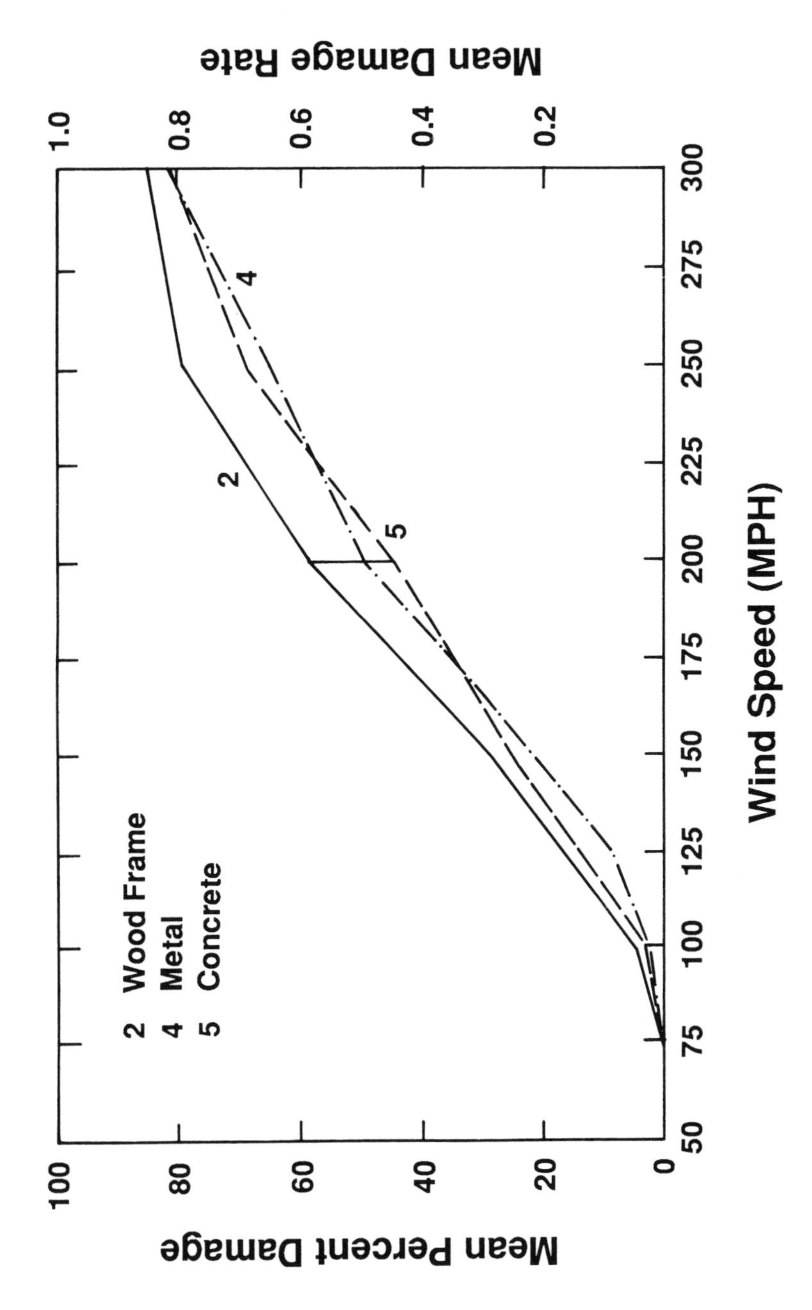

Figure 2.4.4 Mean Damage Versus Wind Speed (One to Three Story Commercial or Industrial) [Hart, 1976]

This clearly demonstrates a need for the increase in the importance factor for hurricanes. On the other hand, tornado damage is only a factor of two less than hurricane, which is consistent with reported losses where, for example, the actual damage loss in 1970 due to tornados and hurricanes was $880 million and $690 million, respectively [Petrak and Anderson, 1982]. However, because of the relatively high cost to tornado proof conventional structures, the benefit gained by requiring additional capacity to resist tornados would not be cost effective.

2.5 COMBINATION OF LOADS

2.5.1 Characteristics of Multiple Loads

A critical decision in the evaluation of design procedures is the selection of a load combination format - a set of equations which specify the way in which loads of different origin are added together.

As explained earlier, individual loads such as dead, live, snow, wind and earthquake are assigned nominal design values. Except for dead loads, these design values correspond to unlikely events with annual probabilities of being exceeded ranging from 0.03 to 0.002.

Basic combination problems involve permanent (or dead) loads acting with only one variable load such as wind load. Permanent loads are random in the sense that their initial magnitudes are uncertain but, once established, their magnitudes do not change with time. For variable loads, there is a random number of significant changes in magnitude during any time period. Each time a load changes, a new magnitude is chosen at random. The probability distribution of the maximum during the lifetime of any variable load is thus a function of the design lifetime.

To establish load combination formats involving two or more variable loads such as wind and earthquake, load-time dependence must be considered explicitly. Simple models such as those shown in Fig. 2.5.1(a,b) may be used. It is evident that large values of different variable loads usually occur at different times.

In general, the maximum of any load during a design lifetime can be expected to occur when more average values of other loads are acting. As shown in Fig. 2.5.1 (c), the maximum value of the sum of two loads generally does not equal the sum of their maximum values. For this reason, a simple addition of the design values of several variable loads would be unreasonably conservative.

A number of theoretical approaches to load combination analysis have been developed [e.g., Turkstra and Madsen, 1980]. A treatise of the subject is given in Wen (1990). Studies of linear combinations of the load processes commonly found in structural design indicate that the probability distribution of a sum of loads depends on the relative magnitudes of the loads involved as well as the total time period considered.

A practical approach for design is to establish combination factors that are risk neutral. That is, design load combinations are established to have probabilities of exceedance during the design lifetime similar to those for any individual variable load acting alone. Resulting load combination factors do not then refer to specific materials or limit states.

To indicate the importance of relative load magnitudes, consider the case involving two basic load types. The first is a design floor load which itself is made up of two parts - a sustained part due to normal usage which changes on average about once very seven years and a transient part due to occasional crowds, parties, or furniture storage during renovations, etc. [Corotis, et al., 1981]. The second load type is a wind or seismic load. If the total live load and each of the other variable load have design values of 100.0, theoretical values of combined design loads corresponding to a probability of exceedance of approximately 0.001 in a lifetime of 50 years are shown in Table 2.5.1 [Turkstra, 1985]. Also shown are the total loads recommended by ASCE (1988) for all three types of live load.

The dependence of combined loads on design load ratios is extremely important because conventional code formats permit a very limited number of simple rules that must cover all possible cases. For example, in any one structure the ratio of live to dead load may be quite small in one element and quite large in another element. However, both elements are designed by the same rules.

Table 2.5.1 Variable Loads

Factored Design Loads				Theoretical Total Design Loads			ANSI ASCE Combination Rules
Case	Total Live	Wind	Seismic	All Transient Live Load	Half Transient Live Load	All Sustained Live Load	
(1)	100	100		100	130	170	131
(2)	100		100	100	130	150	131

The primary consequence of simplicity constraints in design codes is that practical load combination analysis is inherently approximate. In many cases, the effects of one load are significantly greater than the effects of all other variable loads so inaccurate combination rules are not too important. In other cases, however, when two variable loads have significant effects or when one load category is used for substantially different cases (e.g., live load for warehouses and offices in one group) simplification can have important consequences.

A realistic code approach to combinations of variable loads is to consider a set of hazard scenarios where one variable load reaches its lifetime maximum while other "companion" loads have more usual values. As expected, numerical values of the reduced load factors for companion forces depend on the return periods of the primary design load in a combination. Values appropriate to many design situations have been developed [Turkstra and Madsen, 1980].

In design, many cases such as, for example, combined wind and earthquake may simply be ignored. In most situations, more than two variable loads do not have to be considered acting together.

2.5.2 Load Combination Formats in Current Practice

In traditional design practice, load combinations are given for "working stress" or "allowable strength" design. Basic combinations involving dead loads with floor live, live, wind, rain, snow, and earthquake loads and their nominal values D, L, L_r, W, R, S, and E are specified in formats such as:

$$D \qquad (2.5.1a)$$

$$D + L + (L_r \text{ or } S \text{ or } R) \qquad (2.5.1b)$$

(a)

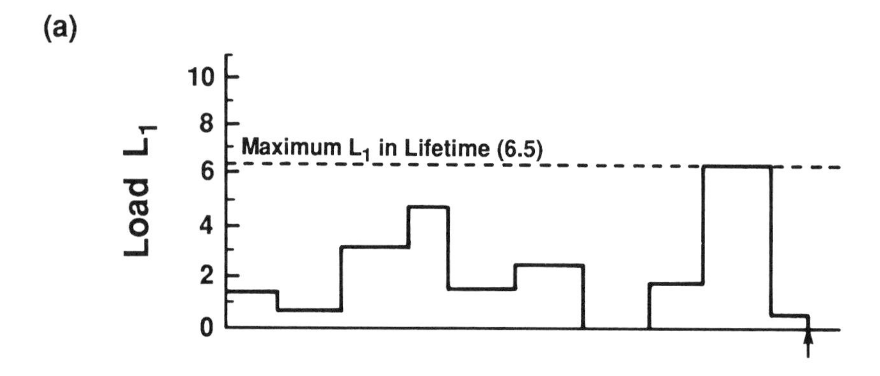

(b)

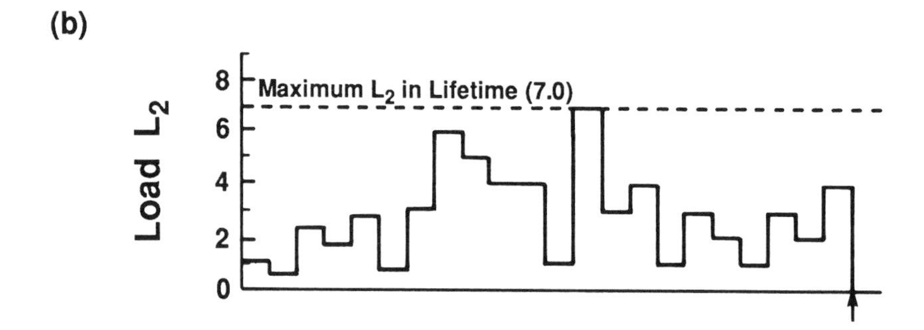

(c)

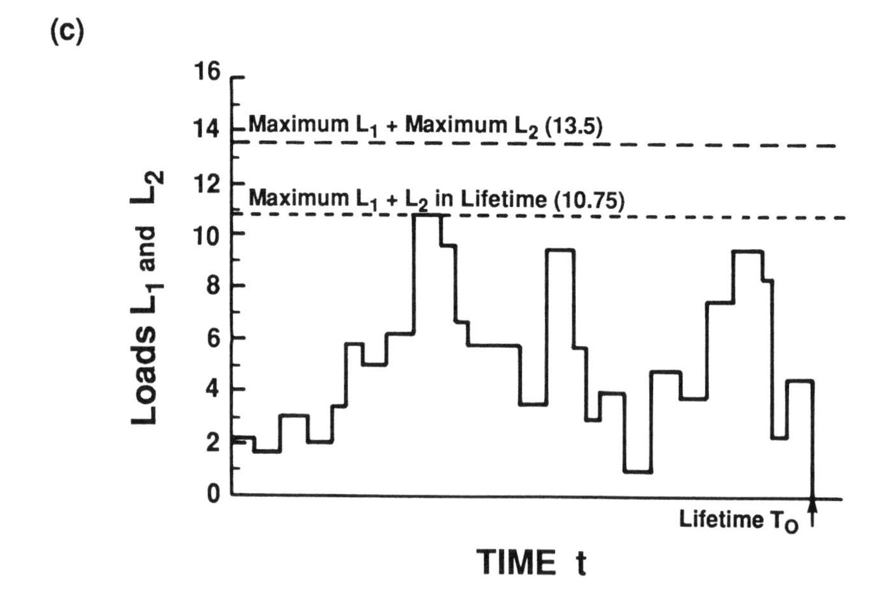

Figure 2.5.1 Variable Loads

$$D + (W \text{ or } E) \tag{2.5.1c}$$
$$D + L + (L_r \text{ or } S \text{ or } R) + (W \text{ or } E) \tag{2.5.1d}$$

The probability of exceedance of such combinations during a design lifetime depends on the relative values of the design loads (e.g., L/D). Wind and earthquakes are not combined because large values are extremely unlikely to occur at the same time.

In traditional practice, combinations involving two or more variable loads are often considered in the form, for example:

$$D + 0.75\,(L + W) \tag{2.5.2}$$

The reduction factor of 0.75 recognizes the fact that large loads can occur simultaneously but the coincidence of two extreme loads is very unlikely.

Another format which is sometimes used in the United States is of the form:

$$0.75\,(D + L + W) \tag{2.5.3}$$

A reduction in dead load in load combinations cannot be justified by any rational analysis. Such a format can lead to the erroneous conclusion that the design value of the sum of two loads is less than the design value of either load acting alone.

In modern "limit states", "load and resistance factor" or "strength" design codes, combinations of nominal values in the form of Equations 2.5.1 are used to check some serviceability states such as deflection. For "ultimate" states, increased or "factored" loads are used. The American Society of Civil Engineers (ASCE) recommends the following: [ASCE 7-88, 1988]

$$1.4\,D \tag{2.5.4a}$$
$$1.2\,D + (1.6L \text{ or } 1.6S \text{ or } 1.6L_r \text{ or } 1.6R \tag{2.5.4b}$$
$$1.2\,D + (1.3W \text{ or } 1.5E) \tag{2.5.4c}$$

To account for cases where dead loads counteract the effects of other forces, other combinations such as

$$0.9\,D - (1.3W \text{ or } 1.5E) \tag{2.5.4d}$$

are specified. The load factors 1.2, 1.3, 1.6, etc. are established on the basis of a safety index analysis over the range of design cases such as L/D ratios. For details, readers are referred to Ellingwood, et al. (1980).

For combinations of variable loads, sets of factored loads involving one variable load at its lifetime maximum and other loads at more usual values is used [ASCE 7-88, 1988]:

$$1.2D + 1.6L + 0.5\,(L_r \text{ or } S \text{ or } R) \tag{2.5.5a}$$
$$1.2D + 1.6\,(L_r \text{ or } S \text{ or } R) + (0.5L \text{ or } 0.8W) \tag{2.5.5b}$$
$$1.2D + 1.3W + 0.5\,(L_r \text{ or } S \text{ or } R) \tag{2.5.5c}$$
$$1.2D + 1.5E + (0.5L \text{ or } 1.2S) \tag{2.5.5d}$$

The reduced load of the form 0.5L, for example, results from the factored value of an average point in time load in the form $1.6L_{apt} = 0.5L$. It is pointed out that in Equations 2.5.4 and 2.5.5, E, the earthquake load, is for allowable stress design. If

E is determined directly from a strength or limit state design (e.g., the NEHRP) procedure, the load factor should be 1.0 instead of 1.5.

Numerical studies have shown that the accuracy of a simple approach such as Equations 2.5.4 and 2.5.5 is quite reasonable for normal design cases. To simplify design and recognize the approximate nature of load analysis, one can devise a rule such as "the effect of any load that is less than 20% of the effects of any other load acting in the same sense may be ignored". Among the existing official documents on load combination, the ANSI (ASCE) recommendations are based more on sound theoretical consideration and careful reliability-based calibration, and therefore can be used as guidelines in design of low-rise buildings.

2.5.3 Resistance Factor

The load combination formulas as described in the foregoing allow one to take into consideration the uncertainties in the loadings. To consider the uncertainty in the resistance, e.g., due to the variability in the material and inaccuracy in the determination of the structural member capacity against a limit state, etc., a resistance factor ϕ can be used such that the resultant resistance is larger than the combined load according to the following:

$$\Phi R \geq \sum_{i=1}^{n} \gamma_i Q_i$$

in which R is the design resistance, Q_i and γ_i are respectively the design load and load factor. Φ depends on the material, its uncertainty, and the prescribed level of reliability. It is less than unity and smaller for a structural member with a larger material uncertainty and a higher reliability (hence, resulting in a larger required member size). For more details of the concept and determination of the resistance factor, readers are again referred to Ellingwood, et al. (1980). Specific values of ϕ for a given construction material are given in the appropriate codes and standards.

3 Wind Forces

3.1 INTRODUCTION

The designer of low-rise buildings should be conversant with the nature of extreme winds that can affect the specific structure being designed. Characteristics of extreme winds, their frequency and location of occurrence and certain special features must be well understood by the designer. This chapter discusses characteristics of extreme windstorms, fundamental wind loading concepts, wind-structure interaction, and wind loading criteria.

3.2 CHARACTERISTICS OF EXTREME WINDSTORMS

In discussing strong winds, three types of winds are usually identified: straight (or extreme), hurricane, and tornado winds. Hurricanes and tornadoes obviously involve rotating winds. Straight winds include winds emanating from cyclones and cold fronts, downslope winds, and downbursts.

3.2.1 Cyclones and Cold Fronts

Widespread strong winds may be experienced with the passage of intense cyclones and cold fronts. The strength of the winds is roughly proportional to the pressure gradient which exists in the vicinity of the front associated with the region of relatively low pressure. Wind gustiness may be enhanced by the sinking of colder air from aloft as a part of the three-dimensional storm structure. The duration of strong winds at a specific location with frontal passages may be relatively short compared to those with a low. Preferred regions of low formation (cyclogenesis) and intensification exist in the lee (regions downwind) of the Rockies and along the Gulf and Atlantic Coasts. These areas may experience prolonged windiness with gradual intensification before the low moves away to the east and north. Preferred cyclone tracks commonly lead the storm into the Great Lakes/Ohio Valley area or along the Eastern Seaboard.

3.2.2 Downslope Winds

Strong, gusty localized windstorms are observed in and near mountainous regions in many parts of the world. These storms are accompanied by a characteristic temperature fall or rise. Warm, descending winds are termed *foehns* and cold, descending winds are *boras*. In the United States, the most destructive and best documented examples of foehn winds occur near Boulder, Colorado.

Severe windstorms occur about once per year in Boulder, although a total of 20 storms were observed in a three-year period, and there is a large interannual variability. Most storms occur in January, but they may occur between September and March. Peak gusts of 100–120 mph have been recorded in populated areas of Boulder. A storm may last a day or more. The Boulder area is a special wind region that requires attention beyond that provided by model codes or standards. Local authorities should be consulted in any mountainous area or any area where downslope winds are a possibility.

3.2.3 Thunderstorms

Wind gusts in excess of 60 mph (fastest-mile wind speed) sometimes accompany severe thunderstorms. Quite often, these winds strike before the onset of precipitation. Their advance may be foreshadowed by an ominous horizontal roll cloud or a wall of dust. These winds are an integral part of the thunderstorm outflow, which develops from descending cold air striking the ground and spreading in front of the storm. The leading edge of these strong winds is called the gust front.

With the availability of satellite imagery, it has been recognized that during the warm season, thunderstorms often become organized into large, persistent assemblages. These mesoscale convective complexes (MCCs) carry a high probability of high winds. During a typical summer, several dozen MCCs affect the central regions of the United States.

3.2.4 Downbursts

Another straight wind phenomena is the downburst. Fujita (1971) defines it as "a strong downdraft inducing an outward burst of damaging winds on or near the ground." Downbursts range in size from one fourth of a square mile (microburst) to 40 or 50 square miles. Wind speeds range from 40 to 200 mph, although the most frequent case is 70 mph or less. The vertically downward force on the roof of a structure produced by a downburst is small, compared to design roof live or snow loads. Hence, only the horizontal component of downburst wind is of concern in building design.

3.2.5 Hurricanes

Hurricanes that ultimately affect the Atlantic or Gulf Coasts of the United States are found in the Atlantic Ocean, the Caribbean Sea or the Gulf of Mexico. Tropical cyclones that form in the Pacific Ocean are called *typhoons*; in Australia they are called *cyclones*. A few *typhoons* cross the west coast of Mexico, but they do not range as far north as the California coast because of the cold ocean currents along the California coast.

A tropical cyclone (which may or may not develop into a hurricane) derives its energy from latent heat available in large quantities over the warm waters in the tropics. A hundred or more hurricane "seedlings" are noted each year. In about ten percent of the disturbances, rain showers release sufficient heat to warm the surrounding environment and cause a surface pressure reduction. Air then begins to accelerate toward the low pressure center. A feedback cycle develops, leading to intensification of the disturbance and circulation (tropical disturbance). Further intensification may lead to formation of a tropical depression (winds less than 30 mph), a tropical storm (winds 31–74 mph), and eventually a hurricane (winds greater than 74 mph). Because of the earth's rotation and the resulting Coriolis effect, the inflowing air follows a counterclockwise spiral toward the low pressure center in the northern hemisphere.

Most hurricanes that affect the Atlantic and Gulf Coasts occur in August, September and October, with the six month period from June through November considered the Atlantic/Gulf Coast hurricane season. Over the last 50 years, the average of about six Atlantic/Gulf Coast hurricanes were reported per year. Of these, only about 25 percent approach close enough to the U.S. coastline to produce hurricane winds over land areas. On the average, only one or two hurricanes reach the coastline per year. However, there are significant deviations from this average. In 1986, six hurricanes crossed the U.S. coastline.

Maximum wind speeds in hurricanes generally occur to the right of the eye (looking along the direction of travel), because of the vector addition of the translation and counterclockwise rotating wind components. The direction of the wind is typically inclined 20°–30° toward the center of the hurricane except near the edge of the wall cloud (towering clouds surrounding the eye) where there is no longer any inflow. The nature of the hurricane wind field is a function of the radius of maximum winds, the central pressure difference, the forward speed, and the geographical location of the storm. The strongest winds occur along the edge of the eye near the wall cloud. Hurricane winds typically reach speeds of 100 to 140 mph (fastest-mile at 33 ft elevation) over open water. Wind speeds of up to 200 mph (e.g. Hurricane Camille, 1969) have been reported in the news media, but these estimated peak gust values correspond to considerably lower design values (see Section 3.3.1). A maximum wind speed of 145 mph was reported by the news media in Hurricane Frederic (1979).

Subsequent evaluation revealed it to be equivalent to a 106 mph design wind speed [Mehta et al., 1983]. Thus, caution is advised in interpreting maximum hurricane wind speeds. Intensities of hurricanes are rated by the National Weather Service according to the Saffir-Simpson Potential of Damage Scale [Saffir, 1974]. The scale places hurricanes into one of five intensity categories ranging from 1 (minimal) to 5 (catastrophic). The scale is used primarily for forecasting.

Hurricane wind speed diminishes rapidly as the storm moves inland. The absence of a heat source from the ocean waters and surface friction is thought to contribute to the rapid decay of wind speed. Normally, by the time a storm has moved 100 miles inland, the winds are well below hurricane force. Hurricane Hugo in 1989 was an exception. As the storm moved across the state of South Carolina, winds were still above hurricane level 200 miles inland at Charlotte, North Carolina. The fact that the storm was intensifying and had an unusually high translational speed (estimated at 25 mph) are factors that contributed to the sustained winds so far inland.

Along the coastline, the storm surge and waves resulting from the combination of low atmospheric pressure in the hurricane vortex and winds pushing water on shore cause as much or more damage than the hurricane winds.

Because hurricane coastline crossings are relatively rare, the records of coastal stations do not adequately reflect the hurricane threat over the relatively short periods of record. Hence, most model codes and standards make special provisions for hurricane risks in the 100-mile zone adjacent to the Atlantic and Gulf Coasts.

3.2.6 Tornadoes

Tornadoes are spawned by severe thunderstorms. The tornado is an atmospheric vortex that extends from cloud base to the ground. The vortex is sometimes narrow and smooth in appearance, but more often is broad and turbulent. A rotating cloud of condensed moisture (funnel cloud) may be visible within the vortex, but the vortex is much larger in dimension and extends well beyond the area of the visible funnel.

Most tornado activity occurs east of the Rocky Mountains. The Texas Panhandle and Central Oklahoma have the highest frequency of tornadoes. Other areas of high incidence are centered in the Midwest (Illinois, Indiana, Ohio and Michigan), and in the Southeast (Mississippi, Georgia and Alabama). Tornadoes have been recorded in every state.

Because tornado wind speeds are virtually impossible to measure with an anemometer, indirect methods are used to estimate maximum tornado wind speeds. Several indirect methods are used:

(1) Structural analysis of damaged structures
(2) Photogrammetric analyses of tornado movies
(3) Analysis of ground marks
(4) Doppler radar

These methods, when applied to the most intense tornadoes observed in recent years, indicate that the maximum tornado wind speeds are in the range of 250–300 mph.

The relative intensities of tornadoes are routinely estimated by the National Weather Service through use of the F-scale [Fujita, 1971]. Classification into one of the six F-scale categories, which range from F0 (weak) to F5 (most intense), is based on appearance of damage. For purposes of risk assessment, those tornadoes that occurred prior to 1971 have been assigned F-scale ratings based on descriptions of the damage.

Table 3.2.1 lists the number of tornadoes by F-scale in the period 1916–1979 [Tecson et al, 1980]. The wind speeds are interpreted as peak gusts at 33 ft above ground. Several important observations can be made from the table. More than half of all tornadoes have a maximum wind speed less than 100 mph. Roughly 85 percent have wind speeds less than 150 mph, while approximately 0.5 percent have wind speeds in excess of 260 mph. The significance of these figures is that while model building codes and standards do not specifically account for the threat of tornadoes, low-rise buildings designed to withstand winds in the 70 to 110 mph range should offer a significant degree of protection against weak tornadoes or near misses by strong ones. Even in regions of high tornado occurrence, the probability of experiencing straight winds in the 70 to 110 mph range is greater than the probability of experiencing the same intensity tornado winds. Thus, ordinary low-rise buildings are not designed to resist tornadoes. Facilities that contain hazardous materials or

TABLE 3.2.1 TORNADO FREQUENCIES BY F-SCALE (1916-1978)

F-scale (Wind Speed Range)*	**Number of Tornadoes**	**Percentage**	**Cumulative Percentage**
F0 (40-72 mph)	6,718	22.9	22.9
F1 (73-112 mph)	8,645	34.7	57.6
F2 (113-157 mph)	7,102	28.5	86.1
F3 (158-206 mph)	2,665	10.7	96.8
F4 (207-260 mph)	673	2.7	99.5
F5 (261-318 mph)	127	0.5	100.0
TOTAL	24,930	100.0	

* Wind speeds are interpreted as peak gusts at 10 m above ground.

processes that pose a threat to people or the environment may require special tornado considerations, which are beyond the scope of this document.

A relatively low pressure exists in the center of a tornado vortex. Prior to 1970, the pressure deficit was thought to be the principal damage mechanism. However, engineering analysis of tornado damage revealed that pressure effects are often overestimated. Wind forces and debris impact ar the main causes of damage.

Waterspouts and dust devils are weak "clear sky" vortices that are generally not of concern relative to damage.

3.3 FUNDAMENTAL WIND LOADING CONCEPTS

Certain fundamental wind loading concepts must be understood before a designer can intelligently determine wind forces for the design of buildings or structures. Background material needed to understand provisions of various standards and model building codes is presented in this section.

The frame of reference in which wind speeds are measured is discussed in this section. Averaging time, terrain roughness and height above ground all affect the magnitude of wind speed measured by an anemometer. Gustiness of the wind produces dynamic effects on structures. However, if properly handled, wind forces may be treated as quasi-static forces for most ordinary structures. Aerodynamic wind pressures develop when the free flow of wind is interrupted by a building or structure.

3.3.1 Frames of Reference for Wind Speeds

The movement of air near the surface of the earth is described in terms of a wind velocity vector having both magnitude and direction. The scalar quantity used to describe wind speed must be defined with respect to averaging time, ground terrain roughness and height above ground, as the wind speed value is a function of these quantities.

Definitions

Wind speed can be described in terms of peak wind gust, mean wind (with a specified averaging time), or fastest-mile wind. Annual extreme, fastest-mile values are used for wind hazard assessment. Each wind speed term has a unique meaning and serves to describe a particular aspect of wind speed:

Peak wind speed is the maximum instantaneous value of the wind speed recorded by an anemometer. Most anemometers have response times of one to two seconds. Hence, peak wind speeds correspond to a one- or two-second gust.

Mean wind speed is the mean value of a wind speed record taken over some time interval. Wind gusts result from fluctuations about a mean value. It is necessary to refer to mean wind speed in terms of an averaging time such as one-minute, ten-minute or one-hour.

Fastest-mile wind speed is defined as the average speed of any one mile of air passing an anemometer during the time period considered. A fastest-mile wind speed of V mph means that a "mile of wind" passed the anemometer in

3,600/v seconds. Thus, the fastest-mile wind speed has a variable averaging time. The National Weather Service and all U.S. codes and standards use fastest-mile wind speed as the reference wind speed.

Annual extreme fastest-mile wind speed is the highest fastest-mile wind recorded at a weather station in a given year. The series of annual extreme fastest-mile winds at a station is used to determine various probabilities of occurrence of wind speeds at that location.

The definition of wind speed has major implications in the determination of wind forces. As illustrated in Figure 3.3.1, the same wind record provides lower wind speed values as the averaging time is increased. National standards from countries around the world use different averaging times. The National Building Code of Canada [National Research Council of Canada (NRCC), 1985] uses mean hourly wind speed, while the British Standard [BSI, 1984] and the Australian Standard [Standards Association of Australia (SAA), 1989] utilize a two-second gust speed.

Inasmuch as wind speed magnitudes are a function of averaging time, there is an obvious need for relationships between wind speeds averaged over various periods of time. This requirement has, to some extent, been met by the work of Durst (1960) and Hollister (1979). On the basis of statistical analyses of open-country wind records from Cardington, England, and Ann Arbor, Michigan, Durst obtained the results summarized in Table 3.3.1. It is important to bear in mind that these are average values which apply to level sites in open country.

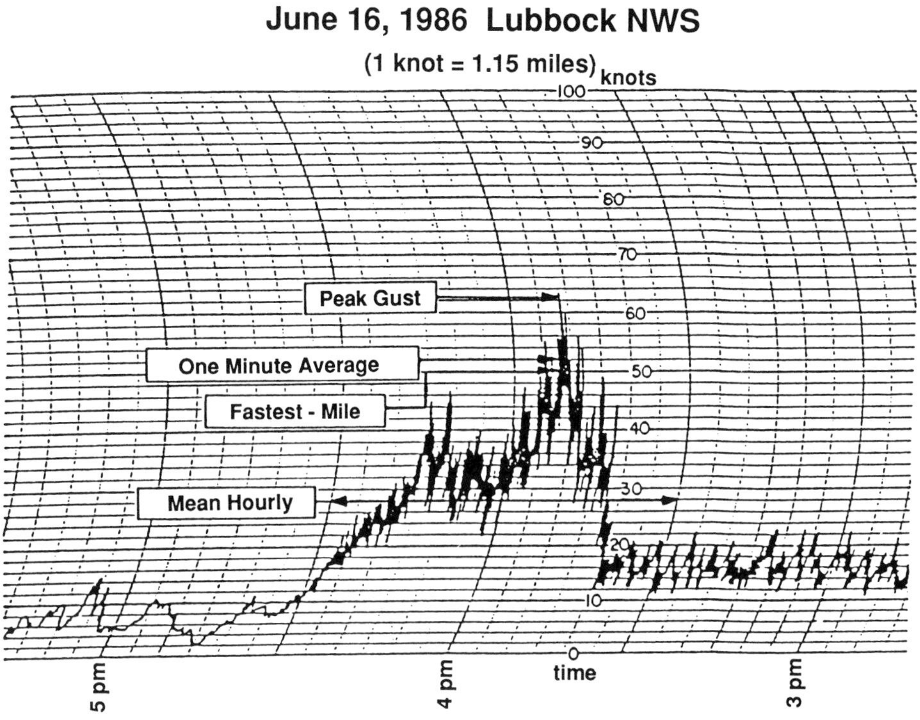

June 16, 1986 Lubbock NWS

(1 knot = 1.15 miles)

Figure 3.3.1 Wind speed recording in a thunderstorm

TABLE 3.3.1 Ratio of Wind Speed for Averaging Time T to Mean Hourly Speed (Durst, 1960)

Averaging Time, T	1 hr	10 min	1 min	30 sec	20 sec	10 sec	2 sec
	1.0	1.06	1.24	1.33	1.36	1.43	1.59

Wind Speed Profiles

Natural and man-made obstructions retard movement of air near the ground surface. This surface roughness causes a reduction of wind speed near the surface within the planetary boundary layer. The height above ground to where the movement of air is no longer affected by ground obstructions is termed the "gradient height," and the associated speed is denoted "gradient wind speed." Above the gradient height, the gradient wind speed is essentially constant. The variation of wind speed with height below gradient height is strongly influenced by terrain roughness.

For engineering purposes, a power law wind speed profile is usually considered acceptable [Davenport, 1960]. The wind speed at any height above ground may be expressed as:

$$V_z = V_g \left(\frac{z}{z_g} \right)^{1/\alpha} \quad (0 \leq z \leq z_g) \tag{3.3.1}$$

where V_z = wind speed at any height, mph
V_g = gradient wind speed, mph
z = height above ground, ft
z_g = gradient height, ft
α = exponential coefficient

The values of gradient height, z_g, and the exponential coefficient, α, depend on the ground surface roughness. Surface roughness is the cumulative drag effect of obstructions to the wind. The roughness is characterized by the spacing, size, and height of buildings, trees, vegetation, rocks, etc. on the ground. Surface roughness is a minimum over large bodies of water and a maximum over centers of large cities. Although the wind profiles for hurricane, thunderstorm and tornado winds can vary significantly from the exponential shape, it is considered the best model for design purposes.

Davenport (1960) examined wind data from 19 different locations and determined the power law exponent a at each location. The variation of α at the different locations was attributed solely to the variation in terrain roughness. The values ranged from 10.5 for coastal waters to 1.6 at the center of a large city. Davenport also found that gradient height, z_g, varied from 885 ft over flat, open country to 2,020 ft over a large city. Typical profiles and associated gradient heights are shown in Figure 3.3.2 for four degrees of surface roughness and a constant gradient wind speed.

To simplify the selection of α and the expression of wind speed as a function of height, four exposure categories are generally accepted by model codes and standards. They account for variations in ground surface roughness that arise from natural topography and vegetation as well as from constructed features. The four exposure categories as defined in ASCE 7-88 (1988) are:

Exposure A: Large city centers with at least 50% of the buildings having a height in excess of 70 ft. Use of this exposure category shall be limited to those areas for which terrain representative of Exposure A prevails in the upwind direction for a distance of at least one-half mile or 10 times the height of the building or structure, whichever is greater.

Exposure B: Urban and suburban areas, wooded areas, or other terrain with numerous closely spaced obstructions having the size of single family dwellings or larger. Use of this exposure category shall be limited to those areas for which terrain representative of Exposure B prevails in the upwind direction a distance of 1,500 ft or 10 times the height of the building or structure, whichever is greater.

Exposure C: Open terrain with scattered obstructions, having heights generally less than 30 ft. This category includes flat, open country and grasslands.

Exposure D: Flat, unobstructed coastal areas directly exposed to wind flowing over a large body of water. This exposure shall be used for those areas representative of Exposure D extending inland from the shoreline a distance of 1,500 ft or 10 times the height of the building or structure, whichever is greater.

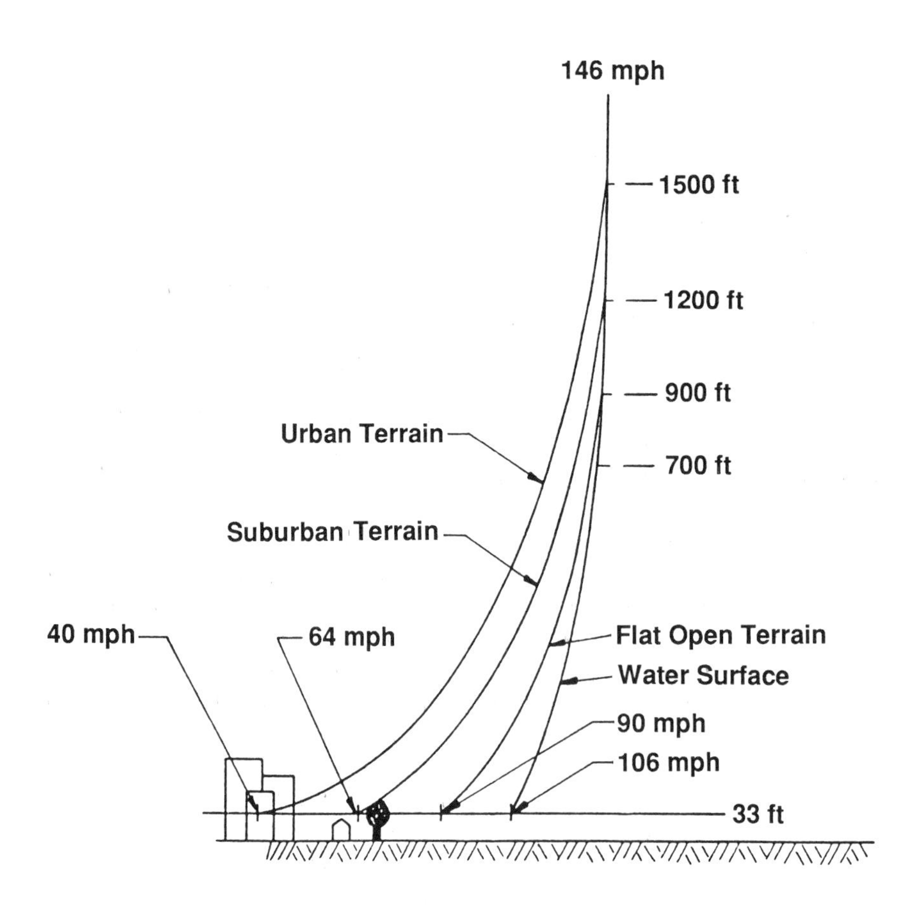

Figure 3.3.2 Typical profiles and gradient heights

3.3.2 Gust Related Effects

Examination of the wind record in Figure 3.3.1 shows that wind speed is continually fluctuating. Wind speed over a given time interval consists of a mean component and a fluctuating component. The mean wind speed (based on, for example, a 10-minute record) increases with height above the ground, but the amplitude of the fluctuating component remains essentially constant. There is, however, a tendency for the amplitude of the fluctuation to be larger near the ground over rough terrain. Turbulence induced by the interaction of moving air with obstacles is referred to as mechanical turbulence. Local turbulence caused by meso-meteorological conditions (e.g. thunderstorms) is called meteorological turbulence.

Since wind speed fluctuates randomly, its fluctuating properties should be considered in statistical terms. Random fluctuations can be shown as a frequency spectrum (see Fig. 3.3.3 as an example). The figure shows that wind speed fluctuates at all frequencies between 0.002 and 0.5 cycles per second (Hz) or periods between 500 to 2 seconds. There is more power (area under the spectrum curve) in the spectrum at a frequency of 0.05 Hz than at a frequency of 0.5 Hz. The power for frequencies larger than 1.0 Hz (periods smaller than 1.0 second) is negligible.

In dynamic analysis of a structure subject to gust loading, significant dynamic amplification of the response occurs at the resonance frequency, i.e. when natural frequencies of vibration of the structure and frequencies of the wind match. If the structure has a frequency of vibration of 0.1 Hz (fundamental period of 10 seconds), there would be significant dynamic amplification of the response, because the fluctuating component of wind has a fair amount of power at that frequency, as is shown in Figure 3.3.3. On the other hand, if the natural frequency of vibration of the structure or its component is higher than 1 Hz (fundamental period of less than one second), the dynamic amplification of the response would be negligible because power in the wind speed spectrum at these frequencies (as shown in Figure 3.3.3) is extremely small. This consideration of natural frequencies justifies application of wind loads as quasi-static loads rather than dynamic loads on most structures, including low-rise buildings. The wind speed spectrum illustrated in Figure 3.3.3 was obtained from wind measured in an open field (Exposure C) in Lubbock, Texas. Its general shape is typical of the winds measured at other open exposure locations.

A structure will not respond fully to the impact of a gust which is only a small fraction of the size of the structure. A gust, to be fully effective, must have sufficient spatial extent to envelop both the structure itself and the flow patterns on the windward and leeward sides, which are responsible for the maximum pressure on the structure. To account for gust-size effects, a correlation function is defined which accounts for the fact the wind gusts are not likely to act simultaneously over the full extent of the structure [Vellozzi et al., 1968]. The gust correlation function varies from unity for complete correlation to zero for no correlation.

Wind pressures on a structure are derived from a mean wind speed plus the effects of fluctuating wind speeds. The response of a structure depends on the mean wind speed, the correlation between gust size and structure size and the correlation between gust frequencies and structural frequencies of vibration. These effects are incorporated in *gust response factors* in standards and modern building codes.

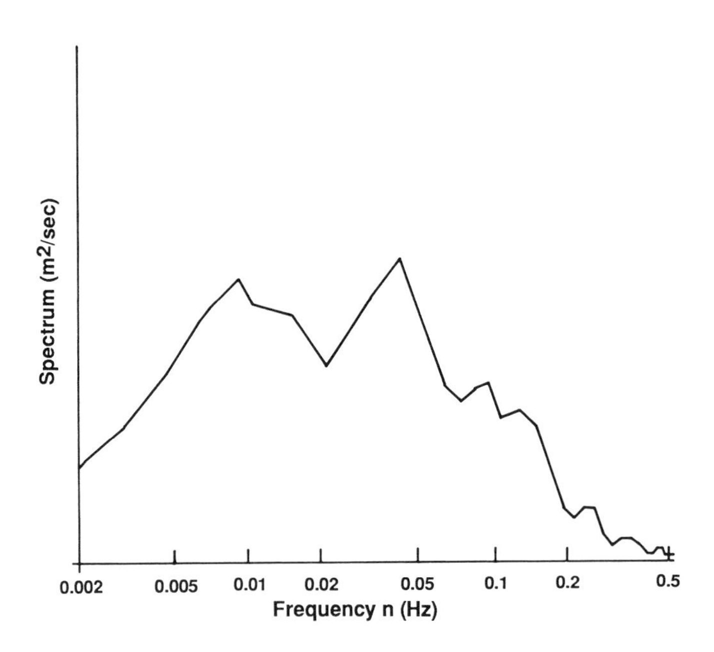

Figure 3.3.3 Typical wind speed spectrum (from the data collected in Lubbock, Texas)

3.3.3 Wind Pressures

When wind approaches and envelops a structure, the direction of air flow is changed and several phenomena are apparent near the surface of the building. The wind phenomena and resulting pressures can differ considerably adjacent to the various building surfaces (windward wall, side walls, leeward wall, roof, wall corners, eaves, roof corners, etc.). Additional wind effects take place if openings occur in the walls or roof of an enclosed building.

Bernoulli Effect

The over-all pattern of air flow produces inward-acting pressure on the windward wall and outward acting pressure on the two side walls, the leeward wall and the roof. Outward pressures relate to acceleration of the air flow as it travels the longer distance around or over the structure, which is the well known *Bernoulli effect.*

The equation which characterizes this fluid flow embodies a complete statement of energy principle, i.e. a balance between kinetic energy and potential energy over every part of a stream line in steady fluid flow. This equation is known as the *Bernoulli theorem* [Rouse, 1938]:

$$0.5\rho V^2 + p + \gamma h = \text{Constant} \tag{3.3.2}$$

where

ρ = mass density of air, slugs/cu ft

V = velocity, ft/sec
p = pressure intensity, psf
γ = specific weight, lb/cu ft

By examining interrelations between the three terms in the equation, we can see that the pressure intensity will vary along a stream line with change in air velocity and with change in elevation. Where air flow over and around a low-rise building is concerned, effects of changes in elevation of a streamline are negligible; hence, it can be seen that regions of accelerated flow (e.g. flow over a roof or along a wall) will be regions of lower pressure, relative to pressure in the free stream in advance of the building. If the building is enclosed and contains air at ambient pressure, p, an upward or outward acting pressure differential will act across the roof and side walls. It may also be observed that a change in velocity to zero along a stream line will cause the pressure intensity to increase by $0.5\rho V^2$, which by definition is the stagnation pressure. Substituting a value for mass density of air at standard conditions (59° F and 29.992 in. of mercury) and factors to convert mph to ft/sec, the expression for stagnation pressure becomes

$$p = 0.000256\rho V^2 \tag{3.3.3}$$

The stagnation pressure has units of psf, if V is expressed in mph.

Flow Separation

The smooth flow of air cannot negotiate the sharp corners at wall corners, eaves and roof corners (flow over "bluff" bodies). As a result, a separation of flow from the boundary surface takes place [(Davenport, 1960]. The point at which separation takes place is a point of stagnation on that stream line which divides the oncoming flow from the reverse flow in the region of discontinuity. The region of discontinuity downstream from the flow separation point is termed the *wake region*. The wake region is characterized by energy dissipation that takes place due to the action of viscosity. In view of this energy loss, it is no longer possible to use the Bernoulli equation to relate pressure and velocity. The pressure in the wake may be expressed as a dimensionless pressure coefficient

$$C_p = \frac{p_w - p_0}{0.5\rho V_0^2} \tag{3.3.4}$$

where p_w = pressure in wake, psf
p_0 = free stream pressure, psf
V_0 = free stream velocity, ft/sec

In the absence of theoretical models for determining the distribution of pressures on building surfaces in the wake region, the structural designer must rely on experimental evidence from wind tunnel studies or full-scale measurements in the field. For very sharp corners and certain angles of wind incidence, C_p values can be very large -- ranging up to 5 or more. Thus, the upward acting pressures at a roof corner can be as much as 5 times the stagnation pressure.

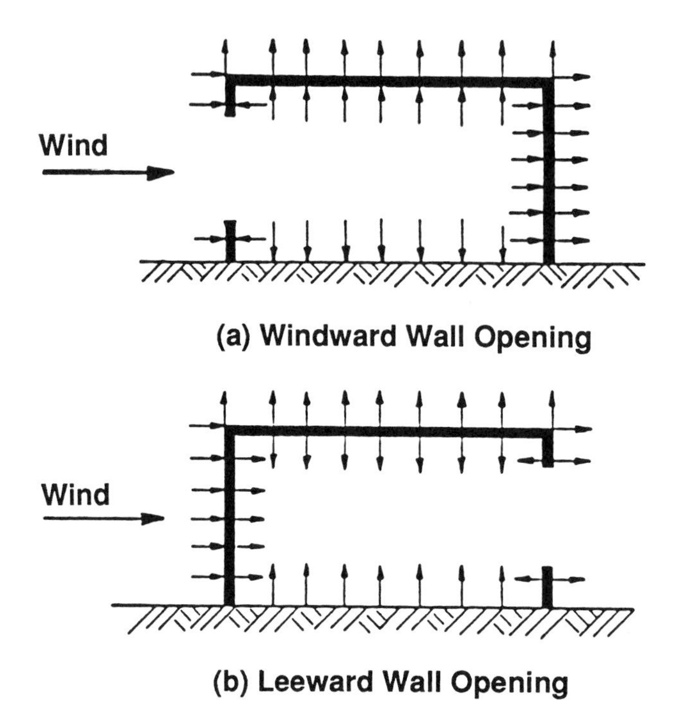

Figure 3.3.4 Effect of combined external and internal pressure

Internal Pressure

In addition to the Bernoulli effect and flow separation phenomena discussed above, additional pressure differences across the roofs and walls develop when openings occur in the structure, either by design or through component (windows, cladding, etc.) failure. Windward wall openings cause an increase in pressure within the building (see Figure 3.3.4a). The increase in internal pressure combines with the outward pressures already acting across the roof, the leeward wall, and the side walls to intensify the net, outward-acting pressures across these surfaces. Conversely, an opening in a side wall or the leeward wall (see Fig. 3.3.4b) causes a decrease in pressure inside the building, because air is drawn out. The decrease in internal pressure combines with the inward-acting pressure across the windward wall to produce a larger inward-acting pressure across this surface. The decrease in internal pressure tends to relieve outward-acting pressures across the roof, side walls and leeward walls.

In view of the above discussion, three important observations can be made: (1) it is clear that wind can produce outward-acting pressures across all building surfaces except the windward surface, (2) the aerodynamically poor character of conventional rectangular buildings results in relatively large, outward-acting pressures near roof corners, eaves and wall corners, and (3) openings in buildings may compound external pressures acting on the building surface.

3.4 WIND-STRUCTURE INTERACTION

Extensive examinations of wind-induced failures support the general observations identified in the previous section. Over-all wind effects tend to push windward walls inward, pull side and leeward walls outward, and lift roofs upward.

Local wind pressures tend to break windows at building corners and peel roofing at eaves and roof corners. Photographs and diagrams presented in this section illustrate wind-structure interaction and depict the corresponding failure modes.

3.4.1 Over-all Wind Effects

Figure 3.4.1 illustrates over-all effects of air flow around a building. Wind pressure acts to lift the roof off this house and to pull the side and leeward walls outward, while pushing the windward wall inward. This mode of failure is not uncommon in residences and other types of low-rise buildings. Engineering analysis often reveals that the roof is lifted first and the walls, being no longer supported at the top, topple inward or outward. Although this particular house was damaged by tornado winds, houses damaged by hurricanes and straight winds exhibit the same failure mode.

The windward wall of the gymnasium in Figure 3.4.2 was blown inward. This failure was the result of inward acting pressure on the large vertical surface presented to the wind. In this case, weak roof-to-wall connections may have been a factor in the failure. The roof probably lifted first, because of local, upward acting pressures along the eaves, the wall was left free-standing between the two vertical control joints, which offered little or no support. Collapse occurred at a wind speed of less than 65 mph.

3.4.2 Local Wind Pressures

The eaves of the building shown in Figure 3.4.3 are severely damaged by the high local pressures created by flow separation at the sharp edge between the wall and roof. Corners of roofs are especially vulnerable to local wind effects. The school building shown in Figure 3.4.4 sustained damage at the roof corner, but no general roof damage. The flow of air approached the building from an angle directed into the corner, which produced large uplift pressures at the roof corner. The local uplift pressures were compounded by the overhang and by the internal pressures resulting from the broken windows. The combined upward pressures were sufficient to uplift the precast double tee roof elements.

3.4.3 Internal Pressures

The effects of a general wind-structure interaction condition can be seen in Figure 3.4.5. The windward wall collapsed inward, the side walls collapsed outward, and the roof was lifted upward. There was evidence that failure of an overhead door in the windward wall allowed pressure to build inside the structure. Thus, outward forces on the walls and roof caused by air flowing over and around the building were combined with internal pressure inside the building. This failure mode is common when large plate glass windows or overhead doors in the windward wall fail or are left open.

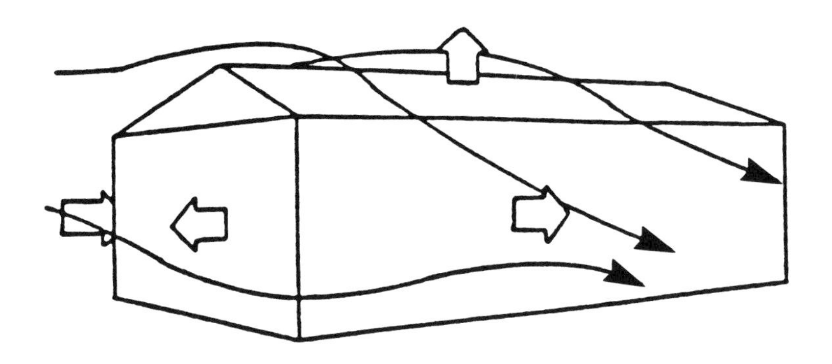

Figure 3.4.1 Over-all effects of an air flow around a building

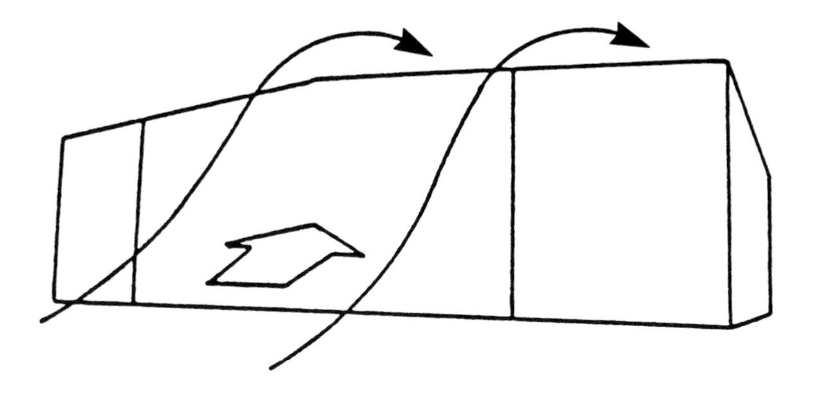

Figure 3.4.2 Failure of a windward wall

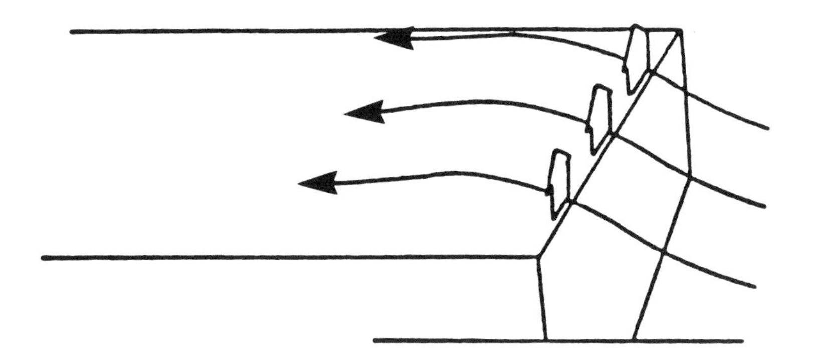

Figure 3.4.3 Local roof damage induced by air flow over eaves

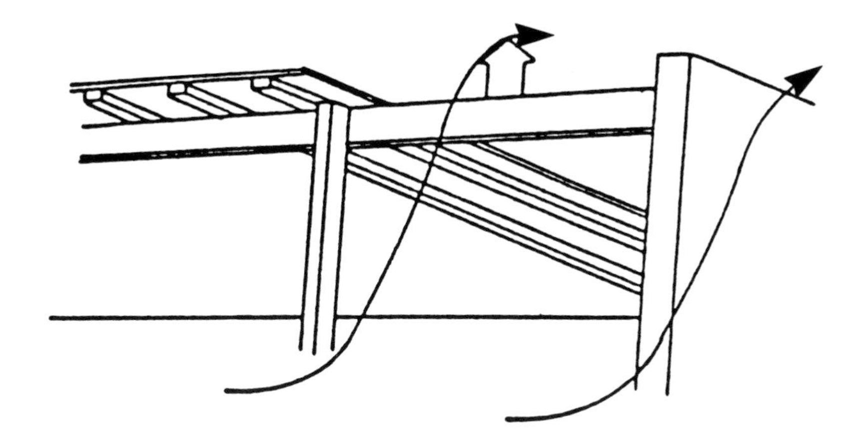

Figure 3.4.4 Damage to corner of roof

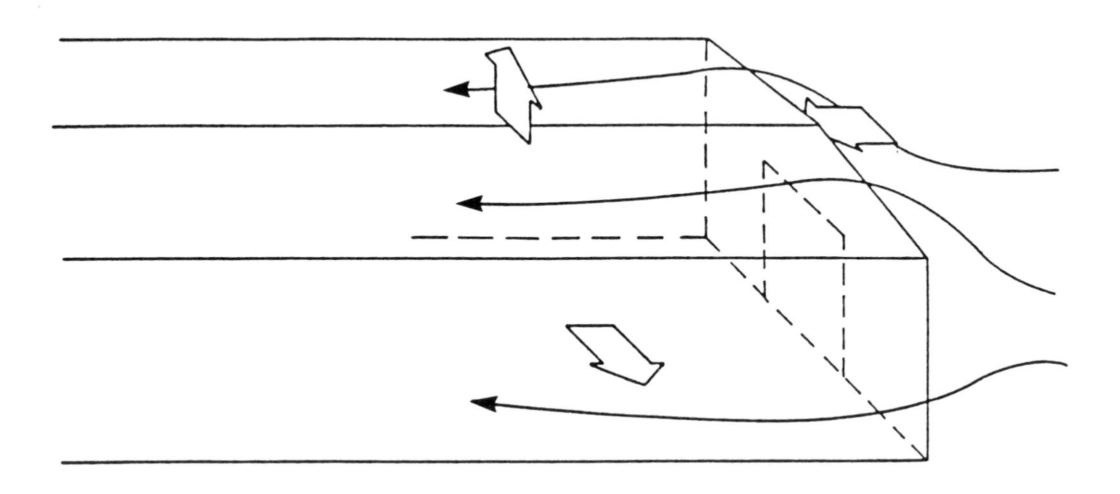

Figure 3.4.5 General effects of air flow which induces failure in windward wall

3.5 WIND LOADING CRITERIA

In this section, the various codes and standards available for establishing wind load criteria are discussed.

3.5.1 Standards

Building practice in the United States is governed by the collective activities of architects, engineers and building officials who are engaged in the design and construction of structures. These collective activities constitute, in effect, the "standard of practice," and building codes and standards in the United States reflect these standards of practice.

Standards reflect current practice, past experience, and research knowledge. Standards are developed by consensus groups (e.g. ASTM, American National Standards Institute (ANSI), by trade associations (e.g., American Institute of Steel Construction (AISC), Metal Building Manufacturers Association (MBMA)), or government agencies (e.g. Department of Housing and Urban Development (DHUD)). Consensus standards carry more prestige than other types of standards, because they are supported by a wider range of interested parties.

3.5.2 U.S. Model Building Codes

A "model" building code is also a consensus document, much like a standard, but is written in language that can be adopted by governmental entities (city, county, state) as a legal document. Model codes include information from many standards and recommended practices.

A jurisdictional building code is a legal document enacted and enforced by a government entity. It comprises a collection of rules and regulations, many of which are adopted from model building codes. When enforced, it provides for safe use and maintenance of buildings and structures. The purpose of a building code is to provide a legal basis for protecting life, health, and public welfare. Building code regulations represent minimum requirements that will assure adequate safety under most conditions. As such, they provide criteria for wind force design.

Three organizations promulgate model building codes in the U.S. The code names, parent organizations and addresses are listed in Table 3.5.1.

TABLE 3.5.1 MAJOR U.S. MODEL BUILDING CODES

Standard Building Code	**1988 Edition**
Southern Building Code Congress International 900 Montclair Road, Birmingham, AL 35213	
Uniform Building Code	**1988 Edition**
International Conference of Building Officials 5360 South Workman Mill Road, Whittier, CA 90601	
Basic/National Building Code	**1987 Edition**
Building Officials and Code Administrators International Inc. 17926 South Halstead St., Homewood, IL 60430	

3.5.3 Other Codes and Standards

Several other codes and standards are available as alternates to the model building codes for determining wind forces.

ASCE 7-88

The ASCE 7-88 is an updated version of ANSI Standard A58.1-1982, Minimum Design Loads for Buildings and Other Structures. Maintenance and supervision of the standard was taken over by the American Society of Civil Engineers in 1986. ASCE 7-88 contains provisions for determining gravity loads, snow loads, rain loads, and earthquake forces in addition to wind forces. The document is a standard rather than a model building code. As such, it reflects the latest research available to the profession. A thorough understanding of the ASCE 7-88 wind load provisions is helpful in interpreting provisions of other model building codes.

In determining wind forces, the ASCE 7-88 standard distinguishes between ordinary buildings, other structures (towers and signs), flexible buildings (tall, narrow buildings), and flexible other structures. A further distinction is made between main wind-force resisting systems and components and cladding. In dealing with components and cladding, a further distinction is made between buildings less than 60 ft. and those taller than 60 ft. Discussion of the ASCE 7-88 standard is limited to those provisions applicable to low-rise buildings.

A five-step procedure is followed to determine wind loads for low-rise buildings:

(1) Define exposure category (based on surrounding terrain) and building occupancy category.

(2) Obtain basic wind speed V from map of wind speed contours for a 50-year mean recurrence interval.

(3) Calculate velocity pressure q, which is a function of basic wind speed, exposure category, height above ground, and building occupancy category.

(4) Use appropriate equations to determine design wind pressures for the main wind-force resisting system.

(5) Use appropriate equations to determine design wind pressures for components and cladding.

Details for accomplishing the above steps and the rationale behind them are presented in Section 3.6. Example problems in Chapter 5 further illustrate application of the ASCE 7-88 standard wind load provisions.

Low-Rise Building Systems Manual

This manual [MBMA, 1986] contains criteria and procedures primarily intended for the design of light-frame, steel low-rise buildings. The wind force criteria reflects results from a series of wind tunnel tests conducted at the University of Western Ontario [Davenport et al., 1977 and 1978].

South Florida Building Code

The South Florida Building Code [SFBC, 1979] has gained recognition in the U.S. as having the most stringent wind force provisions of all U.S. codes. It is a legal instrument in Dade County, Florida (Miami area). The code contains numerous prescriptive requirements to assure that a structure can withstand the prescribed wind forces.

3.5.4 Procedures Beyond Code Provisions

Proper design of structures subject to wind forces sometimes requires the consideration of design situations and the use of design criteria which are not specified in the governing building code or design standard. Good engineering practice requires that the designer understand the intent and limitations of the building codes and standards and how they relate to any specific design problem. The engineer must know when a special procedure, such as a wind tunnel test, is advisable and when serviceability criteria should control design.

The primary goal of all design professionals is to "meet the code." A common belief among designers is that if the design conforms to the appropriate building code, they have met their legal responsibility of proper design. This may well be true for most smaller structures in typical settings. However, there are many structures and design situations for which the code provisions are inadequate and unsafe. This results from localized and other special conditions which cannot be adequately addressed in a usable code of reasonable length. A designer who does not understand or recognize special conditions may produce a design that is not serviceable or is hazardous at certain times. Designers must be able to recognize the need for procedures beyond code requirements.

Three situations that require procedures beyond building code requirements are (1) irregularly shaped buildings, (2) special wind areas, and (3) windborne debris potential. To deal with these problems, the designer may be forced to rely on (1) wind tunnel tests, (2) advice from local officials, (3) other codes and standards, including those from foreign countries, (4) information found in the technical literature, or (5) an expert in wind engineering.

Modern wind tunnel facilities today are capable of modeling terrain and structures to obtain an accurate description of wind pressures and structural response. Because such facilities pay particular attention to modeling the surrounding surface roughness and the wind profile, they are referred to as *boundary layer wind tunnels*. However, the scope of individual low-rise building projects seldom justifies the cost of extensive wind tunnel testing. Today, most wind tunnel testing is limited to high-rise projects.

Local officials may be able to give advice and guidance on the basis of previous experience, especially in areas with unique wind situations such as channeling through a valley or gorge or with unique wind characteristics such as downslope winds.

National codes or standards in Canada, Great Britain and Australia (to mention a few) are sometimes more diverse and provide more in-depth guidance on specific situations than do U.S. codes. They can be very useful if they deal with a specific situation not covered by U.S. codes and standards. The designer must exercise caution in applying provisions of foreign standards to be sure wind speeds are expressed in the proper frame of reference. For example, pressure coefficients from the Canadian code, which are based on mean hourly wind speeds, should not be used directly with fastest-mile wind speeds.

A great deal of technical literature is available to the practitioner in the form of journals, periodicals, proceedings of national and international conferences, commentaries and supplements to foreign codes and standards, design guidelines and textbooks. The list of references and selected bibliography at the end of this chapter contains many publications of this nature.

Irregularly Shaped Structures

Irregularly shaped structures pose a particularly difficult problem for the designer of low-rise buildings. The project may not be large enough to justify the expense of a wind tunnel test. Yet, building codes do not provide pressure coefficients for irregularly shaped structures such as tetrahedrons, truncated pyramids, etc. These shapes are frequently incorporated in modern architecture. Unusual offsets and appurtenances also are not treated in typical codes and standards. The technical literature gives information for some shapes not addressed by building codes.

Even when coefficients can be found for specific shapes, the designer must be cautious. Coefficients from wind tunnel tests that are more than 10 years old likely were not obtained from a boundary layer wind tunnel. Most older tunnels provided non-turbulent smooth flow, which gives significantly different pressure distributions than those obtained from a boundary layer wind tunnel. Similarity of the model and prototype should be verified along with the time frame over which wind speeds are averaged. When no information is available from the technical literature, the only alternative is to take a conservative approach using any relevant trends noted for the wind load provisions of even remotely similar buildings and the designer's intuitive understanding of wind-structure interaction.

Special Wind Areas

Geographical regions affected by downslope winds often merit special attention in determining wind loads. Local areas affected by downslope winds in several areas of the western United States may require advice from local officials or site specific wind hazard assessments. Reliance on wind speed maps [ASCE 7-88] could result in wind forces that do not represent local conditions. The ASCE 7-88 map identifies certain special wind areas, but does not provide advice on how wind forces should be treated in these areas.

Topographical features of an area may require special consideration. Channelling along valleys and gorges may result in higher wind speeds than those estimated from nearby flat open locations. Higher winds are also experienced at and near the tops of hills. Likewise, shielding by hills and promontories may result in lower wind speeds in the wake of these features. In some locations, certain directions may require special consideration of exposure or topographical features. A site located on the shore of a large body of water (e.g. Lake Michigan) will experience different wind characteristics, depending on wind direction.

Windborne Debris

Potential wind generated debris ranges in size from roof gravel to automobiles. From the smallest to the largest pieces of debris, all have potential for causing damage. Flying roof gravel is a significant cause of window glass breakage in windstorms [Minor, 1974]. While windborne debris is usually associated with tornadoes, it can pose a danger in severe windstorms.

The types of debris generated by wind are determined by the availability of potential material in the path of the wind. The four parameters which most significantly define the character of windborne debris are: (1) weight, (2) exposed surface area, (3) aerodynamic drag coefficient, and (4) degree of restraint, if any. In intense tornadoes, items such as steel beams, roof joists and steel pipe may become windborne debris. However, in the vast majority of windstorms, including hurricanes, windborne debris consists of roof gravel, pieces of wood, corrugated sheet metal, and similar lightweight objects.

Debris impact has detrimental structural effects, in addition to the danger of injury or death to people. Perforation of the building envelope allows wind and water to get inside the building. Internal pressure may develop which, in extreme cases, can double pressures acting on building walls and roof.

In view of the above, it seems logical for a designer to choose one of two options in designing cladding in buildings susceptible to impacts from windborne debris: (1) design the cladding system to reject windborne debris without producing openings in the building envelope, or (2) assume that at least some cladding will be perforated and design for internal pressurization.

3.5.5 Recommended Approach for Wind Forces

The jurisdictional building code should always be the starting point in determining wind loads. All requirements of the local code must be met before building permits and occupancy certificates can be issued. The local code likely uses one of the model codes with modifications to meet special requirements or conditions.

It is recommended that whenever local requirements permit, wind resistant designs for low-rise buildings should be based on the provisions of ASCE 7-88, the primary reason being that the provisions attempt to address wind damage modes observed in the field. Detailed descriptions of the ASCE 7-88 wind force provisions are presented in Section 3.6. An understanding of these provisions will assist a designer in interpreting provisions of the model building codes. Application of the ASCE 7-88 Standard provisions are further illustrated by example problems in Chapter 5.

3.6 WIND PRESSURES ON LOW-RISE BUILDINGS USING ASCE 7-88

ASCE 7-88 is recommended for determining wind forces as an alternative to the local building code, if acceptable to the local building official. The ASCE 7-88 wind force provisions reflect the best available data and technical knowledge as of 1988.

In dealing with low-rise buildings, a designer is normally concerned with main wind-force resisting systems or components and cladding for ordinary buildings. With components and cladding, a further concern is if the building is less than or equal to 60 ft or greater than 60 ft. The appropriate equations for design wind pressures are given in ASCE 7-88 Table 4. The general form of the equation for design pressure p in psf is

$$p = q\,G\,C_p - q\,(GC_{pi}) \tag{3.6.1}$$

where

q	= velocity pressure, psf
G	= gust response factor
Cp	= external pressure coefficient
(GC_{pi})	= internal pressure combined with gust response factor

3.6.1 Velocity Pressure

The velocity pressure q (q_z or q_h) is given by a modified form of the stagnation pressure equation (see Equation 3.3.3).

$$q_z = 0.00256\ K_z\,(I\ V)^2 \qquad (3.6.2)$$

where

V = basic wind speed, mph
I = importance factor
K_z = velocity pressure exposure coefficient evaluated at height z

Basic Wind Speed

Basic wind speed values have been established from data collected at 129 National Weather Service stations scattered throughout the United States which have acceptable long term weather records. The wind speed data, which consist of sets of annual extreme fastest-mile wind speeds, were fit to Fisher-Tippett Type I (Gumbel) extreme value distributions. Wind speed contours were drawn for the 50-year mean recurrence interval (0.02 annual exceedance probability) to produce the ASCE 7-88 basic wind speed map (see Fig. 2.4.1). Values on the map are fastest-mile wind speeds for open terrain (Exposure C) at a height of 33 ft above ground. The basic wind speed V used in Equation 3.6.2 is obtained from the ASCE 7-88 map for the geographical location under consideration. Linear interpolation between contour lines is permissible.

Importance Factor

The importance factor accounts for the nature of building occupancy. It adjusts the basic wind speeds for different mean recurrence intervals and also accounts for differences between hurricane and straight wind hazards. The importance factor, in effect, modifies the 50-year mean recurrence interval wind speed values to other mean recurrence interval values.

The probability distribution of hurricane winds obtained through numerical simulation [Batts, 1980] are different from distributions obtained for straight winds at inland stations [Simiu et al, 1979]. In general, higher wind speeds are obtained for hurricane winds than for straight winds. To account for this difference, values of the importance factor greater than one are specified near hurricane prone coastlines. Effects of hurricane winds reduce as the distance from coastline increases. At a distance of 100 miles inland, the effects of hurricane winds are judged to be negligible. Straight line interpolation of the importance factor for locations between coastline and 100 miles inland is acceptable. Values of importance factor are given in ASCE 7-88 Table 5.

Velocity Pressure Exposure Coefficient

The velocity pressure exposure coefficient, K_z, takes into account changes in wind speed with height above ground and with the roughness of the surrounding terrain. Wind speed varies with height above ground; the variation is a function of ground roughness. Four exposure categories (A, B, C, D) are defined in ASCE 7-88. (See section 3.3.1 for descriptions of the four categories). Values of K_z are tabulated in ASCE 7-88 Table 6. Below the height of 15 feet, the value of K_z is assumed to be constant at the 15 ft value because of increased turbulence near the ground surface.

3.6.2 Gust Response Factor

The gust response factor G accounts for the additional loading effects of wind turbulence over the fastest-mile wind and dynamic amplification of structures. For ordinary buildings, the gust response factor is evaluated either at some height z above ground or at mean roof height.

G_z: is used for determining pressures on components and cladding; its value depends on location of component or cladding above ground.

G_h: is used for main wind-force resisting system it has single value determined at mean roof height.

Values of G_z and G_h are tabulated in ASCE 7-88 Table 8.

3.6.3 External Pressure Coefficients

External pressure coefficients specified for enclosed buildings in the ASCE standard are based on wind tunnel tests conducted under turbulent boundary layer conditions. The signs of pressure coefficients should be carefully noted. The sign convention for design pressure p is

+ (plus sign) means pressure acts *toward* the surface
– (minus sign) means pressure acts away from the surface

External pressure coefficients are tabulated in ASCE 7-88 Figures 2 through 4 and Tables 9 through 11. If a plus and minus sign is specified, both positive and negative values should be checked to obtain the controlling design pressure.

In the case of components and cladding, it is impossible to separate the pressure coefficients and the gust response factor in the data obtained from wind tunnel tests. Interaction of wind with building surfaces causes turbulence which is different than the gustiness in the free field wind; the additional turbulence caused by this interaction is of a small scale and is expected to lose its effect on a large tributary area. Thus, combined values of GCp and GCpi are tabulated in ASCE 7-88 for components and cladding. The values are a function of tributary area and location on the building surface. Local pressures at wall corners, eaves, roof corners, and ridges are taken into account.

3.6.4 Determination of Wind Pressures

ASCE 7-88 Table 4 specifies appropriate equations for determination of wind pressures applicable to low-rise buildings.

Main Wind-Force Resisting System

$$p = q\, G_h Cp - q_h\, (GC_{pi}) \tag{3.6.3}$$

q: Use q_z for windward wall at height z above ground
Use q_h for leeward wall and roof evaluated at mean roof height

G_h: Use value determined at mean roof height

C_p: Use values shown in ASCE 7-88 Figure 2

h: Mean roof height, except that eave height may be used for roof slope less than 10 degrees

q_h: Use value determined at mean roof height

GC_{pi}: Use values in ASCE 7-88 Table 9

Components and Cladding: Building Height Less than or Equal to 60 ft

$$P = q_h\,(GC_p) - q_h\,(GC_{pi}) \tag{3.6.4}$$

q_h: Use value determined at mean roof height for terrain Exposure C, regardless of actual terrain roughness. Wind tunnel measurements indicate that for rougher terrain (e.g. Exposure B), the wind speed reduces (lower q_h) but the pressure coefficient increases (higher GC_p), resulting in pressures similar to ones for Exposure C.

GC_p: Use values specified in ASCE 7-88 Figures 3a and 3b. Values depend on location on building surface and tributary area. If tributary area extends into two or more delineated areas on the building surface, a weighted average may be used. When specified, both positive and negative values should be considered in determining critical design pressures.

GC_{pi}: Use values specified in ASCE 7-88 Table 9. Both positive and negative values should be considered in determining critical design pressures.

Components and Cladding: Buildings Between 60 and 90 ft in Height

In the design of components and cladding for buildings having mean roof height between 60 and 90 ft, GC_p values of ASCE 7-88 Figure 3 may be used, provided q is taken as q_h and Exposure C is used for all terrain exposures.

$$p = q_h\,(GC_p) - q_h\,(GC_{pi}) \tag{3.6.5}$$

See ASCE 7-88 Table 4 for components and cladding pressures at heights above 90 ft.

4 Earthquake Resistant Design

4.1 INTRODUCTION

A summary of the historical development of the U.S. and international building codes for earthquake-resistant design is given by Berg (1983). The major thrust to incorporate mandatory requirements for seismic design came after the 1925 Santa Barbara and the 1933 Long Beach earthquakes in California. The insurance industry established earthquake-resistant design standards by which they rated buildings for earthquake insurance coverage. Typically, the earthquake effects are defined in terms of an equivalent lateral force, and a static analysis of the building is performed. In recent years, the building codes have adopted more and more features of the formal dynamic structural analysis, while retaining their original formats. Perhaps the most popular and relatively rigorous building code presently in use in the profession is the Uniform Building Code (UBC, 1991). The seismic design and analysis requirements of the UBC are based on the procedures developed by the Structural Engineers Association of California [SEAOC, 1988]. The developments of the UBC and the SEAOC requirements have been gradual and evolutionary in nature. In a somewhat different approach, the Applied Technology Council, a research and development organ of the SEAOC, undertook a model seismic code development in 1974 under the sponsorship of the National Science Foundation. The effort was coordinated by the National Bureau of Standards. Nearly a hundred scientists and engineers contributed to the project report, popularly known as ATC3-06, or simply as ATC3 (1978). The report was revised in 1985 and again in 1988 by the Building Seismic Safety council (BSSC) for the Federal Emergency Management Agency (FEMA) under the National Earthquake Hazards Reduction Program (NEHRP). The new report is called "NEHRP Recommended Provisions for the Development of Seismic Regulations for New Buildings" (NEHRP, 1988) or, in short, "NEHRP Provisions."

A discussion of the code design methods is presented in this chapter. In general, the use of these codes lead to good building designs. There are certain behavior characteristics that are unique to the low-rise buildings. Additional measures beyond those required by the codes should be taken in those cases. These measures are discussed later in this chapter.

4.2 CHARACTERISTICS OF THE EARTHQUAKE MOTIONS

Earthquakes cause three dimensional random vibratory ground motions at the base of buildings and the buildings respond to these motions. The ground motions consist of horizontal, vertical, and rotational components that occur simultaneously in all directions. The vertical component is not usually considered in design, since the normal design for gravity loads can accommodate the dynamic effects of the vertical component. Also, the rotational component is not usually considered in design, since its effect on typical buildings is not significant when compared to the other components. Consequently, the horizontal components are the main consideration in the design of buildings.

Chapter 2 describes the probabilistic and risk implications associated with earthquake design of low-rise buildings, and recommends seismic ground motion risk maps for use in seismic zoning. These maps give contours of peak acceleration and peak velocity that are used to define the earthquake intensity at the various geographical locations in the United States. This information along with the building site location and site soil type is generally sufficient for the purpose of designing low-rise buildings. An understanding of the probabilistic nature of earthquakes and the implications on structural response is also important and is given in this section.

4.2.1 Ground Motion Parameters

The dynamic response of a given building depends on the characteristics of the ground motion. For example, a near fault ground motion could be characterized by one or two large pulses, while a distant earthquake ground motion could be characterized by a more systematic vibratory motion. Obviously, it is difficult to characterize both of these ground motions by the same parameters, and the response of a low-rise building to these ground motions will be quite different. Therefore, it is highly desirable to have a probabilistic and a quantitative description of the ground motion that might occur at the site of the building during a major earthquake. Unfortunately, there is no one description that fits all the ground motions that might occur at any particular site. The characteristics of the ground motion are dependent on several parameters; namely,

- The magnitude of the source earthquake (i.e., energy released).
- Distance from the source of the earthquake (vertical depth as well as horizontal distance).
- Distance from the surface faulting (this may or may not be the same as the horizontal distance from the source).
- The nature of the geological formations between the source of the earthquake and the building.
- The fault mechanism such as compressional or extensional (thrust, normal or strike-slip faulting).
- The nature of the soil in the vicinity of the building site (e.g., hard rock or alluvium).

Although a fully accurate prediction of site ground motion parameters is not possible, the art of ground motion prediction has progressed in recent years such that probable intensity criteria have been established in areas where historical earthquake records and geological information are available.

4.2.2 Earthquake Intensity Maps

Algermissen et al. (1982, 1987) have compiled and mapped seismic intensity parameters in the 48 contiguous states. These intensity parameters used in these maps are maximum acceleration and maximum velocity in rock for three exposure periods or average return periods. The intensities mapped are predicted as a 90 percent probability of non-exceedence (or a 10% probability of exceedence) in 10, 50, and 250 year periods. The Algermissen et. al. maps update those used for development of current seismic design provisions (such as ATC3, 1978 and NEHRP, 1988).

Although these maps represent the state of the art, the intensity contours and intensity values are subject to change. In this regard, it should be noted that it is common practice to use tentative documents for building design codes, since the definition of natural hazards such as wind and earthquake intensities are dependent on accumulated historical data. Codes such as the UBC are revised annually and reprinted in their entirety every three years. Chapter 3 of the Tri-Services manual (Army, 1986) also includes a comprehensive presentation of the present state of the art of prediction of ground motion at a specific site.

Seismic zoning is described in at least two ways, by areas of constant intensity enclosed by boundaries and by contours of constant seismic intensity. A Seismic zone that is defined by boundaries implies a step function of seismic intensity across the boundaries, and obviously there is a variation of intensity across these boundaries. The value of seismic intensity that is assigned to that seismic zone is the maximum intensity that is probable in that zone and is only appropriate at a boundary with a higher intensity.

All other sites in that zone have lesser intensities that grade to the intensity at the next lower zone at the other boundaries. Clearly, the use of intensity contours as presented in Chapter 2 is more rational. When a seismic zone contour does not contain a higher contour value within the closed contour, the intensity values within the closed contour are considered to be equal to the contour value. When the contour is parallel to the coast line and no lesser value appears in the coastal zone, special seismic studies should be made to determine the minimum intensity values at the coast line.

Comparison of the seismic mapping presented in Chapter 2 (based on Algermissen et al., 1987) with those in other codes [UBC, 1991 and NEHRP, 1988] indicate very substantial differences in intensity values. The commentary of NEHRP (1988) states that there may be locations inside of the highest acceleration contour where higher values may be appropriate, but contouring such small areas would amount to microzoning. Recorded ground motions are the basis of the values presented by Algermissen et. al. (1987) but observation of building damage also supports the professional opinion given in the NEHRP commentary (1988) that buildings designed by recent codes for lesser intensity values generally provide adequate earthquake resistance when exposed to larger motions.

4.2.3 Implications on Building Response

Recent studies of the relationship of building damage to the ratio of velocity to acceleration (V/A ratio) of ground motion [Heidebrecht et al, 1988; Anderson and Naeim, 1987; Kariotis and El-Mustapha, 1989] indicate that this ratio has a significant effect on the possible damage to and the stability of some yielding structures. Studies of nonlinear stiffness degrading structural elements [Kariotis and El-Mustapha, 1989] with pinched hysteretic behavior indicate that initially stiff structures, such as reinforced shear walls, have effective fundamental periods of about six times the initial fundamental period. Figure 4.2.1 illustrates that the structural stiffness of the shear walls on reloading and unloading cycles is very small. Hysteretic behavior of tension only braced frames, chevron braced frames, and concentric braced frames is also pinched due to tensile elongation of the brace members. The pinching of tension only bracing systems is quite pronounced and the use of such systems should only be considered when wind loading (rather than

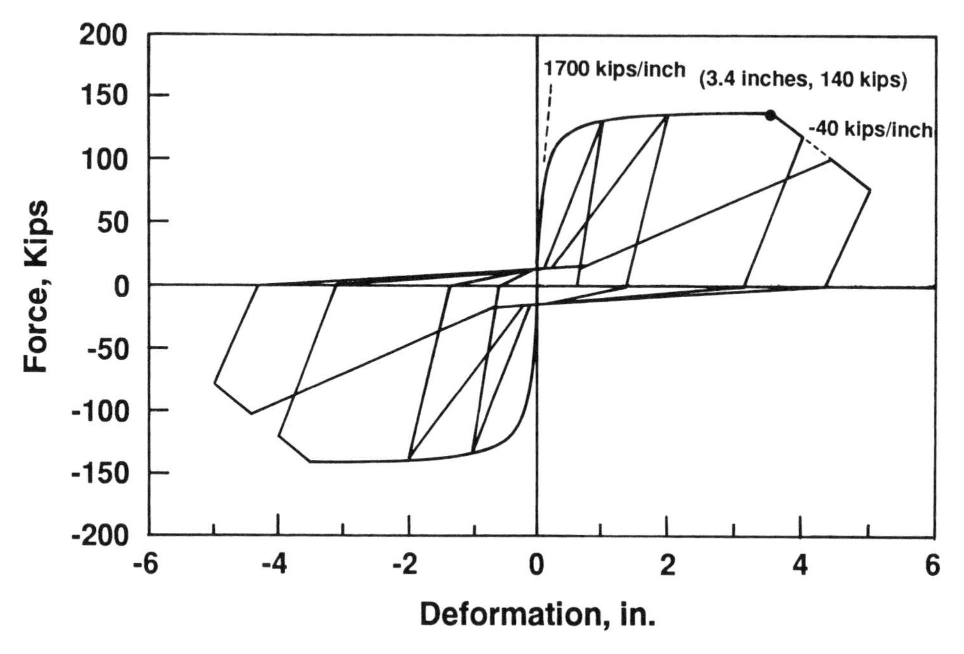

Figure 4.2.1 Force-deformation loops for a shear wall

earthquake) controls the magnitude of the required strength and the detailing standards.

The V/A ratios in surface soils may be up to 2½ times the V/A ratios that are anticipated in rock (Seed and Idriss, 1982). Use of these V/A relationships and the recommendations for the construction of ground surface spectra given by Idriss (1985) indicate that the V/A ratios of the code specified standard response spectra are generally less than the ratios recommended by the referenced techniques and the ratios that have been recorded in the United States.

The study of stiffness degrading systems [Kariotis and El-Mustapha, 1989] indicates that velocity is a better measure of damage than acceleration is. Figures 4.2.2(a) and (b) are taken from this study, and show the spectral accelerations for one of the components of four ground motions that were recorded in an array within 14 km of the fault in the Imperial Valley in 1979. The "Zero Period Acceleration" (ZPA) of these records of ground motion have been linearly scaled to about 0.37g. These figures show that the spectral amplification values of these ground motions have a wide variation in the 0.5 to 1.0 second period range. The scaled velocities of the pairs of ground motion vary by less than five percent, while the spectral acceleration values in the 0.5 to 1.5 second period range vary by more than a factor of two. The non-linear relative displacement of the top of a shear wall shaken by these ground motions will vary by factors of three to four. This study also suggests that the description of the ground surface seismic intensities should be by spectra that are appropriate for the source magnitude, surface soils, distance to the fault system, and the randomness of energy content that is attributed to arrival time of energy at the site.

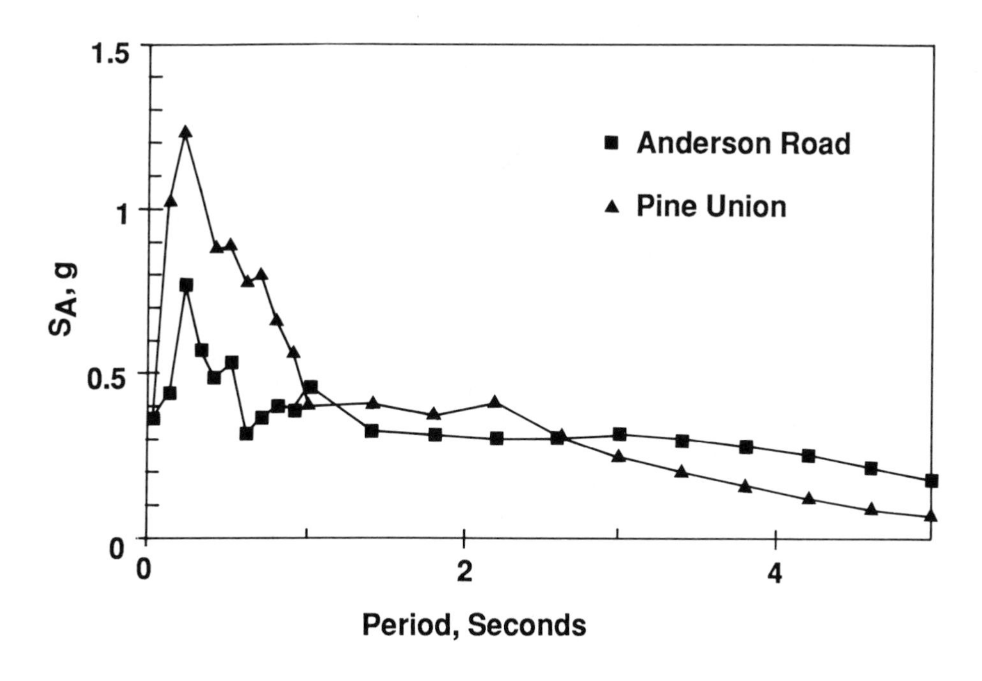

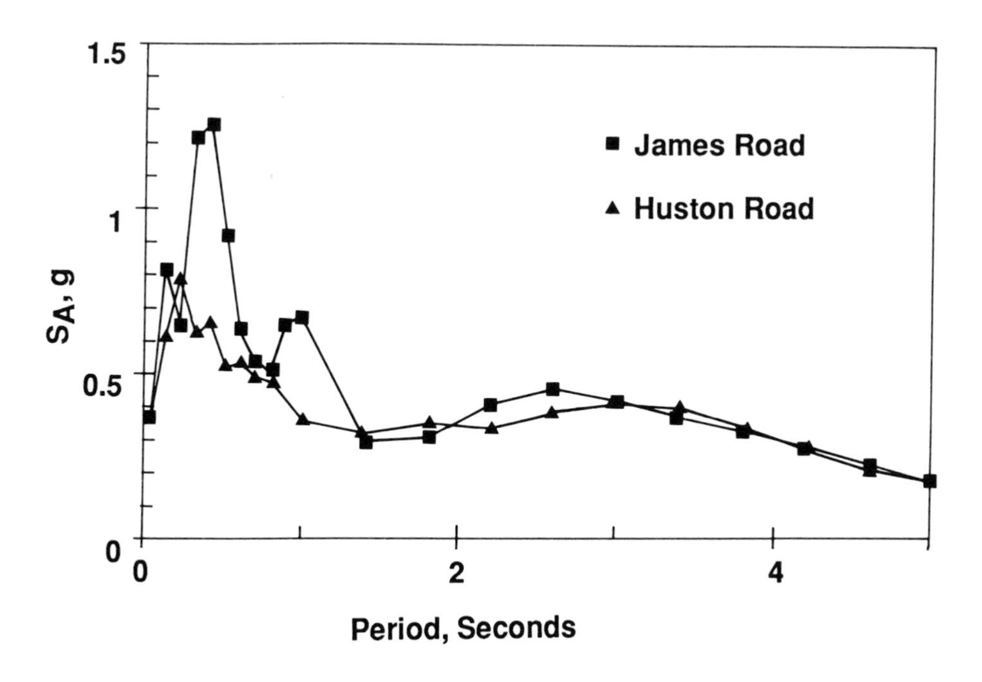

Figure 4.2.2 Comparisons of spectral accelerations for shear walls

4.3 EARTHQUAKE RESPONSE SPECTRUM

As is explained in section 4.4, the earthquake response of a multi-degree-of-freedom, (MDOF) system can be represented as a combination of the responses from its various modes of vibration. In turn, the response in any mode of vibration can be determined in terms of the response of an equivalent single-degree-of-freedom (SDOF) system. In the earthquake response calculations, therefore, it is important to study the behavior of SDOF systems [Chopra, 1981].

For design purposes, it is important to calculate the force in the elastic resistance member (spring) of the SDOF system. This force can be easily calculated by multiplying the spring stiffness (k) by the relative displacement (u) of the mass (m). A quantity that is of special interest is the maximum value of the relative displacement. For a given earthquake motion history one can obtain the maximum value of the relative displacement for various SDOF systems with varying natural frequencies (f, Hertz; or ω, radian/sec) or natural periods ($1/f = T$, sec.) for any damping ratio ζ (i.e. the ratio of the actual damping to the critical value). A plot between the maximum relative displacement as a function of natural frequency or period is called displacement response spectrum. Different damping ratios will give different displacement response spectrum curves. Any ordinate of the response spectrum curve can be denoted by S_D (T, ζ) or simply by S_D. The spectral values can also be represented in velocity and acceleration units: Spectral velocity, $S_v = \omega S_D$, spectral acceleration $S_A = \omega^2 S_D$. The spectral velocity so calculated is not a real physical quantity; therefore, it is called pseudo velocity. It can be shown that in the intermediate frequency range pseudo velocity is quite close to the actual relative velocity of the SDOF mass. Similarly, the spectral acceleration is also a pseudo acceleration. However, the pseudo acceleration is numerically indistinguishable from the actual total acceleration of the SDOF mass. The maximum spring force, $f_s = kS_D = mS_A$.

Figure 4.3.1 shows the ground acceleration history of the north-south component of the 1940 El Centro earthquake, along with the integrated ground velocity and displacement histories. The maximum ground acceleration, velocity and displacements are 0.35 g., 13.2 in./sec., and 4.9 in. respectively. Figure 4.3.2 shows a tripartite response spectra, drawn on a logarithmic scale for several damping ratios. The vertical axis in the figure represents spectral velocity. The two $\pm 45°$ lines represent spectral displacement and acceleration, respectively. We observe in the figure that $S_D = 4.9$ in. in the high period range is the same as the maximum ground displacement; and $S_A = 0.35$ g. in the low period range is equal to the maximum ground acceleration. The high period is characterized by relatively flexible springs or low stiffness. In such a situation, the spring remains practically stationary, thus making the maximum relative displacement of the mass equal to the maximum ground displacement. On the other hand, the spring stiffness of a low period SDOF is high, thus making the mass move practically with the ground with very little relative displacement. The total acceleration in the mass in this situation is almost equal to the ground acceleration, the maximum value of the former being indistinguishable from the maximum pseudo acceleration or S_A.

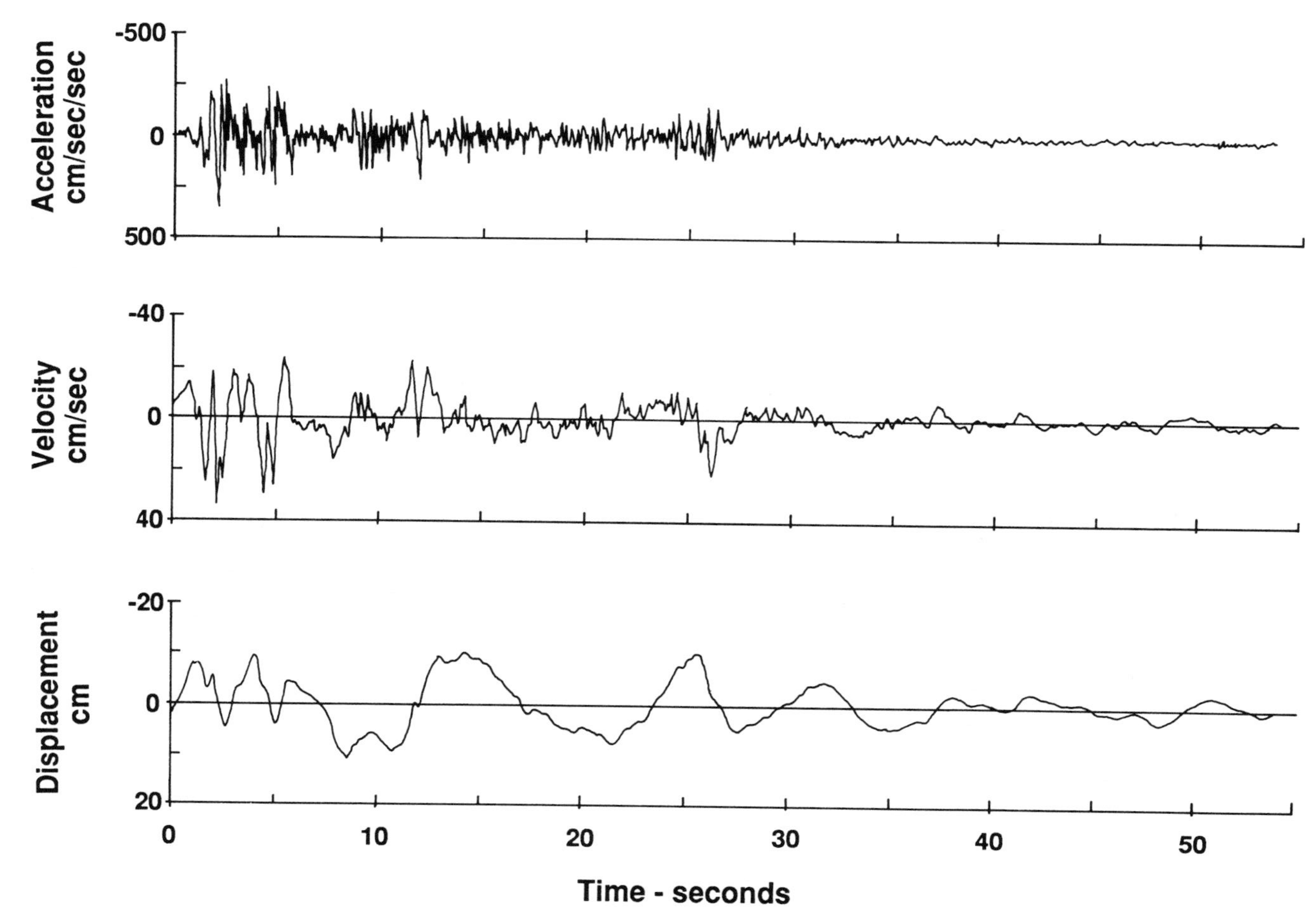

Figure 4.3.1 Ground acceleration and integrated ground velocity and displacement curves for a typical earthquake [Hudson, 1979]

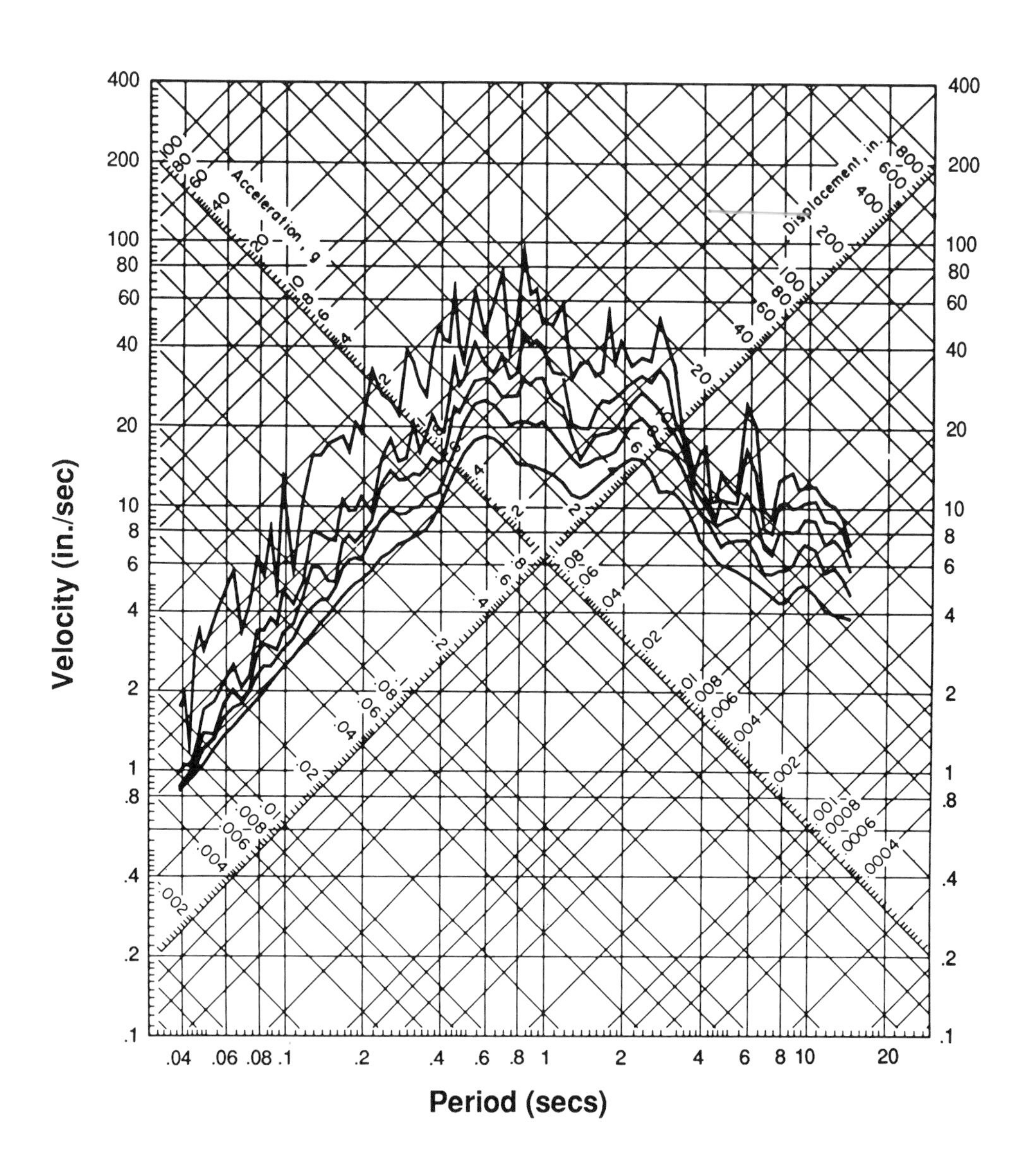

Figure 4.3.2 Typical earthquake response spectrum curves - tripartite logarithmic plot [Hudson, 1979]

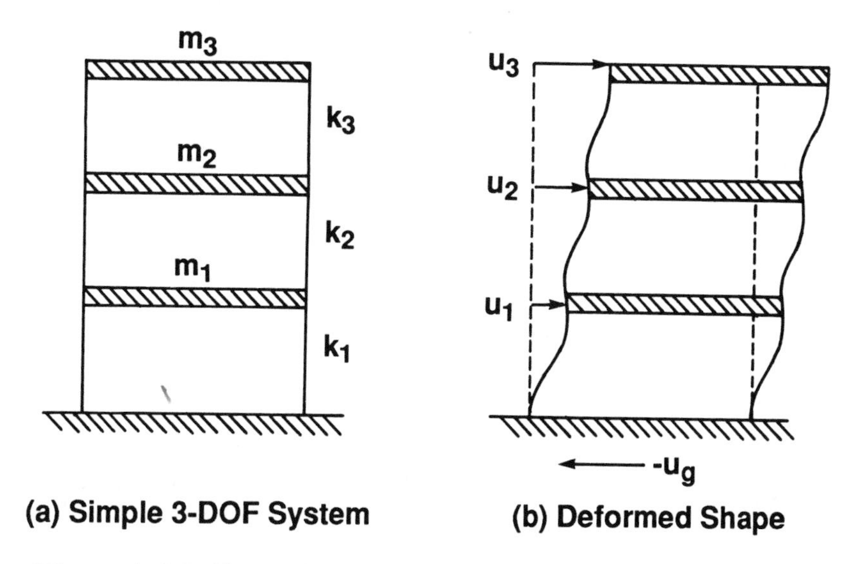

Figure 4.4.1 Example of a multi-degree of freedom system

4.4 STORY FORCES IN A BUILDING

Figure 4.4.1 shows a 3-degree-of-freedom (3-DOF) building which is a simple example of MDOF systems. The relative displacement vector of this building is denoted by $\mathbf{U} = [u_1\ u_2\ u_3]$. We assume here that **U** denotes the relative displacement vector of any MDOF building. The maximum relative displacement vector, $\mathbf{U}_i$, in mode i is given by [Chopra, 1981; Clough and Penzien, 1975]:

$$\mathbf{U}_i = \gamma_i \phi_i S_{Di} = \gamma_i \phi_i S_{Ai}/\omega_i^2 \qquad \gamma_i = \frac{\boldsymbol{M}^T \phi_i I}{\phi^T \boldsymbol{M} \phi_i} \tag{4.4.1}$$

in which ϕ_i is the ith modal vector; $\mathbf{S}_{Di}$ and $\mathbf{S}_{Ai}$, the spectral displacements and accelerations at the circular frequency ω_i ($= 2\pi\, \mathbf{f}_i$), γ_i the modal participation factor, **M** the mass matrix of the building assumed to be diagonal, **I** a vector consisting of unity terms for each horizontal degree of freedom in the direction of the earthquake.

The pseudo-static force vector, $\mathbf{F}_i$, that will cause the displacements $\mathbf{U}_i$ can be calculated from

$$\mathbf{F}_i = KU_i = K\gamma_i \phi_i S_{Ai}/\omega_i^2 = \gamma_i \boldsymbol{M} \phi_i S_{Ai} \tag{4.4.2}$$

where **K** is the building stiffness matrix, and the modal vector property $\mathbf{K}\phi_i = \omega_i^2 \mathbf{M}\phi_i$ has been used. The product $\gamma_i\phi_i$ is dimensionless, and the term S_{Ai} in the above equation has the dimension of acceleration. Therefore, we can look upon the expression $\gamma_i\phi_i S_{Ai}$ as a pseudo-acceleration vector. Thus, the pseudo-static force vector $\mathbf{F}_i$ is expressed here as a product of the mass matrix and the pseudo-acceleration vector. Consequently, $\mathbf{F}_i$ is commonly called the (pseudo-) inertia force vector.

The inertia force for the rth story of the building is $F_{lr} = \gamma_l \frac{W_r}{g}\phi_{lr}S_{Al}$, in which w_r is the weight lumped at the rth story in the building model. The total base shear force, V_l, can be calculated by adding all the story inertia forces. Equations 4.4.1 and 4.4.2 give

$$V_l = C_{sl}W_l \,, \qquad C_{sl} = \frac{S_{Al}}{g} \,,$$

$$W_l = \frac{\sum_r (W_r\phi_{lr})^2}{\sum_r W_r(\phi_{lr})^2} \,, \quad F_{lr} = \frac{W_r\phi_{lr}}{\sum_s W_s\phi_{ls}}V_l \tag{4.4.3}$$

The term C_{sl} is called the seismic coefficient, and W_l is modal weight. The above equations are the basis of the response spectrum analysis of buildings and are included in the NEHRP provisions.

We shall now present the equivalent lateral force (ELF) procedure recommended both by NEHRP and UBC for buildings that are essentially regular in plan and elevation in terms of mass and stiffness distributions. Most of the seismic response of such buildings is derived from the fundamental mode of vibration, which is the only mode considered in the ELF procedure. The fundamental mode shape of a fixed-base, low-rise building can be approximated by a straight line, or $\phi_{lr} \propto h_r$, where h_r is the story height above the base of the building. The force distribution and the total base shear force are given by

$$F_r = \frac{W_r h_r}{\sum_r W_r h_r}V \,, \quad V = C_s W \,, \quad C_s = \frac{S_A}{g} \tag{4.4.4}$$

The fundamental mode in taller buildings may not be a straight line as assumed above. Variations of the above equations are recommended in NEHRP for $T > 0.5$ seconds, and in UBC for $T > 0.7$ seconds, where T is the fundamental period of the building. In NEHRP, the fundamental mode shape is taken as $\phi_{lr} \propto h_r^n$ where n is an exponent whose value varies between 1.0 and 2.0 for $0.5 \leq T \leq 2.5$ seconds. In the case of UBC, an additional concentrated force of $F_t = 0.07TV$ is applied to the top of the structure if T is greater than 0.7 seconds.

4.5 DESIGN SPECTRUM AND SEISMIC COEFFICIENT

The seismic design coefficient, $C_s = S_A/g$, used in Equation 4.4.4 is a non-dimensionalized representation of the spectral acceleration. The coefficient given by NEHRP and UBC can be evaluated from a Newmark-type design spectrum [Newmark and Hall, 1982]. Figure 4.5.1 schematically shows a part of a design spectrum on logarithmic scales. The part AB is the amplified acceleration region with a constant acceleration S_a. The spectral acceleration diminishes and the spectral velocity remains constant (S_v) along BC. The equivalent lateral load procedure in NEHRP and UBC gives seismic coefficients for AB and BC regions that will be presented here.

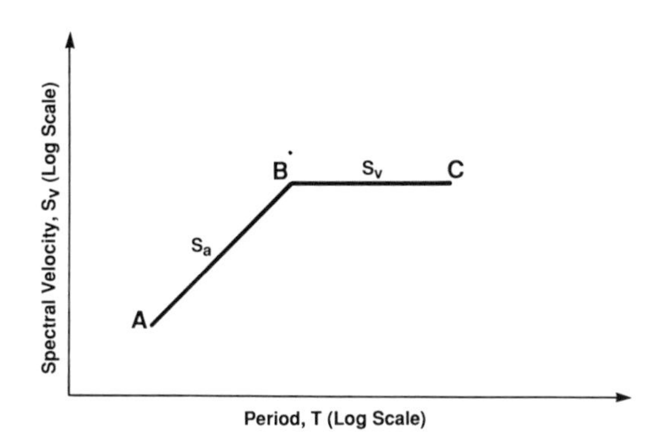

Figure 4.5.1 Schematic plot of a response spectrum.

4.5.1 NEHRP Recommendations

The original ATC3 recommendations and also those included in the NEHRP provisions use an acceleration coefficient (A_a) for AB and a velocity-related acceleration coefficient (A_v) for BC part of the design spectrum. The value of A_v was obtained indirectly from the acceleration data. Recently, USGS has published contour maps with maximum ground accelerations (a) and velocities (v) attenuated independently (see Chapter 2). The 1988 NEHRP provisions give equations for seismic design coefficients (C_s) that can be calculated directly from the values of a and v obtained from the maps. These are based on study performed by Wu and Hanson (1987). Using acceleration and velocity amplification factors of 2.5 and 2.2, respectively, the spectral region ABC in Figure 4.5.1 can be described by

$$S_A = \frac{(2.2v)2\pi}{T} \leq 2.5a$$

where S_A, v and a have the same length unit. When S_A is in cm/s^2, v in cm/s and a in g (981 cm/s^2) units, we can rewrite the above equation as

$$S_A = \frac{0.014v}{T}g \leq 2.5ag \tag{4.5.1}$$

NEHRP has modified the above spectral acceleration equation as follows:

1. In Equation 4.5.1 the spectral acceleration (therefore, the inertia force a building is designed for) is inversely proportional to the period of the building beyond a certain period, point B in Figure 4.5.1. The taller a building is, the longer is its period, and the lower the spectral acceleration. Since most low-rise buildings would not fall into that category, we can briefly state that the NEHRP (and UBC) recommendations have changed the term T in the denominator to $T^{2/3}$ to increase conservatism in the design of tall buildings and account for various uncertainties.

2. NEHRP considers four types of soil profiles that are described in Table 4.5.1. To account for the effect of soil profile on the amplification factor, a coefficient S is introduced in the numerator of the S_A expression for the amplified velocity region (BC in Figure 4.5.1). For the soil profiles S_3 and S_4 the amplification factor for the amplified acceleration region (AB in Figure 4.5.1) is reduced from 2.5 to 2 for relatively stronger earthquakes (a $\geq$ 0.3) to be consistent with the observed phenomenon.

TABLE 4.5.1 Soil Profiles and Coefficient (NEHRP)

Type	**Description**	**S Factor**
S_1	1. Rock of any characteristic, either shale-like or crystalline in nature. Such material may be characterized by a shear wave velocity greater than 2,500 feet per second or by other appropriate means of classification, or 2. Stiff soil conditions where the soil depth is less than 200 feet and the soil types overlying rock are stable deposits of sands, gravels, or stiff clays.	1.0
S_2	Soil Profile Type S_2 is a profile with deep cohesionless or stiff clay conditions, including sites where the soil depth exceeds 200 feet and the soil types overlying rock are stable deposits of sands, gravels, or stiff clays.	1.2
S_3	Soil Profile Types S_3 is a profile with soft- to medium-stiff clays and sands, characterized by 30 feet or more of soft- to medium-stiff clays with or without intervening layers of sand or other cohesionless soils.	1.5
S_4	Soil Profile Type S_4 is a profile with more than 70 feet of soft clays or silts characterized by a shear wave velocity less than 400 feet per second.	2.0

3. To account for inelastic deformation, Veletsos and Newmark (1960) have suggested that the spectral quantities in the BC region (Figure 4.5.1) should be divided by the ductility ratio μ and those in the AB regions by $\sqrt{2\mu - 1}$. NEHRP has replaced both the μ and $\sqrt{2\mu - 1}$ factors by a single response modification factor R, that is given in Table 4.5.2. Noting that the seismic coefficient $C_s = S_A/g$, and incorporating the three items listed above, we get from Equation 4.5.1

$$C_S = \frac{0.014vS}{T^{2/3}} \leq \frac{2.5a}{R} \qquad (4.5.2)$$

$$\text{or} \leq \frac{2a}{R} \text{ for soil profiles } S_3 \text{ and } S_4 \text{ when } a \geq 0.3$$

TABLE 4.5.2 Response Coefficients

BEARING WALL SYSTEM			
	NEHRP		**UBC**
Seismic resisting system	**R**	C_d	R_w
Light framed wall with plywood shear panels, up to 3 stories	6 1/2	4	8
Reinforced concrete shear walls	4 1/2	4	6
Reinforced masonry shear walls	3 1/2	3	6
Concentrically braced frames	4	3 1/2	4
Unreinforced masonry shear walls	1 1/4	1 1/4	--
BUILDING FRAME SYSTEM			
Seismic resisting system			
Eccentrically braced frames, moment resisting connections at columns away from link	8	4	10
Light framed walls with plywood shear panels, up to 3 stories	7	4 1/2	9
Concentrically braced frames	5	4 1/2	8
Reinforced concrete shear walls	5 1/2	5	8
Reinforced masonry shear walls	4 1/2	4	8
Unreinforced masonry shear walls	1 1/2	1 1/2	--
MOMENT RESISTING FRAME SYSTEM			
Seismic resisting system			
Special moment resisting space frame (SMRSF), steel	8	5 1/2	12
SMRSF, concrete	8	5 1/2	12
Intermediate moment resisting space frames (IMRSF), concrete	4	3 1/2	8
Ordinary moment resisting space frames (OMRSF), steel	4 1/2	4	6
OMRSF, concrete	2	2	5

DUAL SYSTEM WITH A SPECIAL MOMENT FRAME CAPABLE OF RESISTING AT LEAST 25 PERCENT OF THE PRESCRIBED SEISMIC FORCES			
Complementary seismic resisting system			
Eccentrically braced frames, moment resisting connections at columns away from link	8	4	12
Concentrically braced frames- steel with steel concrete with concrete	6 6	5 5	10 9
Reinforced concrete shear walls	8	6 1/2	12
Reinforced masonry shear walls	6 1/2	5 1/2	8
Wood sheathed shear panels	8	5	--
DUAL SYSTEM WITH AN INTERMEDIATE MOMENT FRAME OF REINFORCED CONCRETE OR AN ORDINARY MOMENT FRAME OF STEEL CAPABLE OF RESISTING AT LEAST 25 PERCENT OF THE PRESCRIBED SEISMIC FORCES			
Complementary seismic resisting system			
Concentrically braced frames, concrete with concrete	5	4 1/2	6
Reinforced concrete shear walls	6	5	9
Reinforced masonry shear walls	5	4 1/2	7
Wood sheathed shear panels	7	4 1/2	--

4.5.2 UBC Recommendations

The forms of UBC and NEHRP equations are slightly different. The base shear equation (Equation 4.4.4) is written in the NEHRP format. Therefore, to be consistent, we shall adapt the UBC equations to the NEHRP format. The modified UBC equation for the seismic coefficient is given by

$$C_S = \frac{1.25\,ZIS}{T^{2/3}R_w} \le \frac{2.75\,ZI}{R_w} \qquad (4.5.3)$$

in which Z is the seismic zone factor, specified for zones 1, 2A, 2B, 3 and 4 as 0.075, 0.15, 0.20, 0.30 and 0.40 respectively; I, an importance factor that has a value of 1.0 for standard and special occupancy buildings, and 1.25 for hazardous and essential facilities; S, the site coefficient similar to the S-coefficient given in Table 4.5.1 for NEHRP with some differences with site types S_3 and S_4; and R_w, a response modification coefficient given in Table 4.5.2 along with R-values for the NEHRP. The 1.25ZI expression in the UBC equation replaces the 0.014v expression in the NEHRP equation, and the 2.75ZI term in UBC is equivalent to 2.5a or 2a term in the NEHRP provisions. The NEHRP provisions are to be used in conjunction with the ultimate strength method, and UBC with the working stress method.

4.5.3 Building Period and Frequency

As shown in the previous sections, the seismic coefficient (design spectrum) is a function of the building period. Most low-rise buildings would have low periods for which the seismic coefficients become independent of the period. When necessary, a reasonably accurate elastic period can be calculated using Rayleigh's method. Let ϕ represent an approximately estimated fundamental modal vector. The maximum strain energy associated with the mode is $\frac{1}{2}\phi^T\mathbf{K}\phi$, or $\phi^T\mathbf{F}$, where $\mathbf{F}$ is the static force vector that would cause the ϕ displacement vector. The maximum kinetic energy in the mode is $\frac{1}{2}\omega^2\phi^T\mathbf{M}\phi$, in which ω is the circular frequency of the fundamental mode under consideration. Equating the maximum strain and kinetic energies we get

$$\omega^2 = \frac{\phi^T F}{\phi^T \mathbf{M} \phi} = \frac{g \sum_r F_r \phi_r}{\sum_r W_r (\phi_r)^2}, \quad T = \frac{2\pi}{\omega}, \quad f = \frac{1}{T} \tag{4.5.4}$$

in which the subscript r refers to the story level. To apply the above equation, a reasonable inertia force vector, $\mathbf{F}$, consistent with the fundamental mode shape is assumed and is subjected to the building that gives the estimated mode shape ϕ.

NEHRP gives empirical expressions for evaluating approximate period (T_a) of buildings. The following equations apply to an N-story building of height H (from base to the highest level) and length L (in the direction of earthquake motion).

1. For buildings in which the lateral force resisting system consists of moment resisting frames capable of resisting 100 percent of the required lateral force and the frames are not enclosed or restrained by more rigid components.

$$T_a = C_T H^{3/4}, \quad C_T = 0.035 \text{ (steel frames)} \quad \text{or} = 0.030 \text{ (concrete frames)} \tag{4.5.5}$$

2. For concrete and steel frame buildings, described in the paragraph (1) above, that have 12 or less number of stories (N), none of which have a height of less than 10 feet, Equation 4.5.5 may be replaced by

$$T_a = 0.1N \tag{4.5.6}$$

3. For buildings not conforming to the description in the paragraph (1) above

$$T_a = 0.05H/\sqrt{L} \tag{4.5.7}$$

The seismic lateral forces are inversely proportional to the building period. Buildings with higher period will be designed for a smaller lateral force than those with lower period. To control the minimum design lateral force, therefore, NEHRP recommends that the period used in the lateral force calculations do not exceed C_aT_a, where T_a is given by Equations 4.5.5, 4.5.6 or 4.5.7, and C_a is given in Table 4.5.3.

UBC also recommends Equation 4.5.5 for calculating the approximate building period. The UBC-values of C_T are identical to those of NEHRP for the steel and concrete moment resisting frames. UBC allows use of the same formula for all other buildings with a C_T value of 0.20. Alternatively, for buildings with concrete or masonry shear walls,

$$\frac{1}{C_T} = 10\sqrt{\sum_{walls} A_e[0.2 + (L_e/H)^2]} \tag{4.5.8}$$

in which A_e is the minimum cross-sectional area (ft^2) in any horizontal plane of the shear wall, length L_e, in the direction of the earthquake. As in NEHRP, UBC avoids underestimation of the base shear by requiring that the seismic coefficient calculated using the period from Rayleigh's method, Equation 4.5.4, would not be less than 80 percent of that calculated using the approximate building period formulas given above. This requirement is equivalent to specifying $C_a = 1.4$.

4.5.4 Review of the UBC and NEHRP Requirements

Following the San Fernando Valley earthquake of 1971, the SEAOC seismic design requirements underwent a major revision which were reflected in the 1976 edition of UBC (now replaced by the 1991 edition). The seismic coefficients in the 1976 UBC are based on the response spectra calculated for several ground motions in the Los Angeles basin, assuming a 10 percent damping ratio. Format of the ATC3 seismic requirements was modified to make it consistent with the response spectrum method of analysis. As stated in section 4.1, the UBC is an evolutionary code that contains a wealth of experience about the performance of buildings in past earthquakes. Therefore, the ATC3 committee "benchmarked" the R-values for various structure types such that the base shear forces calculated from the ATC3 equations would be nearly equal to those calculated from the 1976 UBC equations in most cases. The ATC3 requirements are now included in the NEHRP provisions. The base shear forces calculated using UBC will be different from those calculated using ATC3 and NEHRP due to the singular definition of ground motion (zone factor, Z) used in UBC as opposed to the dual definition (v and a) used in ATC3 and NEHRP and the differences in the R_w and R factors. In the high seismicity areas (zone 4, a=0.4) the lateral force resisting system requirements are, for all practical purposes, the same whether determined by UBC requirements or ATC3 and NEHRP provisions. The UBC base shear when multiplied by the appropriate factor of safety (used in working stress design) and the resistance factor (the capacity reduction factor in strength design) and divided by 1.33 (33 percent stress increase for combination of loads) will be approximately equal to the ATC3 and NEHRP base shear. The different contour mapping characteristics used in the UBC and NEHRP documents will lead to geographical differences, sometimes significant, in the base shears.

TABLE 4.5.3 Coefficient (C_a) for Upper Limit on Calculated Period (NEHRP)

v (cm/s) Range		Coefficient C_a
Upper	Lower	
--	30	1.2
30	20	1.3
20	15	1.4
15	7.5	1.5
7.5	--	1.7

Table 4.5.4 compares the base shear forces from the 1976 UBC, the NEHRP provisions and the 1988 UBC, in a high seismicity region. The site related coefficients for all the three are assumed to be 1.5, and the importance factors for the two UBC documents are taken as unity. The Table shows that base shear forces are almost equal for all the three cases when the building system consists of shear walls, concentric steel braced frames, or special moment resisting structural steel frames. Both NEHRP and the 1988 UBC require higher shear capacity for the ordinary movement resisting steel frames than does the 1976 UBC.

4.6 FORCE DISTRIBUTION AND DRIFT

4.6.1 Forces Tributary to Each Level

The design force determination in both NEHRP and UBC is based on the dead loads of the building and on some limited portions of live or transient loads. The starting point for seismic design is the same as for gravity load design, that is, determining the weight of the building.

Seismic forces are directly related to the weight of the building as it vibrates during an earthquake, and every weight element which is attached to the structure will affect the total weight. NEHRP has the most complete definition of dead load and of gravity load applicable to seismic force determination. Dead load is defined in

TABLE 4.5.4 Design Base Shear Forces

		DESIGN BASE SHEAR FORCES		
Building Type	**Fundamental Period Sec.**	**1976 UBC**	**ATC 3-06 NEHRP***	**1991 UBC**
Bearing Walls of Reinforced Concrete	0.2	0.19 W	0.17 W	0.18 W
Complete Frame of Structural Steel w/ Concentric Braced Frame	0.5	0.14 W	0.16 W	0.14 W
Special Moment Frame of Structural Steel	1.0	0.07 W	0.07 W	0.06 W
Ordinary Moment Frame of Structural Steel	0.7	0.12 W	0.16 W	0.16 W

* Equivalent working stress design values.

Section 2.1 of NEHRP as "the gravity load due to the weight of all permanent structural and nonstructural components of a building, such as walls, floors, roofs and fixed service equipment." The load, W, to be used in the earthquake force equation is defined in Section 4.2 of NEHRP as "the total gravity load of the building. W shall be taken equal to the total weight of the structure and applicable portions of other components including, but not limited to, the following:

1. Partitions and permanent equipment including operating contents.
2. For storage and warehouse structures, a minimum of 25 percent of the floor live load.
3. The effective snow loads defined in Section 2.1."

The latest UBC provisions state that the seismic load, W, should be based on "... the total dead load and applicable portions of other loads listed below.

A. In storage and warehouse occupancies, a minimum of 25 percent of the floor line load shall be included.
B. Where partition load is used in the floor design, a load of not less than 10 psf shall be used.
C. Where the snow load is greater than 30 psf, the snow load shall be included. When considerations of siting, configuration and load duration warrant, the snow load may be reduced up to 75 percent when approved by the building official.
D. The weight of permanent equipment shall be included."

The total loads W_r at, or tributary to, each floor level of the building must be determined for seismic force computation, rather than loads at a particular location, such as column loads, usually determined for gravity load design. Determination of the unit loads for each of the components of the structure and gathering them in one location near the beginning of the design calculations is a good practice. The unit loads can then be readily referred to, and modified when necessary, during the design process.

Wall partition loads in any story are typically distributed equally to the levels above and below the story. Where partitions are cantilevered from a floor, the partition load should be distributed only to the level on which it rests. There are many different types of partitions that might be used in an office building. Often the type and location of partitions is not known at the structural design stage. Therefore, it is difficult to quantify the partition weight that should be considered when determining the seismic loads. Unless specified by the code, this should be left to the judgment of the engineer with a lower limit of perhaps 10 pounds per square foot, or 30 to 50% of the vertical load allowance. For offices where movable partitions are used, SEAOC recommends using 50% of the vertical load allowance.

The exterior walls extending one-half of the story height above and below a floor are considered to be acting at that floor level. The exterior walls of the top story which are extended as parapets above the roof are typically considered as having the weight of one-half of the story height, plus the full parapet extension tributary to the roof. This may be unconservative for many low-rise buildings. A more accurate weight contribution may be found by using the reactions from the walls acting as a continuous vertical beam.

The amount of live load that would be expected to be acting on each floor of an office or apartment building at the time of the earthquake is difficult to predict. Surveys of in-service buildings have shown great variability of actual live load, but the average live load on a floor was generally considerably lower than the design gravity loads require. It is not zero, however, so some live load should be included. Several buildings collapsed during the 1985 Mexico Earthquake because of excess "live" load. The judgment of the design engineer should be used in determining how much live load to use when computing the seismic design forces. However, a value of 20% to 25% of the vertical live load is not unreasonable, if no required amount is specified in the applicable design code. Another possibility would be to use the entire vertical load allowance for partitions instead of including some of the live load. It should be noted that fixed, or permanent, file systems should be considered as dead load for the purpose of calculating seismic forces.

In many cases, some portion of the snow load should be considered as effective under seismic loadings. A great deal of judgment should be used in this determination. Gabled and similar roof shapes, where there are no impediments at the eave to prevent snow from sliding from the roof, might call for 25% of the snow load as seismically active. Snow melting and freezing can cause "ice dams" at the eaves, so we would not consider a reduction to zero load. There have been a number of failures of eave cantilevers due to these ice loadings. Where the height of ground snow or the positioning of the building with respect to topography or other conditions would not allow the snow to dislodge from the roof, more than 25% of the snow load should be used. Where roof configuration would cause snow to be trapped on the roof, or where drifting can occur, 75% might be used for the portions where entrapment can occur. The ANSI A58.1 commentary discusses the problems of intermittent heating and ice formation as well as drifting. All of these factors suggest that reduction of snow load in combination with earthquake by 90%, as recommended by some researchers based on statistics for ground snow only, is unsafe.

4.6.2 Horizontal Shear Force Distribution

The seismic design story shear in any story, V_x, is determined by adding the story forces (F_r, calculated from equation 4.4.4) at and above the story r.

$$V_x = \sum_{r=x}^{N} F_r \qquad (4.6.1)$$

in which N is the total number of stories. The shear force V_x is distributed to the various vertical elements of the lateral load resisting system in the story under consideration, with due thought given to the relative stiffness of the diaphragm and the vertical resisting elements.

When the diaphragm is flexible, the distribution would be on the basis of the tributary loading, taking into account the displacement of the mass. A diaphragm can be considered flexible if the maximum deflection of the diaphragm between two of the lateral load resisting systems (at the level of the diaphragm) is more than twice the average story drift of the associated story.

TABLE 4.6.1 Plan Structural Irregularities (NEHRP)

	Irregularity Type and Definition
A	Torsional Irregularity - to be considered when diaphragms are not flexible Torsional irregularity shall be considered to exist when the maximum story drift, computed including accidental torsion, at one end of the structure transverse to an axis is more than 1.2 times the average of the story drifts at the two ends of the structure.
B	Re-entrant Corners Plan configurations of a structure and its lateral force-resisting system contain re-entrant corners, where both projections of the structure beyond a re-entrant corner are greater than 15 percent of the plan dimension of the structure in the given direction.
C	Diaphragm Discontinuity Diaphragms with abrupt discontinuities or variations in stiffness, including those having cutout or open areas greater than 50 percent of the gross enclosed area diaphragm, or changes in effective diaphragm stiffness of more than 50 percent from one story to the next.
D	Out-of-Plane Offsets Discontinuities in a lateral force resistance path, such as out-of-plane offsets of the vertical elements.
E	Nonparallel Systems The vertical lateral force-resisting elements are not parallel to or symmetric about the major orthogonal axes of the lateral force-resisting system.

TABLE 4.6.2 Allowable Story Drift Δ_a (NEHRP)

	Seismic Hazard Exposure Group		
Buildings	I	II	III
Single story steel buildings without equipment attached to the structural resisting system and without brittle finishes	No Limit	$0.020h_{sx}$	$0.015h_{sx}$
4 stories or less without brittle finishes	$0.020h_{sx}$	$0.015h_{sx}$	$0.010h_{sx}$
All others	$0.015h_{sx}$	$0.015h_{sx}$	$0.010h_{sx}$

For low-rise buildings with flexible diaphragms the seismic forces at midspan of the diaphragm may exceed those at the vertical resisting elements at the ends by a factor of two or more. The designer should account for this effect.

When the diaphragm at any level is not flexible, as defined above, additional shear forces due to torsion should be considered. Often, the diaphragm is assumed to be infinitely rigid and the forces are distributed in proportion to the rigidities of the individual vertical resisting elements of the elements of the system. When the diaphragm is neither flexible nor rigid, additional consideration must be given to the distribution of forces.

4.6.3 Horizontal Torsional Moment

When the centers of mass and rigidity do not coincide at any floor, the torsional moment on the floor can be calculated by multiplying the story shear force, V_x (acting through the center of mass), by the eccentricity (distance between the two centers perpendicular to the direction of the forces). For design purposes the calculated eccentricity is increased by an accidental eccentricity equal to 5 percent of the dimension of the building perpendicular to the direction of the applied forces.

If a torsional irregularity exists as defined in Table 4.6.1, the accidental torsional moment calculated above should be amplified using a factor A_x determined from the following:

$$A_x = (\delta_{max}/1.2\delta_{average})^2 \qquad (4.6.2)$$

in which δ_{max} is the maximum displacement, and $\delta_{average}$ is the average of the displacements at the extreme points of the structure, all calculated at the level x using Equation 4.6.3 in the next section. The value of A_x need not exceed 3.0.

4.6.4 Drift Limitations

One of the most important aspects in the design of buildings for lateral forces is drift control. This is particularly true for moment frame structures. The following are the recommendations of the NEHRP provisions. The reader should refer to UBC for its requirements that are similar to those of the NEHRP provisions. Story drift, Δ, is defined as the displacement (δ_x) of a floor (x), relative to the displacement (δ_{x-1}) of the floor below it (x–1). If the elastic analysis for the specified earthquake force gives a story displacement δ_{xe}, the corresponding inelastic displacement (δ_x) is calculated from

$$\delta_x = C_d\, \delta_{xe} \qquad (4.6.3)$$

in which C_d is the deflection amplification factor given in Table 4.5.2. NEHRP limits on the story drift (Δ_a) are given in Table 4.6.2. The calculated displacement from Equation 4.6.3 may be amplified due to the P–Δ effect. However, such amplification is not likely in low-rise buildings and can normally be ignored.

4.7 MODELLING

The modelling concepts used for low-rise buildings are not themselves different from those used for medium- and high-rise buildings, however the models used must match the characteristics of the low-rise building that is being designed. While simplifications are required for the design of all buildings, these assumptions should not be used without first ensuring that they are valid. This means that the designer must be aware of the assumptions that are implied by the design requirements.

Buildings are designed for resistance to earthquake shaking by processes that are not necessarily appropriate for all buildings. Current seismic design requirements lead the designer to believe that the damped response of the vertical element of the lateral load resisting system dominates the response of the total structural system and that this will behave as an oscillator having a single value of fundamental period.

However, for low-rise buildings, the roof or floor diaphragm may be the most important oscillator, since the materials commonly used for structural elements do not have a single value of stiffness and therefore fundamental period, and the damping that limits the dynamic response is hysteretic, not viscous.

Current seismic design requirements specify a minimum lateral strength for the structural elements of the system and assume that the probable elastic response of these elements to design level earthquake ground motions will be reduced by yielding in some of the structural elements. The design requirements do not provide the designer with formulae and material strength values that could enable the actual peak strength of the structural elements to be determined. Moreover, the seismic design requirements do not give any warning to the designer that increasing the strength of primary elements above the minimum requirements may cause a secondary element to have a greater response than would be the case if the primary element had been allowed to yield.

Designers of low-rise buildings should be familiar with the probable peak strength of the materials used for low-rise buildings. Although these peak strengths are not used for seismic design, estimates of the peak strength of the various elements can be used by the designer to predict the location of yielding in the response model and to concentrate the design effort on these critical areas. All seismic design codes require a building with a "soft story" to be designed with special consideration given to the nonlinear behavior in that "soft story".

The peak strength of the structural elements cannot be calculated by code formulae or by reference to code allowable stresses for the materials. The estimated peak strength should be taken as the mean probable strength. Extrapolation of the minimum and maximum probable peak strength is best based on the judgement of the design engineer. Strength methods are used to determine the probable peak flexural strength of reinforced concrete systems, as described in ACI-318 (1983). The designer should use all vertical reinforcement, including distributed wall reinforcement, at its probable yield stress (not minimum guaranteed yield stress), for the computation of probable flexural strength. Capacity reduction factors should not be used in determining the peak strength. Peak shear strengths for reinforced concrete and for reinforced masonry should be estimated from reported research rather than by use of empirical code formulae. Peak strength of braced frames constructed of structural steel should be calculated by considering the development of successive yield zones. Peak strengths of wood framed elements, steel deck elements and similar materials can be estimated from published test data or from values given in the Tri-Services manual (Army, 1986).

In the case of braced frames, consideration must be given to the nature of the bracing members. If they are slender and likely to buckle at a low compressive load compared to their tensile strength, then the compression braces should be ignored when determining the period of the structure as well as its peak strength. If the braces have significant compressive strength, they will contribute to the initial strength and stiffness of the structure, but these will be reduced when the compression bracing buckles. In addition, the bracing members will undergo reversals of loading during an earthquake and this could lead to low cycle fatigue failure of the members and their connections.

Modelling concepts for seismic design should not only consider the probable peak strength of the structural elements but should also take into account their nonlinear behavior. Unlike gravity load design where the maximum loading experienced by the structure is not expected to exceed the design values, the demands on a building subjected to a major earthquake are expected to be substantially greater than the demands imposed on the structure by the code-prescribed design forces. Thus, the structure may be required to resist deformations that exceed the elastic deformations of the structural elements.

While it is possible to design buildings to remain elastic under the maximum expected earthquake forces, it is not normally economic to do so. Besides the cost, there is the performance of the structure to take into account; for instance, the floor accelerations within a building could be exceptionally high if the response is completely elastic. It is more desirable for the structure to respond inelastically if it can displace in a stable and ductile manner. As structural elements within the building are displaced into the inelastic range, the periods of vibration of the building are increased and the structure absorbs more energy. This reduces the amount of resonance amplification of the earthquake ground motion and increases the hysteretic damping of the structure. Thus the floor accelerations and interstory shear forces for a structure responding inelastically are generally less than for a similar building responding elastically. Lateral displacements may be greater for inelastic response than for elastic response; however, these displacements are limited due to the cyclic characteristics of the earthquake motion, the maximum displacement of the ground motion and the interaction between building response and earthquake motion.

In summary, special consideration must be given to the following items when modelling low-rise buildings:

- Inclusion of the horizontal diaphragm in the response model, when appropriate.
- Consideration of the peak strength of structural elements in the response model in lieu of code calculated design strengths.
- Estimation of the elastic period from force-displacement relations rather than from arbitrary formulae and limitations that may be imposed on computed elastic periods.
- Foundation models that allow for the separation of the foundation from the soils, when appropriate.
- Correlation of the structural stiffness that is used in the response model with the state of stress that is estimated from the response model. This is of special importance for reinforced concrete and masonry elements as well as wood diaphragms.

4.8 INTERCONNECTION FORCES IN LOW-RISE BUILDINGS

Most of the modelling concepts for low-rise buildings discussed above have assumed that the connections between the elements will be capable of sustaining all probable forces within their elastic strength values and do so without significant displacements taking place. This implies that the peak strength of all moment frames and braced frames can be fully developed without connection failure and that the connections between reinforced masonry and concrete members and other structural elements of the system can develop the peak strength of the adjoining members.

This is not to say that nonlinear distortion of a connection should be avoided at all costs. If the nonlinear displacement is in a single direction such as tension loading where compression buckling of the connection on cyclic loading is not probable, the inelastic behavior will reduce the excitation force that is applied to the connected member. All proposed nonlinear connection behavior should be carefully evaluated to ensure that the connection can develop the strength of the adjoining members.

5 DESIGN

5.1 INTRODUCTION

All buildings are required to be designed for gravity loads and lateral forces. Lateral forces for most buildings are produced by windstorms and earthquakes. Perhaps the most important consideration for design is the difference in the limit states for structural material behavior upon which the design criteria are based. Design for wind forces is based upon conventional linear behavior (stresses remain within elastic limit) while design for earthquake forces is based on the same concept for code level forces but anticipates that the real earthquake forces in a strong ground motion event will result in excursions into the inelastic range of material behavior.

This difference in expected performance means that the energy absorption or dissipation capacity of the structural system (the system, not just the material) is of extreme importance for earthquake resistant design. The seismic resisting structural system must provide stable resistance over repeated cycles of inelastic strain while there is no such requirement for the wind resisting structural system. Redundancy is more critical for seismic resistance than for wind resistance since failure of some part of the seismic resisting system is more likely where the actual response is so close to ultimate capacity.

The common practice, unfortunately, of ignoring the special seismic detailing requirements where design wind forces exceed earthquake forces is incorrect and potentially unsafe. The correct approach is to compare the wind design forces with the earthquake design forces multiplied by a factor representing the reduction for inelastic action. The 1991 UBC requires that the seismic detailing requirements be followed even when wind forces govern the design. It is recommended that the following guidelines be applied to the determination of detailing requirements:

1. When the wind forces are equal to or more than two times the earthquake forces determined using the NEHRP procedures design and detail the lateral force resisting system for the wind forces.

2. When the wind forces are less than two times the earthquake forces design the lateral force resisting system for the wind forces but detail the parts of the system to satisfy the earthquake requirements.

When earthquake forces equal or exceed the wind forces the codes require the lateral force resisting system to be designed and detailed to satisfy the earthquake requirements.

The basic tenet of good lateral force design is, as Newtonian principles tell us, that for every force there must be an equal and opposite reaction. We believe these principles apply to all design. To resist the forces induced in the structure there must be a complete and continuous load path provided from the point of origin of the force to the point of final resistance. Lateral force design, especially design for earthquake forces, has often been construed as increasing the cost of design and construction. Low-rise buildings which are well designed and detailed for gravity loads can achieve the expected level of performance for lateral forces for little or no extra cost when the design adheres to the principles stated above and quality control in construction is exercised.

5.2 FRAMING SYSTEMS

Low-rise buildings have been generally characterized as buildings in which gravity load strength requirements dominate the design of the system as well as the individual members. Historically the typical approach to design for lateral forces was linked to the ratio of height to width of the building. Where this ratio was less than 0.5 it was assumed that wind stresses would be so small they could be ignored in design [McDonald, 1975]. The lateral forces due to earthquake in regions of low seismicity have typically also been ignored. There has been an assumption that internal walls and partitions and unaccounted for redundancies were available to provide resistance to lateral forces and limit lateral displacement, drift. For many buildings this is a false assumption, especially with open warehousing, commercial and manufacturing uses and new construction techniques in office and residential uses. The structural systems must, of course, be adequate to support the gravity loads to which they will be subjected during the useful life of the structure. The structural systems must be integrated with all other systems in the building. Interference between systems has resulted in sacrificing the integrity of the structural system.

For purposes of design it has been common to categorize various structural systems into four groups. The categories are: Bearing Wall Systems; Building Frame Systems; Moment Resisting Frames; and Dual Systems. The characteristics of each of these systems will be discussed below. The above categories are associated with the vertical load and lateral force resistance of the systems. There are, in all buildings, horizontal, or nearly horizontal, elements or systems which transfer lateral forces to the vertical resisting elements.

5.2.1 Bearing Wall Systems

Bearing wall systems are used in structures in which major portions, if not all, of the gravity loads are supported by the walls of the structure, see Figure 5.2.1. Such systems have been labeled as box systems in some codes. These systems may be considered to have vertical and horizontal elements which provide lateral force resistance. The vertical elements of these systems are the walls which act as shear resisting components. The horizontal elements of these systems are the roofs and, where present, the floors of the building which transmit the lateral forces to the vertical elements. See Sections 5.6 and 5.7 for walls acting as vertical resisting elements of lateral force resisting systems.

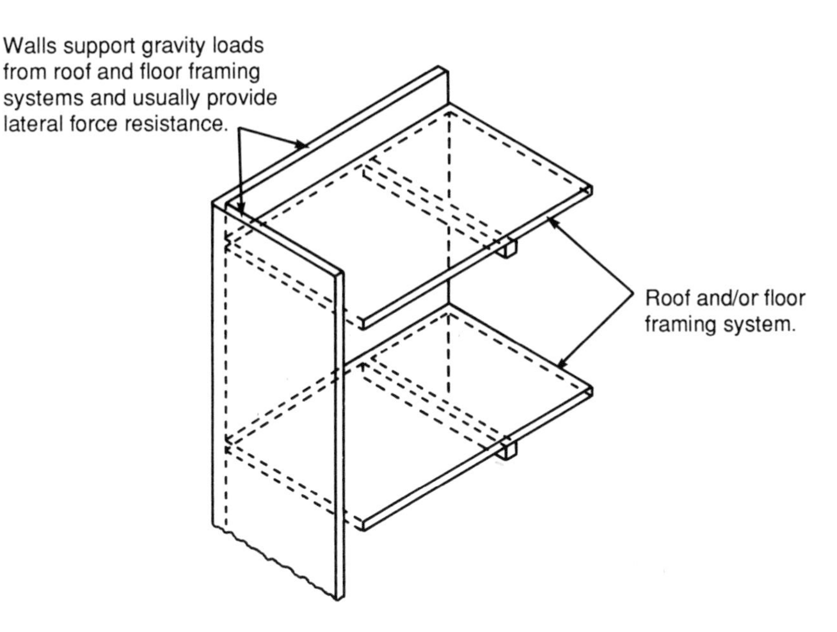

Figure 5.2.1 Bearing wall system

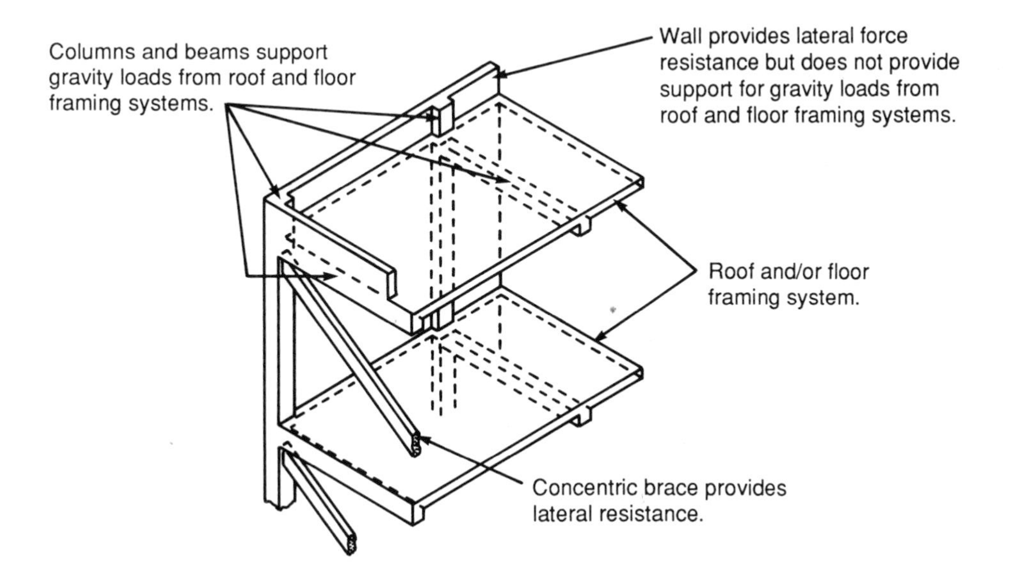

Figure 5.2.2 Building frame system

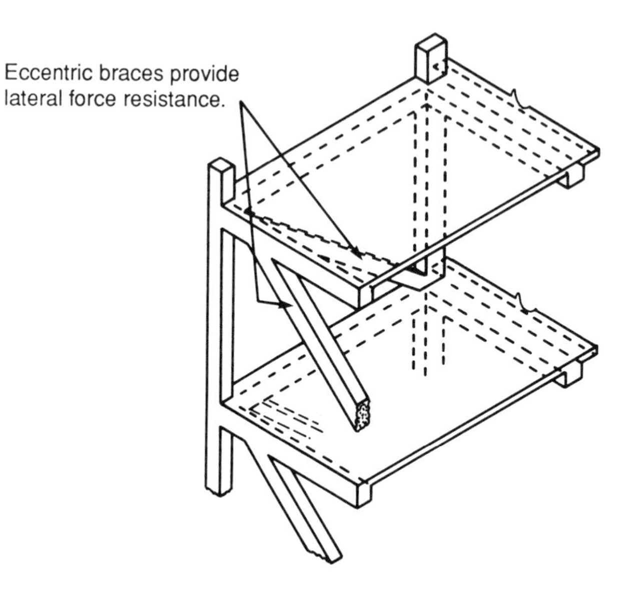

Figure 5.2.3 Building frame system

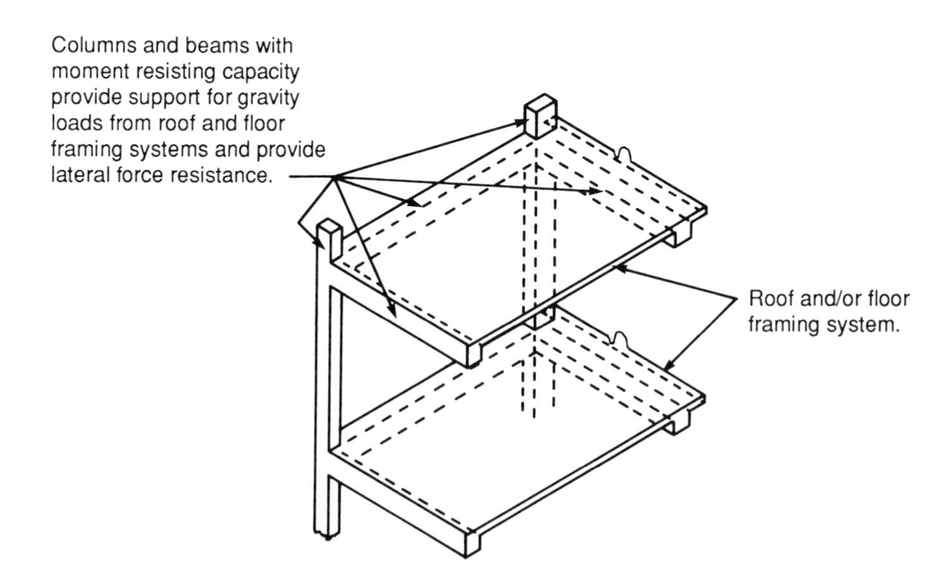

Figure 5.2.4 Moment frame system

5.2.2 Building Frame Systems

Building frame systems are used in structures in which essentially all gravity loads are carried by a space frame, see Figures 5.2.2 and 5.2.3. For this system the connections of the horizontal members to the vertical members of the space frame are normally considered pinned. It is advisable to provide some nominal moment capacity in these connections for lateral force resistant design. The lateral force resistance for this type of system is provided by shear walls or vertical trussing systems and the roof and floor diaphragms. See Sections 5.6, 5.7 and 5.8 for vertical resisting elements used in this category of construction.

The reduction in capacity due to buckling of members in vertical resisting systems during inelastic excursions is recognized in the latest codes. The forces in the bracing members, and their connections, of certain vertical lateral force resisting systems are increased for seismic effects to account for expected inelastic buckling behavior and provide a measure of control on brittle failure.

5.2.3 Moment Resisting Frame Systems

Moment resisting frame systems are types of space frames in which the connections of the horizontal members to the vertical members are designed to develop, to varying degrees, the moments induced at these locations by gravity loads and lateral forces, see Figure 5.2.4. These systems may be only segments of the complete space frames system supporting gravity loads. The lateral design provisions consider two general types of moment resisting frames; ordinary and specially detailed frames. Ordinary moment resisting frames do not have the capability of withstanding more than a few excursions into the inelastic range of the materials. Specially detailed moment resisting frames have limitations placed on aspect ratios, connection capacity and detailing which will permit the frames to experience a number of excursions into the inelastic range without adverse effect on their capacity to sustain loads. See Sections 5.4 and 5.5 for detailed discussion of these systems.

5.2.4 Dual Systems

Dual systems, as the name implies, are systems in which both moment resisting frames and shear walls or vertical trusses are used to resist lateral forces. Figures 5.2.2 and 5.2.3 would represent dual systems if the beam-column connections were made moment resistive. The seismic code provisions require that there be a complete vertical load carrying space frame in this type of system. The moment resisting frames are required to have the capacity to resist at least 25 percent of the code level seismic forces. The older codes required the shear wall or vertical trusses to be designed to resist 100 percent of the code level seismic forces but the later provisions require that the force distribution be made on a relative rigidity basis. Details for the elements of this category can be found in the remaining sections of this chapter.

5.2.5 Horizontal Resisting Elements

Roofs and floors, in addition to providing support for gravity loads, can be and are used to resist lateral forces and transfer such forces to the vertical resisting elements of buildings. The two most commonly used horizontal resisting elements are horizontal trusses and diaphragms.

Horizontal trusses can be analyzed in the same manner as vertical trusses with the loads being the tributary lateral forces due to wind or earthquake which are acting on the members of the truss. Diaphragms are analogous to built up plate girders. The webs of the girders are the wood, metal or concrete roof or floor membrane and the flanges are the framing members at the boundaries. The webs are considered to resist the shears induced by the lateral forces while the flanges resist the stresses due to flexure. For many of these systems, shear governs the action and the displacement or deflection of the diaphragm. Details of these elements are discussed in Sections 5.9 and 5.10.

5.2.6 Selection of Structural System

The selection of the framing system for capacity to support vertical loads is typically the first step in structural design. The usage or occupancy of a building, together with architectural considerations, will often tend to dictate the type of structural system to be used. An industrial structure with a need for wide, high roofed bays will eliminate poured-in-place reinforced concrete from consideration. Buildings which require high-level fire resistance will eliminate wood framing from consideration. Poor foundation conditions may dictate that the lightest framing system available be used. Many other examples could be cited which reduce the number of choices for the design engineer. There are many instances when the choice of the system has been made before the engineer becomes involved.

The structural engineer should be involved in the development of the design of the structural framing system for a project as early as possible. The structural system must not only provide the strength to resist the imposed gravity loads but also provide a serviceability level commensurate with the needs of the intended occupancy or usage. A good structural system is one which avoids discontinuities for gravity load support and lateral force resistance.

Good engineering has often been defined as the art of providing a serviceable product for the least cost. The ingenuity of the structural engineer to meet this goal is frequently taxed by predetermined conditions that do not lend themselves to least-cost solutions. The discussions which follow, together with the design examples, are intended to help guide in the selection of the most effective system for the project under consideration.

All of the commonly used structural systems and materials can be incorporated into a structure having a high degree of lateral force resistance when the principles stated in the introduction to this chapter are followed. The importance of providing continuity in the structural system to assure transfer of forces from their point of origin to the point of final resistance cannot be overemphasized.

Architectural systems which provide environmental protection to make a building habitable have to be accommodated by the structural system. The deformation compatibility of the architectural and structural systems need to be considered when making the selection of the structural system. Deformation control is typically provided by limiting inter-story drift. Story drift limitations accomplish three objectives: 1. Secondary P-Delta effects have a measure of control; 2. Damage to structural and non structural elements is limited; 3. Lateral motion control is provided. Deformation compatibility is more critical in seismic resistant design than it is in the design for other forces.

Current design practice for wind forces has no story drift limitation imposed. An ASCE task committee recently studied drift control in steel buildings and found that the most frequently used drift limit was H/400 [Bjorhovde and Fisher, 1986] with a range of H/667 to H/333, where H = building height. The Canadian Institute of Steel Construction recommends drift limits between H/400 and H/200 for industrial buildings. The American Institute of Steel Construction Specifications [Fisher, 1984] and the American Concrete Institute Code contain no recommendations on drift limit nor do the model codes have recommendations on drift control limits for wind forces. Drift control limits for seismic force design are contained in the model codes and apply to all types of construction. Shear wall construction with concrete, masonry or wood will remain within the prescribed drift limits for low-rise structures.

Moment-resisting frame systems are most susceptible to excessive story drifts which lead to excessive damage to non-structural components and even to structural damage. Normally the structural elements of the lateral force resisting system will be checked for both strength and displacement during the design phase. The vertical deflection of framing members which influence non-structural components is typically checked and consideration is given the compatibility of this deformation problem. The same consideration is too often neglected when considering lateral deformation compatibility of structural and non-structural components.

Only limited data are currently available on the lateral resistance of non-structural wall systems. It has been reported [Fisher, 1984] that industrial buildings with sheet-metal wall systems have performed well with bare frame wind drifts as large as H/100. Research conducted in Mexico and reported by Schneider and Dickey (1980) show masonry shear walls failing at deflections between H/100 and H/200 for various force levels and reinforcement ratios. Experience seems to show that most masonry cannot take a racking shear distortion greater than about H/500 without showing some distress [Council on Tall Buildings and Urban Habitat, 1980].

Tests conducted by John A. Blume and Associates on partition systems of gypsum wall board with wood or metal studs indicate that first damage occurs with a displacement of approximately 1/4 inch (H/400) at the top of an 8'-0" x 8'-0" wall panel. Ultimate capacity for the panel is reached at a displacement of approximately 3/4 inch (H/125) [Freeman, 1977]. The results of tests on additional wall panels which were tested with plaster, plywood and combinations of plywood and gypsum board are reported in other Blume reports. They have also specifically addressed wall panels of low-rise structures in an internal report.

The ASCE task committee on drift control has identified some additional test data for racking of non-structural walls [Allison and Fisher, 1986], data from which is presented in Table 5.2 - A, with drift indices given for first distress and ultimate load, and is beginning a study of this aspect of building drift. First distress was defined as any type of cracking noise or any noticeable signs of distress while ultimate load was defined as failure of any major element or loss of integrity. A review of the data indicates that for drift ratios, or indices, of 0.0025H none of the walls showed distress. Tarpy (1984) presented the results of an extensive study of racking tests, seventeen wall types, in 1984. These walls were constructed with 20-gauge steel studs at 16 and 24 inch centers with gypsum board, stucco and plywood facings. For most cases reported, first distress occurred at deformations less than 0.5 inch or a drift index of approximately 0.005. All of the testing for walls utilized the ASTM racking test procedures, that is they were not cyclic tests. Some recent work subjecting sheathed walls to cyclic displacements found a lower first distress incidence for some materials.

Typical design practice has not investigated the interaction effects between the structural system and the non-structural components of the building. The effects of the deformation of the structural system on the serviceability of, or the distress to, the non-structural components has be considered for some designs. There is a growing awareness and concern for the effects of interaction between the structural frame and the non-structural components such as partitions and exterior cladding [Freeman et al, 1980; Klinger, 1980]. Studies conducted at the University of Western Ontario and reported by the Metal Building Manufacturers Association [MBMA, 1980] verify that actual building drift under wind loadings is not the same as the calculated drift for the bare frame. In the case of the high-rise Dravo Building in Pittsburg the skin was designed to participate with the structural frame to control drift, but not considered in the lateral strength design of the frame. Should drift limits be related to partition type, should drift calculations include the effects of non structural components and, in seismic resistant design should degrading stiffness of non-structural components be recognized -- these are some of the questions now being asked by designers and researchers alike.

The discrepancy in drift control requirements in the codes for wind and earthquake forces should be eliminated. Although there are no formal recommendations for drift control limits for wind forces, a reasonable drift index of 0.0025 would seem appropriate for serviceability for structures where non-structural components are important for continued utilization of the building. The cost-benefit ratios between providing drift control and repairing damage must be considered in conjunction with the probable risk of occurrence of the damaging event.

A major concern for providing resistance to seismic forces is the need for redundancy to preclude instability arising from failure of a single element of the lateral force resisting system when subjected to the dynamic effects of earthquake ground motion. Some designers have limited the lateral force resisting system to a single vertical bay of bracing or moment frame at each face of a structure or of a core area within the structure. It is obvious that failure of a single vertical element will induce large torsional forces as well as increasing the direct forces on the remaining vertical element parallel to the failed element. Where such an arrangement must be used, the elements should be designed for the real seismic forces, not the reduced code level forces. At the least, a reduced R or R_w should be used to compensate for the lack of redundancy in this type of structure.

The selection of the type of lateral force resisting system may be determined before structural design of the system is started. The design engineer should, however, determine the arrangement of the resisting elements that will provide the best system for lateral force resistance and make suggestions for improvement in configuration which will assure better seismic performance.

Buildings with large irregularities can be designed to perform acceptably in an earthquake, but at a premium in cost and complexity. Past earthquakes have consistently shown that buildings with generally regular features perform better than irregular structures when construction and design quality are the same.

5.2.7 Quality Control

The performance of a building subjected to gravity loads and to wind or seismic forces depends on the quality of the design and the quality of the construction. The best design possible will not produce good performance from a building if the design is not faithfully executed.

The prevailing existence of building code use throughout the country has led many people to the belief that buildings will be constructed in conformance with the design and the governing codes. This belief is unfounded, as has been indicated by the building failures reported in the construction press. The prevalent lack of conformance with design drawings and code provisions that were noted following one of the recent earthquakes emphasizes the need for more stringent quality control efforts.

The basic properties of the materials of construction must be verified. Concrete cylinders must be taken and tested; reinforcement identified for grade and heat number or tested if such identification is not possible; masonry units, mortar and grout must be tested; structural steel identified for grade and heat number or tested where such identification is not possible; wood and plywood must be identified for grade and checked visually for conformance with grade. The details of construction must be verified for compliance with the design drawings for reinforcement arrangement and placement in concrete and masonry members; for bolting and critical welds of structural steel members, which may involve non-destructive testing; for connections and fastening of wood members and sheathing.

5.3 LATERAL FORCE DETERMINATION AND DISTRIBUTION

5.3.1 Codes and Background

Building code requirements for lateral forces due to wind and earthquake, and the history of their development, have been covered in Chapters 3 and 4. Low-rise buildings will be designed to conform to the requirements the American National Standards Institute standard [ANSI, 1982] and/or one of the codes promulgated by the model code bodies; Building Officials Conference of America [NBC, 1990], International Conference of Building Officials [UBC, 1991] or Southern Building Code Congress [SBC, no date]. These guidelines were developed to provide the user with the most recent information on design to resist lateral forces. In the case of resistance to seismic ground motions this meant incorporating the latest ground motion mapping from NEHRP [NEHRP, 1988] as indicated in Chapter 2. The use of these maps requires a design procedure somewhat different from, but parallel to, the design procedure in the most recent seismic design recommended source documents [SEAOC, 1990; ATC3, 1978]. The new procedure, and the parallel UBC code procedure, is explained in the Preface to the Design Examples and the design base shears determined by the use of both will be included in the examples.

5.3.2 Calculation of Forces Tributary to Each Level

For wind forces, the lateral force resisting system of a building responds to the total average pressure acting over all of the exposed surface areas of the building. In general, the average pressure is a function of the velocity of the wind which varies with height. The 50-year design wind speeds are given in Figure 2.4.1 of Chapter 2. For wind forces the exposed surface areas tributary to each level and the effective pressures acting on those surfaces at that level will determine the total force acting on the building at that level.

For seismic design the design force determination in each of the three standards cited is based on the applicable gravity loads given in Chapter 4 acting at each level of the building. The normal procedure for gravity load design is determination of the weight tributary to each vertical load-carrying element, the column loads in a frame structure, while seismic design requires the sum of these loads at each level.

5.3.3 Total Lateral Force Determination

For wind forces, the total lateral force is the sum of the forces at each level determined from the pressures and tributary areas at each level.

For seismic forces, the codes recognize two methods for determining the total design force for a building. These are the Equivalent Lateral Force Procedure and the Dynamic (Modal) Force Procedure. Any building may be designed according to the Dynamic Force Procedure, and some buildings must be designed in this manner. The Equivalent Lateral Force Procedure may be used only for the following buildings:

"(1) All structures, regular or irregular, in Seismic Zone No. 1 and in Occupancy Category IV in Seismic Zone No. 2.

(2) Regular structures under 240 feet in height with lateral force resistance provided by systems listed in Table 2 of Chapter 4.

(3) Irregular structures not more than five stories nor 65 feet in height. (See Table(s) 4 (and 5) of Chapter 2 for irregular structures)

(4) Structures having a flexible upper portion supported on a rigid lower portion where both portions of the structure considered separately can be classified as being regular, the average story stiffness of the lower portion is at least 10 times the average story stiffness of the upper portion and the period of the entire structure is not greater than 1.1 times the period of the upper portion considered as a separate structure fixed at the base."

As the vast majority of low-rise buildings fall into one of these categories, only the Equivalent Lateral Force Procedure will be discussed in this chapter since past experience has indicated this to be adequate. See Chapter 4 for a discussion of modeling and dynamic analysis procedures.

Once the total effective weight, W_i, of each level is obtained, the total effective weight, W, of the building may be determined by simply summing the individual weights (W = W_i, i = 1 - n), for the number of levels above the base.

The design philosophy mentioned previously of designing for inelastic behavior is evident from Equation 4.5.2 of Chapter 4. The factor $0.014vS/T^{2/3}$ represents an estimate of the maximum base shear that the building would experience during the design earthquake if it remained elastic. For reinforced concrete and masonry structures the elastic threshold occurs at the displacement level where the steel reinforcement yields. The factor R represents the total reduction in this base shear that is used for design. R depends on the building type as given in Table 4.5.2 of Chapter 4. The value of R specified for each building type accounts for the assumed energy dissipation capacity, damping and excess strength that might be expected, as well as the overall level of risk that is believed to be acceptable. Note that the largest value of R is 8 for special moment resisting space frames of steel or concrete. This is a very large reduction in the expected force levels and is only possible if special detailing requirements are followed and quality construction is assured. Thus, it is important to note that the force provisions of the codes may not be used without also following the material specifications.

5.3.4 Vertical Distribution of Forces

For wind forces the vertical distribution was determined when the tributary forces at each level was calculated.

As previously mentioned, an earthquake induces inertial forces in the building. Low-rise buildings will vibrate primarily in their fundamental mode of vibration. Therefore, the lateral accelerations throughout the structure will be proportional to the amplitude of the mode shape at that location, see Chapter 4.

5.3.5 Drift Limitations

One of the most important aspects in the design of buildings for lateral forces is drift control. This is particularly true for moment frame structures. Story drift is defined as the displacement of a floor relative to the floor below it. Story drift ratio is the story drift divided by the story height. The drift limits in the NEHRP Provisions differ from those in the Uniform Building Code. The NEHRP limits are from 0.010 to 0.020 times the story height for the calculated elastic drift times the deflection amplification factor. The UBC code limits the calculated elastic seismic story drift to the smaller of $0.04/R_w$ or 0.005 times the story height for buildings less than 65 feet in height and $0.03/R_w$ or 0.004 times the story height for buildings more than 65 feet in height. The code allows these limitations to be exceeded if the designer can demonstrate that the greater drift can be tolerated by both the structural elements and the non-structural elements that could affect life safety. Since inelastic behavior during an earthquake may result in unpredictable drifts, design drift limits greater than those specified by the code should not be used. Many engineering firms use more stringent requirements than those specified, with one firm using a value as low as 0.002 times the story height. In most cases the shears and moments resulting from P-delta effects would have to be included in the analysis. The "*Blue Book*" [SEAOC, 1990] requires that P-delta effects be included if the drift at any level under design loads exceeds $0.02/R_w$. The over-all stability of the structure should also be confirmed at the maximum anticipated drifts under inelastic behavior which NEHRP approximates as the deflection amplification factor, C_d, times the calculated elastic drifts. Stability for most low-rise buildings is not a concern. As noted earlier, there are no code limits on drift for wind forces, although there have been attempts to introduce such provisions.

5.3.6 Special Considerations

The location of the center of mass at any level is not precise due to variations between assumed unit weights and actual unit weights and due to equipment location changes or additions. For this reason NEHRP requires that the mass at each level be offset from its calculated location by a distance in either direction equal to five percent of the building dimension perpendicular to the direction of forces under consideration. This offset applies whether or not torsion can be generated.

As noted in Chapter 4, flexible diaphragms in low-rise structures have indicated an amplification of motion at mid-span as recorded by strong motion instruments in several recent earthquakes. This amplification effect had not yet been directly addressed by the 1988 editions of the codes, but is reflected in the wall anchorage recommendations of SEAOC (1990).

If the center of mass and center of rigidity do not coincide at every floor, torsional vibration will occur under earthquake motions. Similarly, torsion will occur if the center of pressure for wind loads does not coincide with the center of stiffness of the lateral force resisting system. For seismic design an additional accidental torsion due to the five-percent offset of the mass, as mentioned in Chapter 4, shall be included. The additional shear resulting from this torsion must be considered in the design. The effects of torsion may be computed using well-known principles of mechanics.

5.4 CONCRETE MOMENT RESISTING FRAMES

This section is concerned with design of monolithic reinforced concrete frames. The discussion refers to buildings with lateral strength provided only by frames. It is also assumed that the frames may be analyzed in their planes and that their distortion is not hindered by nonstructural elements.

Proportioning and design of reinforced concrete moment frames for gravity loading has been extensively covered in many texts and will not be further developed here. Proportioning and design of reinforced concrete moment frames to resist lateral forces, especially those induced by earthquake ground motions, will be explored in the following material.

For a given earthquake intensity or magnitude of wind forces, the design process involves three main decisions which the designer controls. They are decisions related to (1) weight-to-strength ratio, (2) weight-to-stiffness ratio, and (3) toughness. Discussions of the relative importance of these decisions in the current environment for lateral-force-resistant design and a description of a simple procedure for checking adequacy of the stiffness of reinforced concrete frames are presented.

5.4.1 Weight-To-Strength Ratio

It is convenient to refer to the lateral strength of a frame in terms of its base shear strength calculated for a collapse mechanism with lateral forces applied at each story. Ideally, the collapse mechanism chosen is the one that results in the lowest base shear force for the "linear" distribution of story forces described in Figure 5.4.1. Even though the actual distribution of story seismic forces under dynamic excitation is likely to deviate from that assumed, the base shear strength so calculated provides a satisfactory measure of frame strength unless the structure has an unusual distribution of mass or exposure area, stiffness, and/or strength over its height. The base shear strength is normalized by taking its ratio to the weight of the building. The ratio so obtained, the base shear strength coefficient, is likely to be considerably larger than the seismic design base shear coefficient. The base shear strength for wind resistant design will also probably exceed the design wind base shear.

Building code procedures for earthquake-resistant design of reinforced concrete frames focus on strength. Weight-to-strength ratio appears to be the most important, if not the only, issue in design. Usually, this consideration is much less important than it appears to be. Because of factored gravity-load demands, "over-design" produced by proportioning sections for maximum effects and extending the reinforcement to other sections, and code minima on member sizes and reinforce-

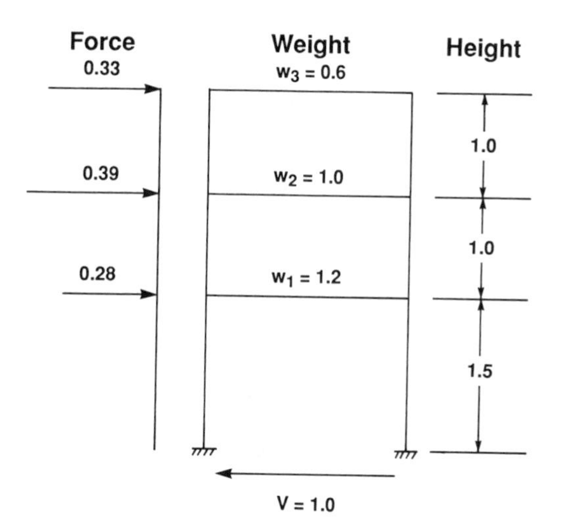

Figure 5.4.1 Three-story frame example

ment, frames often have base shear strengths well in excess of the amount required by the code for earthquake resistance. It is instructive for the designer to check the base shear strength coefficient after the design is completed and compare it with the design base shear coefficient, which is the product of the coefficients from Equations 4.4.4 and 4.5.1 of Chapter 4.

5.4.2 Weight-To-Stiffness Ratio

Section capacities leading to adequate calculated base shear strength may be built into a frame for reasons other than lateral force resistance, but adequate lateral stiffness is not likely to happen accidentally. It has to be carefully planned. Lateral stiffness of a frame is controlled by span lengths, story heights, member sizes, materials, and, in cases of extremely low or extremely high strength, base shear strength. The most important role of structural analysis in lateral-force-resistant design of reinforced concrete frames is in planning for adequate lateral stiffness, given a building weight and lateral forces. A pragmatic measure of weight-to-lateral stiffness ratio is the story drift ratio, which is the ratio of the story drift (relative horizontal displacement of floors bounding the story) to the story height. By limiting the calculated story drift ratio to a tolerable amount, adequate stiffness is assumed to have been attained. It is important to note that the drift ratio is invoked simply to assure a level of stiffness. It is not the intent of design calculations to predict the actual drift in the event of a particular ground motion or wind intensity. As a matter of fact, the tolerable story drift ratio may be totally arbitrary as noted in Section 5.2.6. It may be set to compensate for the level of the fictitious forces used in the calculations for lateral displacement.

5.4.3 Toughness

Of the three main design considerations in earthquake resistant design of reinforced concrete frames, toughness is unquestionably the most important. Cross--sectional strengths that lead to adequate lateral strength may be provided by demands of gravity loads and code minima, but design for gravity loads does not always demand the full transfer of moment at every joint from horizontal to vertical elements and does not demand that such continuity be sustained through repetitions of load reversals in the nonlinear range of response. The degradation of strength during excursions into the inelastic range of the reinforcement has been noted in many tests of concrete joints and members. Toughness is achieved by selection of member sizes and detailing of reinforcement to make certain that the frame elements retain their strength and remain effectively connected during the displacement history that the frame would experience in an earthquake. A frame may survive without adequate stiffness or with substandard strength, but it is very unlikely to survive if it lacks the ability to resist excessive strength degradation, toughness. How tough a frame should be depends on the drift it is expected to experience. The extent of the drift depends on the weight, the ground motion, and the effective stiffness of the reinforced concrete frame. It is difficult to place a categorical limit on drift because the most critical combination of parameters for drift may yet be beyond our most pessimistic prediction. The pragmatic solution is to tune the level or cost of the prescribed actions for toughness to a target drift ratio that is not likely to be exceeded in all but extreme situations. Figure 5.4.2 shows a compilation, from laboratory experiments, of the variation of relative damage to nonstructural walls. It is seen that, according to that interpretation, perceptible damage starts at a drift ratio of approximately one-quarter percent. By the time a drift ratio of two percent is reached, the walls are said to be a total loss. Although there are more than walls to a complete building, it is prudent to assume that if a frame suffers a drift ratio of over two percent, its contents would have to be completely rebuilt unless they were isolated from the motion of the frame.

Ideally then a frame should be stiff enough to limit the drift ratios, preferably to less than one percent and certainly to no more than two percent. It should also be tough enough to accommodate a drift ratio of at least two percent.

5.4.4 Detail requirements for frames

Appendix A of "Building Code Requirements for Reinforced Concrete" of the American Concrete Institute [ACI 318, 1983] and the "Uniform Building Code" of the International Conference of Building Officials [UBC, 1991] have essentially the same requirements for frame details. In both documents, detail requirements differ for frames in zones of high (areas with v greater than 20 c.p.s.) and moderate (areas with v from 15 to 20 c.p.s.) seismicity. For the sake of consistency, references in the following discussion will be made only to Appendix A of ACI 318-83. To simplify reference without creating still another acronym, detail requirements for frames in regions of high and moderate seismicity will be referred to simply as high seismicity and low seismicity details or detail requirements.

In the light of the discussion in section 5.4.3, it appears inconsistent to have different detail requirements in different zones. For an ideal and consistent design process, the maximum drift attained should be the same in all regions.

The detail or toughness requirements should then also be the same. However, the design process is neither ideal nor is it likely to give consistent results as it entails more than a straightforward application of mechanics. The seismicity estimate in regions of low and high seismicity is not expected to be uniform. Frame members have certain threshold properties so that the mean drift ratio for building populations is not expected to vary linearly with ground motion or wind force intensity. More importantly, regions of moderate seismicity or wind force levels suffer damage less frequently than regions of high seismicity or wind force levels. Professions concerned

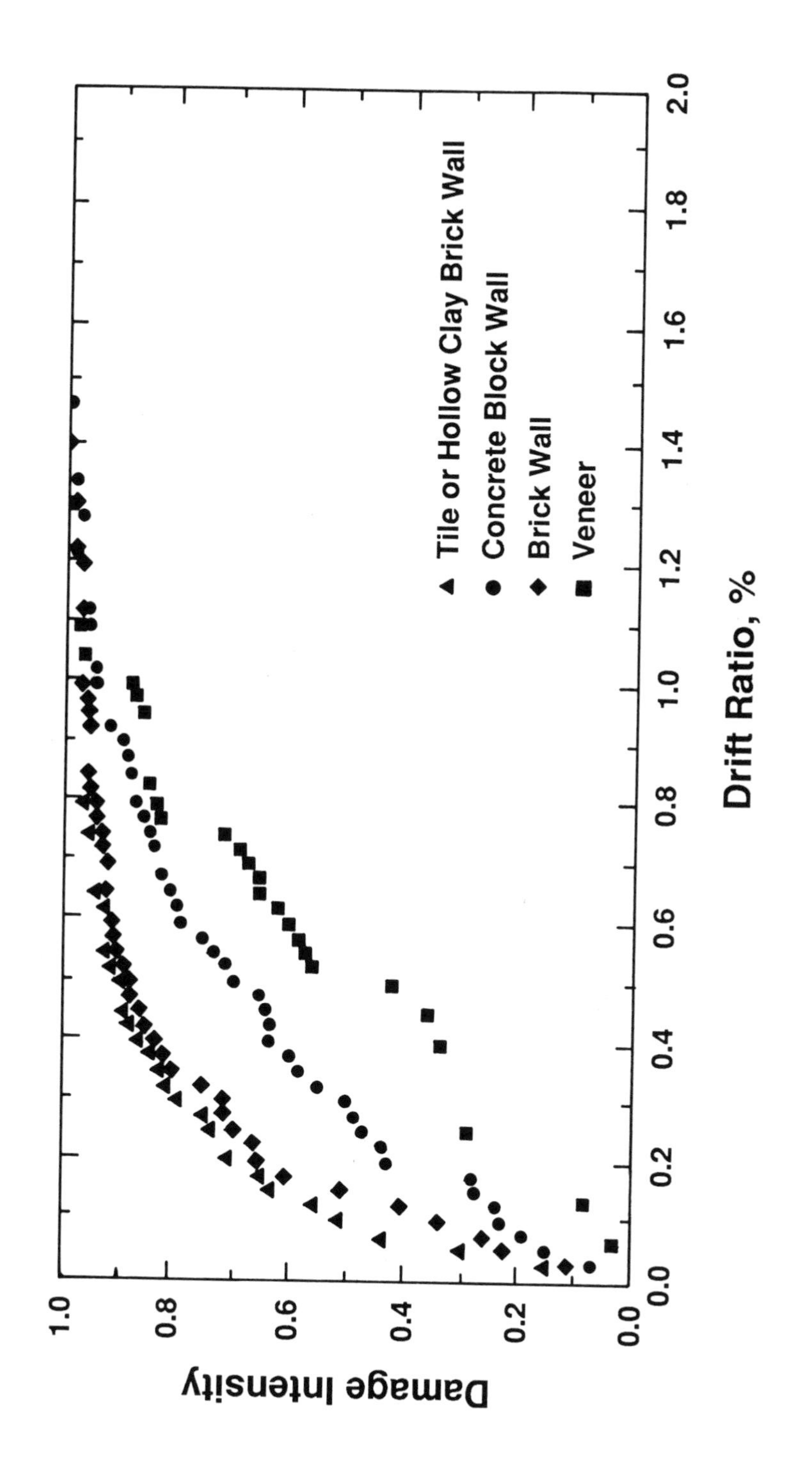

Figure 5.4.2 Variation of damage intensity with damaging distortions as indicated by experimental data on damage to nonstructural partitions

with building and public authorities in regions of moderate seismicity tend to tolerate higher risks. The difference between detail requirements in different zones is due primarily to such non-quantifiable issues.

Field and laboratory evidence have demonstrated consistently that the main sources of brittleness in reinforced concrete frames are shear and bar anchorage-/splice failures. The first and most important step in the design of reinforced concrete frames is make certain that the probability of such failures, within the anticipated performance range of the frame, is virtually zero.

For high seismicity requirements, frame elements are proportioned for the maximum shear force that they are likely to develop based on flexural capacity provided. For beams and girders, this is a straightforward procedure. For columns, the maximum shear force is not uniquely defined in terms of cross-sectional and span properties. It depends also on the axial load and may, therefore, be negotiated up or down. The designer is cautioned to make certain that a critical weakness is not built into the structure for the price of a few hoops. It is, in general, preferable to select member proportions so that the total unit shear stress does not exceed $6/f'_c$ (for normal weight aggregate concrete).

For low seismicity requirements, member shear forces may be based on those related to twice the design base shear. It is not certain that this procedure will guarantee toughness for all possible conditions [ACI 318, 1983, Article A.9.3]. The procedure used to determine member shear forces for high seismicity requirements is likely to satisfy the toughness criterion better.

The salient differences and similarities between high seismicity and low seismicity detail requirements in ACI 318 (1983) are summarized in Table 5.4 - A, for girders, and Table 5.4 - B, for columns. Locations of lap splices are limited to approximately the middle half of all frame members for both requirements, but the confinement requirements are explicitly stated for high seismicity areas. For high seismicity areas explicit rules are given for anchorage of hooked bars [ACI 318, 1983, Article A.6.4] in confined concrete (as defined in ACI 318, Article A.6.4). For low seismicity areas, bars are anchored in accordance with gravity-load requirements recognizing the degree of confinement for the concrete. For girders, the main distinguishing factors between high and low seismicity requirements are the geometric constraints and the stipulation of hoops in hinging regions. The maximum reinforcement ratio of 0.025 for high seismicity areas does not apply to frames in low seismicity areas, but it seldom provides a hardship in design of properly sized elements. Lack of geometric constraints in low seismicity areas permits the use of slabs as components of frames resisting earthquake effects. There are special requirements for reinforcement amounts and details for slabs used in this capacity [ACI 318,1983, Article A.9.6]. However, current building codes do not directly emphasize the inherent flexibility of these elements. Article A.4.4.1 of ACI 318, governing transverse reinforcement in hinging regions, is the main difference between column details for regions of moderate and high seismicity. Columns in high seismicity areas are also required to be stronger than the girders framing into the same joint [ACI 318, 1983, Article A.4.2] and have reasonably compact sections [ACI 318, 1983, Article A.4.12].

Joints of frames in regions of high seismicity have to be detailed according to requirements special to earthquake resistant design [ACI 318, 1983, Article A.6].

TABLE 5.4A
Reinforced Concrete Frames, Basic Requirements for Earthquake Resistance
GIRDERS

	In Regions of High Seismicity (zone 4, 3)	**In Regions of Moderate Seismicity (zone 2)**
Unit (factored) axial-load stress (P/A_g)	Not more than ($f'_c/10$)	Not more than ($f'_c/10$)
Clear span-to-effective depth ratio (l_n/d)	Not less than 4	--
Width-to-effective depth ratio (b_w/d)	Not less than 0.3	--
Width (b_w)	Not less than 10 in.	--
	Not more than ($c_2+0.75d$)	--
Longitudinal reinforcement ratio (applicable to top as well as bottom reinforcement) Minimum*	200 b_wd/f_y	200 b_wd/f_y
Maximum	0.025	(3/4) balanced
Flexural strength at face of joint M_{neg}/M_{pos}	Not less than 1/2	Not less than 1/3
In "hinging" regions*	(Hoops)	(Stirrups)
S_1	Not more than 2 in.	Not more than 2 in.
S_2	Not more than d/4	Not more than d/4
S_2	Not more than 8 * d_{bm}	Not more than 8 * d_{bm}
S_2	Not more than 24 * d_{bh}	Not more than 24 * d_{bh}
S_2	Not more than 12 in.	Not more than 12 in.
In lap splice regions		
S_3	Not more than d/4	
S_3	Not more than 4 in.	
Spacing of stirrups S_4	Not more than d/2	Not more than d/2
Anchorage of reinforcement with standard hooks		
Straight embedment length in confined normal weight aggregate concrete	Not less than 10 * d_b	
	Not less than 6 in.	
	Not less than $(f_yd_b)/(65\sqrt{f'_c})$	

* See Figure 5.4.3

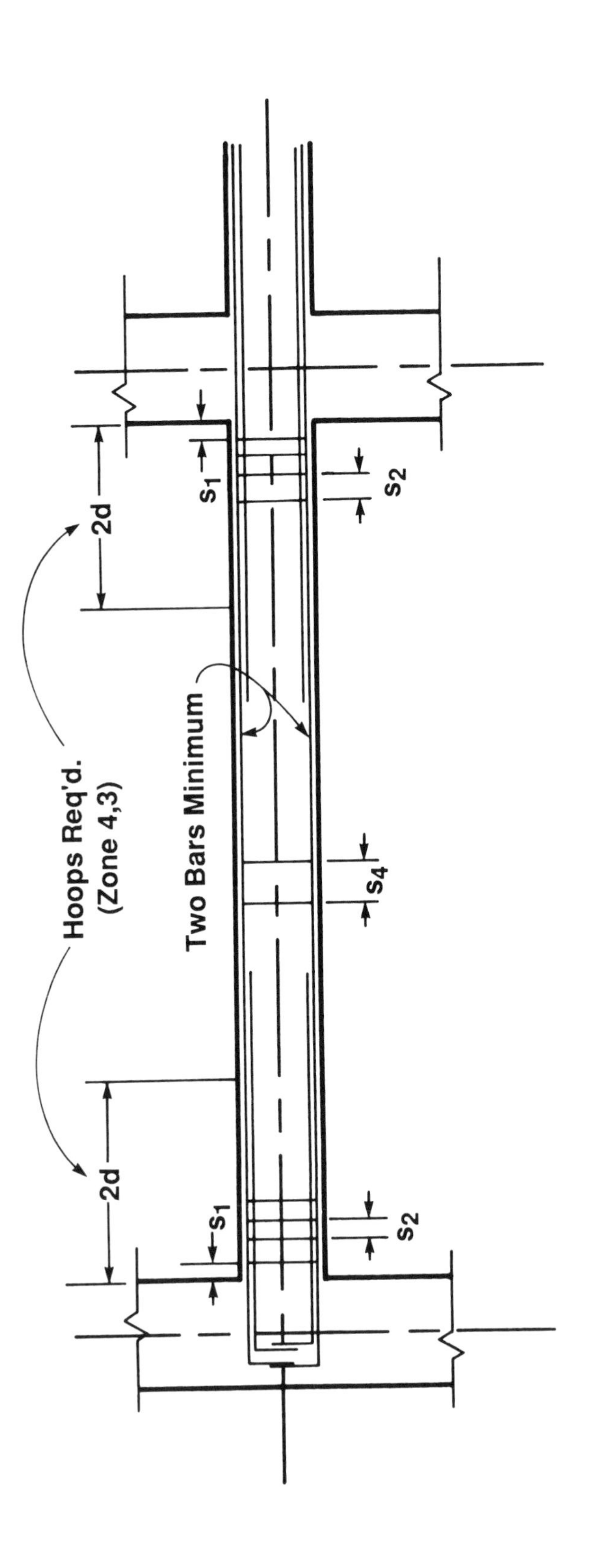

Figure 5.4.3 Girder details

TABLE 5.4 B
Reinforced Concrete Frames, Basic Requirements for Earthquake Resistance
COLUMNS

	In Regions of High Seismicity (zone 4, 3)	**In Regions of Moderate Seismicity (zone 2)**
Shortest cross-sectional dimension	12 in.	
Ratio of shortest to longest cross-sectional dimension	0.4	
Ratio of column to girder flexural strengths framing into a joint	6/5	
Longitudinal reinforcement ratio		
Minimum	0.01	0.01
Maximum	0.06	0.06-0.08
Transverse reinforcement in hinging regions**		
S_5	(Hoops)	(Ties)
S_5	Not more than $h_m/4$	Not more than h/2
S_5	Not more than 4 in.	Not more than 12 in.
S_5		Not more than $8 * d_{bm}$
S_1		Not more than $24 * d_b$
		Not more than $S_5/2$

** See Figure 5.4.4

5.4.5 Drift Control

The drift response of low- and moderate-rise planar frames is dominated by response in the lowest mode. Displacement response of a linear single-degree-of-freedom oscillator, having the same period as the effective period of the frame, provides a simple and good tool for understanding frame response to strong ground motion. It is well known [Blume et al, 1961b] that for strong ground motion experienced on stiff soil, the maximum velocity response of single-degree-of-freedom oscillators with periods ranging from 0.4 to 2 seconds tends to remain essentially the same and may be related linearly to the effective peak acceleration:

$$V_r = A_g * K_a \tag{5.4.1}$$

where:

V_r = velocity response
A_g = effective peak ground acceleration
K_a = constant depending on damping of oscillator and frequency content of ground motion.

Displacement may be related satisfactorily to velocity response by assuming the motion is harmonic. In the period range of Equation (5.4.1), displacement may be expressed as:

$$D_r = A_g * K_a * T \tag{5.4.2}$$

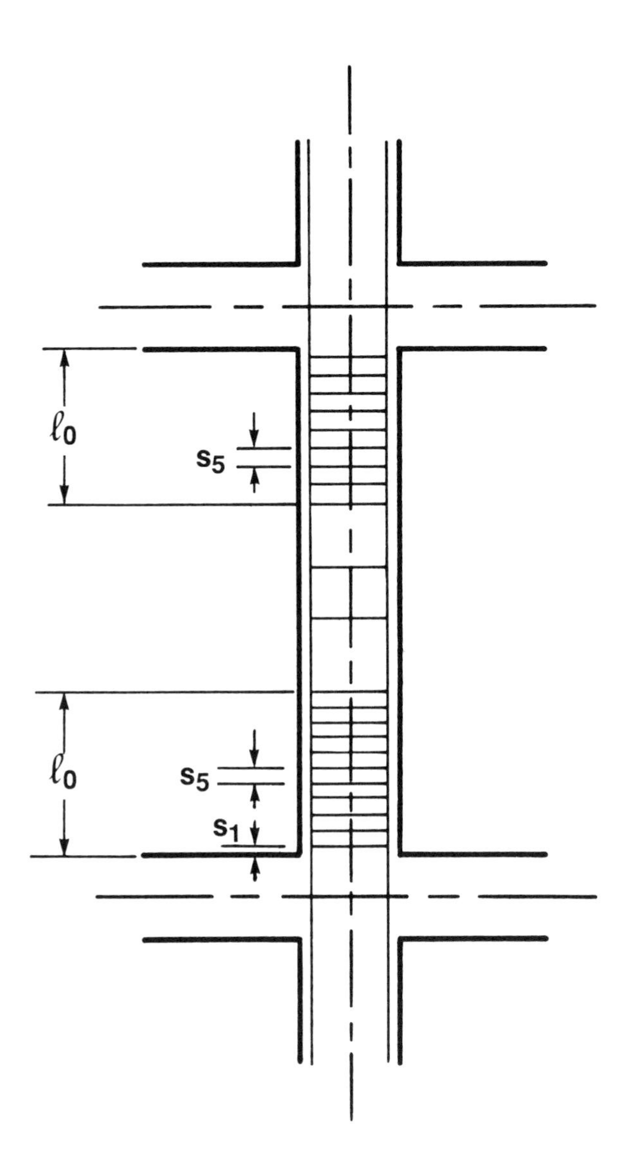

Figure 5.4.4 Column details

Where:

D_r = displacement response
T = oscillator period.

Story displacements of a frame having the same effective period as the oscillator are related to D_r (the spectral displacement) by the following equation based on the deflected shape of the frame for the fundamental mode.

$$D_i = D_r * x_i * \Sigma(m_i * x_i)/\Sigma(mi * x_i^2) \quad (5.4.3)$$

Where: D_i = lateral displacement (drift) at story i
x_i = ratio of lateral displacement at level i to that at top level for the assumed or calculated mode shape
m_i = mass at level i

Equations (5.4.2) and (5.4.3) help put into perspective the important parameters affecting the drift of reinforced concrete frames. It is noted from Equation (5.4.2) that the displacement response is proportional to the term (A_g/K_a) that reflects the properties of the ground motion and the damping of the oscillator. The accuracy of the displacement prediction cannot be better than that of the intensity and frequency content of the ground motion. Properties of the as-built structure also affect the displacement response but not directly. The effective stiffness of the structure, during high-amplitude response, changes the dominant period of response. But the change in displacement response is less than the change in stiffness. In Equation (5.4.2), the capability of the structure to dissipate energy (damping), is included in the coefficient K_a. For reasonably well detailed reinforced concrete structures, it is not likely to be a major source of error in estimating drift.

5.4.6 Check for Required Stiffness

The explicit design criterion for stiffness is the calculated drift. Drift is calculated using the design story forces and by static analysis of a linearly elastic model of the frame. Calculations for drift of a multi-story frame are most appropriately done using the stiffness method considering the rotations and displacements at each joint. The number of equations necessary to obtain a solution will often require the use of a digital computer. As long as a computer is used, it is efficient to use an "off-the- shelf" structural analysis program that takes care of all computations, given the basic input.

On many occasions, there is need for a simple method either to check the results of the structural-analysis program or for preliminary design. The procedure described below provides such an option.

The first step in the procedure is to define or interpret the frame with its many joints resulting in many degrees of freedom at each story as a "shear beam" with a single degree of freedom at each story. This requires the definition of a stiffness for each story that combines the flexibilities of the columns and girders. The equation given below performs that function.

$$k_i = \frac{24E}{h^2} \quad \frac{1}{\Sigma \frac{2}{k_c} + \Sigma \frac{1}{k_{bb}} + \Sigma \frac{1}{k_{bt}}} \quad (5.4.4)$$

Where k_i = equivalent story stiffness
E = Young's modulus
h = column height
k_c = I_c/h
I_c = cross-sectional moment of inertia of column
k_{bb} = I_{bb}/L
L = beam span

I_{bb} = cross-sectional moment of inertia of beam at level below
k_{bt} = I_{bt}/L
I_{bt} = cross-sectional moment of inertia of beam at level above

The story stiffness defined by Equation (5.4.4) will give the same result as an exact solution if the girders are rigid. For flexible girders, the stiffness calculated is approximate. Story drifts calculated using the shear beam model tend to deviate from the exact values as the ratio of column to girder stiffness increases but, for frames of representative proportions, the errors are well within the errors compatible with the definition of section stiffness for reinforced concrete sections.

5.5 STEEL MOMENT RESISTING FRAMES

Steel frames are used extensively [AISC, no date] and some of these systems are designed as moment resisting frames. The AISC Specifications [AISC, 1978] define three types of steel construction: Type 1, designated as "rigid-frame"; Type 2, designated as "simple framing"; Type 3, designated as "semi-rigid framing".

The NEHRP Recommendations [NEHRP, 1988] and the Uniform Building Code [UBC, 1991], recognize two classes of moment resisting frames, special and ordinary. Special moment resisting frames, SMRF, are those conforming to type 1 with additional special detail requirements. All other moment resisting frames are classed as ordinary, OMRF. SEAOC [SEAOC, 1988] and ATC3-06 [ATC, 1978] reference AISC and amend the specification to parallel UBC in establishing the two classes, SMRF and OMRF. The previous edition of UBC did not specifically recognize the OMRF but considered it only as a frame with K = 1.0 for "all framing systems except as hereinafter classified". The AISC Type 2 construction in the UBC would be assigned to the Building Frame System category. Type 1 without the special detailing and Type 3 construction would be considered as an OMRF.

5.5.1 Building Frame System

The gravity loads, defined in ANSI [ANSI, 1982] or UBC, are carried by a space frame which is defined as "a three-dimensional structural system, without bearing walls, composed of members interconnected so as to function as a complete self-contained unit with or without the aid of horizontal diaphragms or floor-bracing systems". The AISC Type 2 construction would meet the requirements of this category. For seismic design it is, however, recommended that the system be provided with a degree of moment resistant capacity in the detailing of the connections. It is known that even simple web connections have some moment capacity, particularly when the floor beams act compositely with the floor system. However, this type does not have the capability of resisting the cyclic forces at yield range stresses which are expected in seismic designs. It would be advisable to consider Type 3 construction for this purpose.

Seismic force resistance in this system is provided by braced frames or shear walls. See Sections 5.8 and 5.6 for information on these elements.

5.5.2 Ordinary Moment Resisting Frames - OMRF

AISC Type 1 construction without the additional details required for a SMRF would fit this category as would Type 3 construction with certain additional requirements. In the more seismically active areas of the country it would be advisable to use

Type 1 construction and, for some structures, the codes require Special Moment Resisting Frame (SMRF) construction. There are special seismic requirements for columns in Chapter 10 of NEHRP and Sect. 2722 of UBC (See discussion in 5.5.3). The R or R_w value for a steel ordinary moment resisting frame is set to provide essentially elastic behavior in a design level earthquake in an area having a value of v = 20 in NEHRP or in UBC Seismic Zone 2, but the probability still exists that inelastic behavior will occur [Montgomery and Hall, 1979]. Previously the UBC required the design base shear for this type of frame (K = 1) to be one and one half times that for an SMRF (K = 0.67). Due to the reduced ductility and the greater uncertainty in the actual capacity that is provided by the OMRF it was decided that this system was not capable of providing the level of performance expected in the code. The force level in the 1988 UBC has been raised to two times that required for a SMRF, R_w = 6 for OMRF compared to R_w = 12 for SMRF. The R values in NEHRP for an OMRF range from 4½ to 5½ while the value for an SMRF IS 8. The increased force level should discourage the use of OMRF for seismic resistance and encourage the use of more reliable connection detailing. The special detailing for SMRF systems should be adhered to where possible even though the codes do not require this. Local buckling of column flanges or webs or torsional buckling of beams or columns will greatly reduce the energy dissipation capacity of the system and will probably lead to unacceptable performance.

5.5.3 Special Moment Resisting Frames - SMRF

Special moment resisting frames are believed to be the most reliable structural system for resisting earthquake ground motions. When designed, detailed and constructed properly [AISC, 1986; ATC, 1978; SEAOC, 1988] they posses excellent ductility and energy dissipating characteristics, and their superior performance has been demonstrated in a number of past earthquakes. SMRF design is predicated on developing the full capacity of the horizontal framing at the connections to the vertical framing members [Butler and Kulak, 1971; Krawinkler, 1978; Popov et al, 1986] thus providing for the development of the full ductility of the material. Special detailing to preclude early inelastic behavior in elements of the members; i.e. local buckling of flanges or webs [Popov and Pinkney, 1969; Roeder and Popov, 1978], or panel zones [Krawinkler, 1978; Popov, 1978] or tearing at connections [Popov and Pinkney, 1969]. The over-all stability of the structure at realistic story drifts, not less than C_d or 3(Rw/8) times the computed elastic drifts for code level forces must be available.

AISC Type 1 construction with the special detailing requirements added in the NEHRP, UBC, SEAOC and ATC documents is required for seismic resistance only in limited applications. SMRF are required for buildings over 160 feet in height in areas with a value of v of 30 or greater in NEHRP or in UBC Seismic Zones 3 and 4 which places this requirement out side our low-rise category. Whenever SMRF is used the special seismic requirements for steel frames given in Section 2722 of UBC must be followed. AISC Type 1 construction may be designed in accordance with either Part 1 or Part 2. The special seismic requirements are tied more closely to Part 2 provisions than to Part 1 provisions or to the new load and resistance factor procedure [AISC, 1986]. The use of SMRF systems for wind resistance and for seismic resistance in low-rise buildings may be self-limiting because of the added costs in comparison with OMRF systems.

It has been observed through studies of instrumented buildings shaken by strong earthquakes and through laboratory testing of structures and sub-assemblages that the probable strength of an SMRF structure may be 300 to 500 percent of the design strength. This excess strength results from yield strengths in excess of those specified, strain hardening, increase in effective yield strength due to cyclic yield reversals, redistribution of forces through inelastic deformation and the participation of non structural elements such as walls, cladding, stairs, etc. This excess strength is essential for acceptable performance if R values as high as 5½ or R_w values as large as 6 for OMRF systems and R values of 8 or R_w of 12 for SMRF systems are used in design and if the story drifts are to be limited to a reasonable level (1.5%) under the design level earthquake ground motions, as defined in Chapter 4.

This excess strength leads to larger connection forces and column load requirements (compression and tension) than would be produced by the code level forces. As a consequence the codes have additional requirements for seismic design of steel frames. Columns must be designed for two additional load combinations in UBC.

A. Axial Compression

$1.0\ P_{DL} + 0.7\ P_{LL} + 3(R_w/8)\ P_E$

B. Axial Tension

$0.85\ P_{DL} + 3(R_w/8)\ P_E$

where P_{DL}, P_{LL} and P_E are the axial forces resulting from dead load, live load and earthquake, respectively. These are special load combinations beyond those cited in Chapter 2 to safeguard these critical gravity load supports against failure when they are subjected to the effects of actual, not design level, seismic forces. The column capacity, computed as $1.7\ F_aA$ must be equal to or exceed the column force. The special column load combinations apply to OMRF and SMRF systems. There are two exceptions to these requirements related to the limit of the force that can be transmitted to the columns by the foundations or the members framing to the column. NEHRP limits the stress level in columns to $0.6\ F_y$.

The basic requirement for SMRF systems is to have strong columns and weak beams, i.e. to force any hinging at inelastic excursions to take place in the beams. This requirement was added to the 1988 edition of the UBC whereas before it was only a recommendation in the "Blue Book" Commentary. There are some exceptions to the requirement and work is underway to further limit the exceptions. Large axial forces in conjunction with plastic hinges at the ends of columns could lead to failure of the column and, possibly, the building. The strong-beam weak-column system could produce a soft story behavior where all of the inelastic deformation occurs at a single level. This has been demonstrated to be unsafe and undesirable. Other special detailing requirements are provided, which will be discussed briefly in Section 5.5.5.

5.5.4 Detailing

The special detailing requirements which are beyond the normal AISC requirements [ATC, 1978; NEHRP, 1988; UBC, 1991] are intended to ensure acceptable performance for many cycles of inelastic deformation. A more complete description of detailing requirements is presented in Chapter 6. The girder-beam to column connection strength is required to develop the strength of either the girder-beam or the panel zone. Connections with flanges having full-penetration welds and webs either welded or high strength bolted are considered to satisfy the requirements without further investigation for members whose flanges provide at least 70 percent of the moment strength. Strength as used here is the plastic moment capacity. Where joint moment capacity is dependent on bolted flange connections there are severe limitations placed on the placement of the bolts.

The panel zone is required to develop a shear strength equal to 80 percent of the sum of strengths of the girder-beam framing to each side of the column. The shear strength of the panel zone is given by Formula 22-1 of Sect. 2722 of UBC or Sect. 10A.7.5 of NEHRP. Trusses are permitted as part of SMRF with strict control on the relative strengths of members and requirements for column bracing. There are also requirements for girder-beam flange bracing and limitation on change of flange area.

5.5.5 Non-structural Damage and Drift Control

Although steel moment frames provide excellent resistance to collapse when subjected to large earthquake ground motions, it is well known that they are quite flexible. They therefore offer little protection for the non-structural elements and cladding. Damage to building contents should be expected when subjected to moderate earthquake ground motions. The non structural damage in a "design earthquake" could lead to losses and replacement costs in excess of the original cost of the building. Some structures have been detailed with separation of non-structural elements from the structural frame and have survived strong ground motions with minimal damage. Such detailing is costly and will not be done unless mandated by the owner or by a code provision. Since building codes are typically minimum requirements for life safety, non-structural damage will continue to be accepted by the codes. For structures in the lower seismic zones it has been suggested that a moment frame system with an added nominal bracing system to limit wind drifts would be a possible solution to the problem. It should be remembered that the effects of such bracing on the moment frames would have to be investigated and proven to be acceptable.

5.6 CONCRETE AND MASONRY SHEAR WALLS

Concrete walls, cast in place and precast, and masonry walls are often found in low-rise structures. Many of these walls are provided to support gravity loads and generally provide enclosure for the structure. These walls can, and often are designed to, act as vertical elements of the lateral system resisting wind and earthquake forces.

5.6.1 Principal stresses

These walls have been designated as "shear" walls since the controlling stress in providing resistance to lateral forces is typically shear, especially in low-rise buildings. For design purposes, these walls are normally considered as being fixed at the base and are thus acting as vertical beams cantilevering from the base. The validity of this assumption will depend on the configuration and stiffness of the foundation or framing below the base and of the soils beneath the foundation. One medium rise shear wall building located on sand was determined to have rocked in a past earthquake based on the observed displacement of the sand at the face of the footing. Similar rocking could occur on isolated shear wall elements of low-rise buildings. This rocking functions as a non-linear mechanism, similar to a non-linear flexural hinge in a shear wall, and absorbs some of the energy imparted by the earthquake.

The height-to-width ratio of shear walls for many low-rise building walls would be one or less and they would be classed as deep beams. These walls should be designed and reinforced as deep beams. For wall segments where the controlling stress will be flexural the typical flexural strength design is appropriate. It should be remembered that lateral forces from wind and earthquake can reverse and that reinforcement may be in tension or compression. Present U. S. seismic codes require confinement of reinforcement in "boundary elements" to prevent buckling. There is evidence of that bars at the ends of shear walls have buckled in the past but there is some question of the necessity for confinement based on tests in this country and abroad.

An isolated shear wall with a height to width ratio of 2.5 or greater resting on a shallow footing will probably exhibit rocking behavior where the flexural strength of wall is greater than the stability moment at the foundation-soil interface. For such walls a rational design procedure is to determine the required out of plane flexural reinforcement and the minimum vertical reinforcement. The minimum vertical reinforcement for walls which are part of the seismic resisting system, shear walls, is twice that required for walls normally. Determine the in plane flexural strength of the wall with the reinforcement determined from above. Where this strength is less than the code required flexural strength increase the vertical reinforcement as required.

There is an implied requirement in the seismic codes that the resisting elements have an adequate energy dissipation capacity. A shear wall would have to behave elastically, have a shear capacity greater than 2.5 times the design shear, or have a shear capacity greater than its probable flexural strength. The probable flexural strength is determined utilizing all of the flexural reinforcement at its probable, not minimum, yield strength.

5.6.2 Panel Construction

The use of precast tilt-up concrete panel or prefabricated masonry panel walls has altered the manner in which many of these walls perform under lateral forces. The manner in which panel wall systems will behave when subjected to lateral forces must be analyzed using well-established principles of mechanics and providing for all forces imposed on each element from the origin of the force to the final point of resistance. Resistance to the vertical components of shear at the panel junctures, control of the deformations of the wall elements and of the horizontal diaphragms must be provided. In addition boundary elements to resist the forces developed in both wall elements and diaphragms will normally be necessary.

5.6.3 Reinforcement

Concrete and masonry walls providing resistance to earthquake forces have typically been reinforced, while many such walls not designed to resist earthquake forces have been unreinforced. The stresses determined from code level earthquake forces in many walls of low-rise buildings in areas of low seismicity would indicate that reinforcement would not be necessary. It is advisable, however, to provide reinforcement since the actual forces in "maximum probable" seismic event would exceed the code level forces by several hundred percent.

The ACI code [ACI 318, 1983] has special requirements in Appendix A for seismic resisting shear walls. The new masonry code [ACI/ASCE 530, 1988] prepared jointly by ACI and ASCE also has provisions for such walls.

5.6.4 Design

The lateral forces from wind are normally delivered to the horizontal diaphragms and transferred to the shear walls while the lateral motions of the ground in an earthquake generate inertial forces in the diaphragms as the walls follow the ground motions. The method of attachment of the diaphragm to the shear wall is a critical element of design.

The forces that will be delivered to the individual wall elements in the system is dependent on the rigidities of the diaphragms at the various levels. The code requires that torsion be provided for when diaphragms are not flexible and provides a definition for a flexible diaphragm.

Where diaphragms are not flexible the distribution of lateral forces at any level will be in proportion to the rigidities of the various vertical elements and torsional effects can be transmitted to these resisting elements. For individual solid wall elements the rigidity is the inverse of the deflection of the cantilever beam from the base to the top of the story under consideration. For wall elements with door and window openings the determination of the rigidity of the wall element is more complex. The relative rigidity of the wall piers in the wall element must be determined in order to assign the forces proportionally and then the rigidity of the wall element can be found. The deflection of piers acting as cantilevers or as vertical beams fixed top and bottom can be readily calculated. The deflection of the total wall element involves the proper combination of the individual pier contributions and the contribution of the spandrel elements connecting them. For one-story buildings, Amrhein (1978) and PCA (no date) are good guides. For taller structures, it is more appropriate to use a computer program. Several available programs permit easy input for complex arrangements of openings in shear walls.

5.6.5 Detailing

The Uniform Building Code (1991) in Section 2625(f) has detailing requirements for concrete "shear walls – serving as parts of the earthquake resisting systems". This section, with some modifications, is taken from ACI 318-89 Appendix A.5. Since ACI 318 deals only with reinforced concrete design there are no provisions in ACI or UBC Sect. 2625(f) for unreinforced concrete. UBC Chapter 24 permits unreinforced masonry in Seismic Zones 0 and 1 as does the new ACI/ASCE 530 standard. As previously mentioned it is recommended that concrete or masonry shear walls used to resist earthquake forces be reinforced. Where concrete or masonry walls are required to be reinforced there are minimum amounts of reinforcement required. There are special anchorage requirements for the longitudinal (horizontal) reinforcement and, for concrete, a requirement for providing "boundary members" at the ends of walls. UBC Sect. 2412 on Strength Design, Reinforced Masonry Shear Wall has a requirement for boundary members.

The physical limitations imposed by a thin, 6 to 8 inch, wall on obtaining anchorage has been recognized for concrete walls by permitting a 90 degree hook at the end of the horizontal reinforcement to engage the "minimum edge reinforcement". For masonry walls there is also a limitation on the maximum size of bar and the maximum percentage of a cell area that the reinforcement can occupy. UBC and the new ACI/ASCE 530 permit 6 percent.

5.7 FRAMED SHEAR PANEL WALLS

Framed shear panel walls may be sheathed with panel products (typically 4' x 8') or lumber boards. Lumber sheathed walls are no longer commonly used.

Building codes and material manufacturers publish lists of panel products that meet "wall bracing requirements". Wall bracing and shear walls are not synonymous. Conventional wall bracing is not engineered for a specific lateral load and is intended to provide adequate lateral strength for simple low buildings located in areas with low wind or seismic forces. Shear walls, specifically designed for the applied loads, are necessary for light framed structures of unusual shape, size, or split level, located in areas with velocity contours of 15 and above (NEHRP) or Seismic Zone 2A or higher (UBC).

5.7.1 Engineered Panel Sheathed Shear Walls

The shear resisting sheathing of a wall may be the siding-grade paneling or it may be the substrate sheathing for the final wall finish, such as stucco or lap siding. In either case, shear wall design starts by choosing the panel required for either its appearance or substrate qualities. Next this panel is checked for its shear resistance.

Shear wall design values are available for most panel products from tables included in Building Codes or from tables published by manufacturers or trade associations representing manufacturers of panel products. Since there are practical limits on how many combinations of sheathing, fasteners, and framing that can be included in a simple table, the designer should carefully check the table headings and footnotes to determine the exact applicability of the table. Also the footnotes often include details on how to adjust the table values for combinations not shown.

Usually, the lateral strength of the fasteners rather than the shear strength of the panel determines the shear capacity of a fastener-sheathing combination. Fastener head size is particularly important in resisting fastener head pull-through. Fastener choice is often limited when the shear-resisting panel also serves as the "finished" siding. Typically, the fastener head must be of minimum size (such as casing nails) for appearance and also the fastener must be corrosion resistant. Fastener strength may be further reduced by decorative grooves, cut into the panel, that reduce the panel thickness at the point of fastening.

The panel thickness at the location of the panel edge nailing determines the "net" panel thickness to be used in determining its lateral shear strength. Should it be possible to nail through the full panel thickness along the edges of a grooved panel, shear through the panel thickness should be checked at the minimum thickness occurring at the maximum-depth groove. If the shear values require nailing through the full panel thickness, the width of the usual combination shiplap-groove detail may make framing wider than 2x nominal necessary. Even with extra wide framing, careful workmanship and supervision is required.

Under controlled conditions, it is possible to place a single row of nails to fasten both the upper and lower lips of a shiplap joint. The thin material at the joint, combined with the extreme care required in proper placement of the fasteners makes this detail unsuitable for normal construction. When the shear load is low enough to allow it, a more practical approach is to nail only along the edge of the panel having the top lip of the shiplap joint. Proper nailing along the next interior framing member makes the wall the equivalent of a series of spaced 32"-wide shear panels (when studs are spaced 16" o.c.).

The design shears of panel-sheathed walls can be doubled by placing the identically same sheathing, using the same fastener schedule, on both sides of the wall. Should the sheathing on the two sides not be identical, the design shear is taken as twice that of the weaker side. In a double-sided shear wall, it is good practice, and often required by code, to offset the panel joints on each side to fall on different framing members.

The framing members at the panel joints and along the edges of the shear wall are very important. The width must be sufficient to avoid any possibility of splitting when the wall requires large and/or closely spaced fasteners. The end members must also be large enough for the framing anchor required to prevent uplift at the end of the wall. The large vertical forces, that develop in a shear wall under lateral load require design consideration. The anchor and hold-down is quite obvious, but sometimes overlooked. Typical hold-downs with bolts in single shear in the wood end members do not perform well under cyclic loading. Equally critical is compression of the lumber plates by the vertical forces from the end stud. Lumber has a lower allowable design stress when loaded perpendicular to the grain than when loaded parallel to the grain. Often the vertical load in the end member of a shear wall is sufficient to require a steel bearing plate to distribute the load. Another method of construction for very high loads, is to stop the plate at the end stud and extend the stud directly to a bearing plate at the foundation (for end grain bearing). This type of column base generally also provides side straps placing bolts in double shear in the wood member giving better uplift resistance.

Occasionally it is necessary to design a shear wall using wood-based panels attached to metal framing. There is very little published information on lateral design values for fasteners suitable to be used to fasten panel sheathing to metal framing. In most cases, it will be necessary for the engineer to test a sufficient number of specimens of the fastener-sheathing-stud combination to develop design shears. At present, there are only two references for metal framing: ICBO Report 2392 for sheet metal screws into 14-gauge and thinner steel and ICBO Report 4144 for steel pins into 11-gauge and thicker steel.

5.8 BRACED FRAMES

Braced frames are vertical trusses with the members loaded axially when resisting the lateral forces from wind or earthquake. Braced frames may be constructed of concrete, steel or wood or combinations thereof but steel is used for most braced frames. Braced frames are economical systems for resisting lateral forces, especially in low-rise buildings.

Braced frames may be divided into two categories, concentric and eccentric and these two categories will be treated separately.

5.8.1 Concentric Braced Frames

Concentric braced frames are defined as having members which are subjected primarily to axial forces. Concentric braced frames may be classed according to their geometry into diagonal, cross or X, K and chevron groupings. A knee-braced column does not qualify as a concentric braced frame, because the column section of the assembly is placed in flexure. These configurations are defined in Sect. 2722 of UBC applying to seismic resisting systems. K bracing is not permitted in UBC Seismic Zones 3 and 4 for buildings over two stories in height unless designed for elastic response, i.e. the bracing must be designed for $3(R_w/8)$ times the seismic forces generally applied to braced frames. They are prohibited for NEHRP Seismic Hazard Exposure Groups D and E. This requirement is imposed since the columns are primary vertical load carrying members and buckling of one leg of the K would impose high flexure on the column.

Chevron braces may attach to the top or bottom of a beam at mid span. Where the brace attaches to the bottom of the beam the beam in seismic resisting systems must be designed for the tributary gravity loads ignoring the presence of the brace. This provision will require a beam with sufficient capacity that buckling of one leg of the brace will probably not cause excessive strength degradation.

Framing members loaded axially into the inelastic range are relatively poor energy dissipators. Diagonal members of braced frames have exhibited cyclic buckling and plastic extension. Tests [Popov and Black, 1981], have demonstrated that the cyclic response of axially load struts is characterized by severely pinched hysteresis loops and progressive reduction in cyclic buckling capacity, Figure 5.8.1. UBC Sect. 2722 for seismic resisting systems provides some control on the bracing members by limiting the L/r of the member and imposing a cyclic reduction factor, related to the proportion of l/r to critical slenderness ratio, on the allowable stress. Tests of cold worked members, such as structural steel tubes [Jain, 1980], revealed that fracture of such members may occur after only a very few cycles of loading. The local buckling of the tube walls imposes critical strains at the corners of the tubes which have previously been strained by the cold working.

Limits are imposed on width-thickness ratios of elements of compression members in both NEHRP and UBC. The slenderness ratio of a part of a built up bracing member must be not greater than 75% of the slenderness ratio of the bracing member. There is an additional control against failure in any line of bracing in a seismic resisting system which requires an approximate balance between the capacity of members acting in tension and those acting in compression.

Bracing members themselves are not always the weak link in the system, as has been noted in past failures. Bracing connections now must be capable of developing the lesser of the full tensile strength of the brace in both NEHRP and UBC or 3(Rw/8) times the design force on the brace in UBC.

It should be noted that the above discussion concerns only steel-braced seismic resisting frames, since all of the information relates to NEHRP Sec. 10.8 or UBC Sect. 2722. The detailing requirements and member configuration limitations do not specifically apply to wind bracing systems but, as noted at the beginning of this chapter, they should be followed to provide the level of performance expected from good design. Braces in concrete frames would be governed by UBC Sect. 2625(e) with the longitudinal reinforcement required to be confined. Wood braced frames are not directly addressed. Wood exhibits high short term strength so the full increase in connection capacity required for steel is not appropriate. It would, however, be advisable to have the connection develop at least 50 percent of the member capacity.

5.8.2 Eccentric Braced Frames

The eccentric braced frame as a lateral force resisting system is relatively new. There have been low, generally one story, knee braced industrial buildings built for a number of years but the normal braced frame in the past has been concentric. The eccentric braced frame recognized by the code is limited to steel construction at this time and has a number of limitations placed on configuration and brace capacity. The criteria for the eccentric brace requirements was developed from a number of tests of frames and sub-assemblages [Kasai and Popov, 1980, 1986a,b; Malley and Popov, 1984], and there have been some buildings designed in accordance with the developing criteria.

The eccentric brace must have at least one end of the brace engaging a horizontal member at a location other than at the intersection of the horizontal and vertical members. The eccentric brace is not permitted to intersect a vertical member (column) at any location except at the intersection with a horizontal member (beam). The segment of the beam between the end of the brace and the column, or between the ends of two braces intersecting near the mid point of the beam, is termed the "link beam". This requirement would eliminate a knee braced column from consideration as part of an eccentric braced system.

This system has demonstrated the favorable combination of the stiffness of the braced frame system and the ductility and energy dissipation capacity of the special moment frame system.

5.9 DIAPHRAGM SHEARS AND DEFORMATIONS

A diaphragm is an approximately horizontal structural system that transfers the lateral forces applied to a structure to vertical structural elements for transfer to the foundation. In typical building design, the lateral forces are from wind or seismic forces. The resisting elements are often shear walls (vertical diaphragms) but may be vertical trusses and, where resisting elements are placed only at the perimeter, may be moment resisting frames.

Often diaphragms are described as acting in a manner analogous to a deep beam, where the floor-roof membrane material acts as the "web", resisting shear, while the framing members at the diaphragm edge perform the function of "flanges", resisting the bending stresses. These flange members act in compression or tension (along the length of the diaphragm) and are traditionally called chords, while the members transferring shear to the vertical resisting element (along the ends of the diaphragm) are called struts. Framing members in the interior of the diaphragm that transfer shear forces into the diaphragm or to vertical structural elements are referred to as collectors or drag struts.

The "horizontal" in the definition of a diaphragm refers to the direction of the forces resisted by the diaphragm and not necessarily to the orientation of the actual diaphragm. Diaphragms can slope, can be curved (roof using bow-string trusses), or can include changes of slope (roof with ridge). Significant curvature or change of slope generates forces perpendicular to the plane of the diaphragm and the design must consider these forces.

Diaphragm design uses the floor and roof as effective and economical elements to provide the lateral force-resisting system for a building. Very often in wood and steel construction the sheathing and framing required to support the vertical design loads will also serve as the lateral force-resisting system by properly specifying the fasteners and/or reinforcing to make it function as the "web." Similarly, by detailing the necessary connections to give continuity and by sizing them for the tension and compression forces, the edge members will become the "flanges." Metal decks with concrete fill and reinforced concrete slabs also act as diaphragms and some of the criteria noted above apply.

Diaphragm design involves the following steps:

1. Calculate the diaphragm shears.
2. Design the shear-resisting element, the "web".
3. Determine required flange (chord) area.
4. Calculate deflection.

5.9.1 Diaphragm Shears

In a simple rectangular building with shear walls at the ends such as Figure 5.10.1, the design diaphragm shear, in pounds per foot, is simply the total lateral force divided by twice (for the two end walls) the width of the building. This shear is maximum at the end walls and uniformly decreases to zero at the mid-length of the building.

The structure must be analyzed both across its width and along its length. Normally the higher shear will be parallel to the least dimension of the structure. Also the design must consider that the wind or seismic loading will reverse and can be from either direction.

5.9.2 Design of Diaphragm Components

The first step in the design of diaphragms is to design the "sheathing" material and its supporting framing members for the required vertical load. Next the material chosen is checked for its shear resistance.

Should the sheathing material chosen on the basis of the vertical loading not have sufficient shear resistance for the lateral loading, it will be necessary to increase its shear resistance. This may be by increased thickness, increased reinforcement (for concrete), or by going to a grade or type having higher shear properties. Upgrading the sheathing may allow such offsetting benefits as allowing the framing members to be spaced further apart. Also to be considered, the required shear resistance of the web (sheathing) is reduced as the shear in the diaphragm decreases.

The design shears of building materials can usually be obtained from tables included in building codes, or published by manufacturers or trade associations representing manufacturers of sheathing materials. It is also possible to calculate design shears by principles of mechanics, using building code-accepted lateral design methods, the shear properties of the sheathing material, the axial properties of the flanges and the capacities of the fasteners.

One caution in using design values from published tables for panelized construction; the common use of these values is for the design of diaphragms used to resist wind or seismic loading and, since most building codes allow an increase for short term wind or seismic load for all building materials, the tables already include a one-third increase in design stresses. Check the table heading and footnotes for confirmation of the type of loading assumed in developing the table. For normal load duration (10 years for wood) use 75% of the published values if they were intended for wind or seismic loading. For wood-based materials, reduce this an additional 10% for permanent loading, such as for resistance of permanent earth pressure.

The diaphragm chords (flanges) are designed for the axial "flange" forces calculated on the basis of the diaphragm acting as a deep beam. The force at any point along the chord can be determined by resolving the moment in the diaphragm at that location into a couple (equal and opposite forces separated by the width of the diaphragm). Since these forces relate to the bending moment, they are maximum at the mid-span of the diaphragm and reduce to zero at the ends.

It should be noted that the "flange" forces may never actually occur where the flange members are walls having shear resistance capacity which would transfer the forces directly to the foundation. The use of the girder analogy is conservative and the provision of "flange" or "chord" members is not a major cost, so it is better to use this procedure and avoid the complexities in design associated with determining when such members are not needed. This is further discussed in Section 5.9.3.

It is common practice to use already existing wall members (such as the top plates or horizontal wall reinforcement) or roof edge members as the chords. The size or number of these members may have to be increased depending on the load in the chord at that location. Also the members will have to be made continuous by splicing or staggering end joints. Each chord must be designed for both tension and compression since the chord forces reverse with the reversal of load on the diaphragm.

5.9.3 Diaphragm Deflection

The important considerations in diaphragm deflection are that the structure does not have any irregularity to restrict deflection of any portion and that the maximum possible deflection will not allow the structure to contact another structure. Note that under seismic loading, it is possible for two adjacent structures to be out of phase or, in other words, they may deflect toward each other.

Anything attached to the structure must be either strong enough to resist the lateral force at the point of its attachment or flexible enough to deflect with the structure. A common oversight is to incorporate one or more of the interior roof supporting columns into the walls of a very rigid enclosure such as a concrete vault. The top of any column, so stiffened, is then no longer free to move with the roof, and failure will occur if the column and its attachment to the roof do not have adequate strength to resist the lateral force at that point.

Exact methods of computing diaphragm deflection do not exist for conventional panelized diaphragms constructed of individual elements. The commonly published equations are based on tests and assume "worst possible" conditions. The equations are useful only to give the designer an indication of the maximum possible deflection that could occur. The actual deflection will usually be much less. The usefulness of this calculated deflection is further reduced by the lack of a meaningful maximum deflection limit.

The test diaphragms used to establish the factors used in the deflection equations have been restrained only by the simulated shear resisting elements, usually at the ends, and were otherwise totally free to deflect. Likewise the flanges of the test diaphragms developed the entire axial forces caused by the bending moment in the diaphragm. In buildings where the flanges are walls having shear capacity, the maximum theoretical chord force does not develop since the force being transferred from the diaphragm to the flange is immediately transferred through the wall to the foundation. This shear transfer also reduces the force on flange splices, if any, and the resulting joint slip contribution to deflection.

Diaphragm chords and their continuity are important to proper structural design, and the fact that they can be shown to be theoretically unnecessary should not be used to lessen or eliminate their use. "Chordless" diaphragm design assumes that the shear transfers from the roof to the wall and that the wall becomes an extension of the "horizontal" roof diaphragm, which makes the foundation or footing the diaphragm chord. Omitting the chord is not conservative design since it allows any construction or engineering oversight the potential of causing major structural weakness. Any opening or change in stiffness of the shear wall will also disrupt the transfer of shear between diaphragm and wall.

5.9.4 Special Diaphragm Considerations

Strong ground motion instrumentation records have shown that the motion at the roof at mid span of flexible diaphragms is two to two and one half times the motion occurring at the end walls. This amplification was anticipated by the research work described in Blume et al (1961). This amplification has not been reflected in the codes in the past and the performance of code complying structures in past earthquakes has been generally good. The notable weakness has been in the wall to diaphragm connections and the transfer of these anchorage forces into the body of the diaphragm. The UBC 1991 has provided for the amplification for the anchorages.

The discussion in the previous sections has been limited to rectangular buildings with shear resisting elements available along all four sides. In actual practice, often there are openings necessary in the diaphragm or the building may not have sufficient wall length along one or more sides to resist the shear that would normally be transferred from the roof or floor diaphragm to the wall. These problems will be covered in the following discussion.

The relative size of the opening in comparison to the over-all size of the diaphragm determines how thoroughly the framing around the opening has to be engineered.

The connections of the framing members at each corner of an opening must be designed for both compression and tension. As the load on the diaphragm reverses, the corners are alternately in tension and compression. Likewise both framing members at a corner of an opening must be extended. (A four-sided opening has eight extensions of framing members.) These framing members become drag struts or the flanges of local or "sub" diaphragms.

Calculations are seldom necessary if the opening is quite small, say 4' by 8', in a diaphragm that is 100' wide by 200' long. Common practice is to extend the framing members at each corner of the opening for a distance equal to the parallel dimension of the opening. The sheathing must be attached to these "extended" framing members using a typical "boundary" fastener schedule. This is particularly important, but easily overlooked in a diaphragm constructed of individual elements when the framing member is not at an edge joint between sheathing panels. Peery (no date) writes: "One of the established principles of mechanics, formulated by Saint Venant, states that the stresses resulting from such a system of forces will be negligible at a distance from the forces. The distance is approximately equal to the width of the opening."

Rectangular buildings with walls on only three sides can be designed to resist lateral forces. Calculation of shear resistance in the direction perpendicular to the open side is exactly the same as for a building walled on all sides. The chord is very important on the edge of the diaphragm which lacks a shear wall. This chord will receive the full theoretically calculated forces since there is no possible transfer to a supporting shear resisting wall.

In resisting forces parallel to the open side, the diaphragm is forced to simulate a cantilever beam extending to the open side from the opposite wall. A cantilever beam must have a fixed end; however, in this case, the top of the wall will not be rigid, but will rotate with the diaphragm. This rotation transfers shear directly from the diaphragm to the two walls perpendicular to the applied load. The shear in the diaphragm is zero at the open end and increases uniformly to its maximum at the opposite end. Likewise, the shear in the walls perpendicular to the open side is zero at the open end and increases uniformly to its maximum at the end of the wall furthest from the open side. At this point, its value is equal to the unit shear in the single wall parallel to the lateral force.

Because of the inherent weakness of an open-sided building, the use of a continuous member (chord) along the wall-roof intersections is again suggested as good engineering practice. Structural symmetry is also recommended to help reduce the eccentricity of the building.

5.10 DIAPHRAGM DESIGN

Diaphragms, as discussed in Section 5.9, can be made of three of the common construction materials; concrete, steel and wood. Masonry has not normally been used in diaphragms. The general considerations and requirements for the three common diaphragm construction materials are discussed below.

5.10.1 Concrete Diaphragms

Reinforced concrete structural slabs are commonly used to support gravity loads. These same slabs can typically function as diaphragms to resist lateral forces. For most building configurations the diaphragm would be considered a deep beam and should be designed as such. Deep beam theory dictates that the flexural reinforcement be distributed over a wide area of the beam and not concentrated in a narrow area at the edge of the beam. Concrete diaphragms will normally be considered rigid but the relative rigidities of vertical elements and diaphragms should be calculated when diaphragm length to width ratios exceed four to one. The provisions of Section 2312(e)6 govern the determination of flexible diaphragm designation.

5.10.2 Steel Diaphragms

Diaphragms utilizing steel are typically formed steel decking of differing configurations and used as forms for concrete or support for rigid material underlying roof membranes. Diaphragm action to resist lateral forces is a secondary purpose and might be considered a bonus for the designer and owner.

Steel decking used as a slab form is often embossed to provide for composite action with the concrete for gravity load carrying capacity. Decking for floors may be double layered with voids formed for use as electrical or mechanical raceways or ducts. Where a structural concrete fill is used over the deck it provides the "web" of the diaphragm but the attachment of the deck to the framing provides for the transfer of forces to and from the diaphragm. The diaphragm action for these decks is the same as for reinforced concrete decks described earlier.

Many metal decks are used for roofs with no concrete fill. These decks act as panelized systems with the connections to the framing members providing for the interaction to resist the web shears and develop the flange forces in the diaphragm. Metal decks are typically classified into three generic types; narrow rib (NR), intermediate rib (IR) and wide rib (WR). Decks are made from light steel sheets from 22 to 14 gauge (0.030 to 0.074 inch thickness) in panel widths of 18 to 36 inches.

Metal deck diaphragm strengths and rigidities have been established by tests and accepted by the model code groups on the basis of the results of such tests. The 1989 addendum to the Cold Formed Steel Specification [American Iron and Steel Institute (AISI), 1989] has a diaphragm test procedure. Most of the tests were monotonic, not cyclic, so their behavior under seismic forces cannot be reliably predicted by the tests. Based on the tests which have been run a semi-empirical design formula has been developed for predicting strength and stiffness of metal deck diaphragms. Generally metal deck diaphragms without concrete fill would be considered flexible but the relative rigidities of vertical resisting elements and diaphragms must be determined to verify the designation as noted in Section 5.10.1.

Fastening of the metal decking to the framing members and to itself along the seams is critical to the behavior of the diaphragm. The initial form of attachment of the deck to the framing members was by burn through puddle welds. Seam connections were by fillet welds or button punching. These are still quite typical methods of connection but there are now approvals for connections using self drilling self tapping screws or powder actuated drive pins.

Detailed design of steel deck diaphragms will not be discussed here and it is recommended that the reader refer to AISI (1989), Ha et al (1979) and Atrek and Nilson (1980).

5.10.3 Shear Panel Diaphragms

Shear panel diaphragms consist of structural wood panel sheathing over framing members. The common wood panel diaphragm is constructed of 4' by 8' plywood panels; however other wood based, structural use, panel products [American Plywood Association (APA), 1988a] are being developed that have approximately equivalent shear strength to plywood. These non-veneer or partial veneer panels are accepted for shear resisting elements by building code officials on the basis of "performance testing" showing their equivalence to previously accepted products [APA, 1988b].

The design of diaphragms constructed of modular elements such as plywood or other wood-based panels requires specific attention to panel layout and fastener size and spacing. The first step in the design of diaphragms constructed of individual panel elements is to design the panel and its supporting framing members for the required vertical load. There are constructions where vertical loads are carried by other sheathing, tongue-and-groove planking, and the panel elements are used only to resist diaphragm shears. Next, the panel chosen is checked for its shear resistance. In panel-type elements, shear resistance may be dependent on such things as:

A. Panel layout relative to adjacent panels.

B. Number and size of fasteners.

C. Framing spacing.

D. Provision for shear transfer at panel edges at right angles to the framing members.
   a. Blocked diaphragms have framing members or other provision along these edges to transfer shear to the adjacent panel.

b. Unblocked diaphragms do not have provision for transfer of shear. Where necessary, the unblocked edge must be properly designed for potential differential vertical loads by use of such details as T & G panel edges or panel clips.

If the panel layout or positioning of the panel joints affects the shear resistance, the desirable design would be to place the stronger layout (case 1 or 2 for panel-type sheathing, Figure 5.10.1) in the direction of greatest shear.

The required fastener size and spacing can usually be obtained from published tables. Fastener spacing is closely related to the choice of sheathing and its layout and often the two are combined in the same table of design values included in building codes or published by manufacturers or trade associations representing manufacturers of sheathing materials. These tables usually show minimum combinations of fastener size and sheathing thickness that will develop the diaphragm shear.

There are practical limits to the number of sheathing grades, fastener sizes and spacings, and lumber species that can be shown in a typical table and the designer should understand how the tabular values apply to other fastener-sheathing combinations. Thicker sheathing can be used with the same fastener for the same design shear, provided the fastener is long enough to meet the minimum penetration into the framing required in the building code. Footnotes to the table will usually describe how the shear values should be adjusted for combinations not shown. If reductions are necessary for both sheathing grade or type and lumber species, only the greater reduction is applied. Reductions are not cumulative.

Typically, the lateral strength of the fasteners, rather than the shear capacity of the sheathing, controls diaphragm design. The size and maximum number of fasteners must be limited so that there is no potential of splitting the lumber framing. Diaphragms with design shears in excess of 750 to 850 plf, require the use of framing members wide enough to allow placement of the fasteners in multiple rows. Staples do not tend to split lumber to the same extent as nails. This allows much closer spacing of fasteners when staples are used.

The number of fasteners can be reduced as the shear decreases towards the center of the diaphragm. However, all sheathing materials have a specified minimum number fasteners required to hold the panels in place and to lessen the chance of panel buckling between widely spaced fasteners.

5.11 DESIGN EXAMPLES

The following design examples were prepared to illustrate some of the principles discussed previously. The presentation is arranged in a manner which might be used for submission to a building department or checking agency when applying for a plan check prior to obtaining a building permit.

A general statement listing the code criteria and the materials properties used are the lead items of information. A listing of the gravity dead loads and live loads for the structure is needed before design can proceed. The next steps in design are the development of member loads and forces and the sizing of members or systems to resist these loads and forces. For lateral forces it is convenient to present the information in a tabularized format.

This handbook has discussed in Chapter 2 the probabalistic basis for the hazards due to wind and earthquake that the codes, and the designer, should consi-

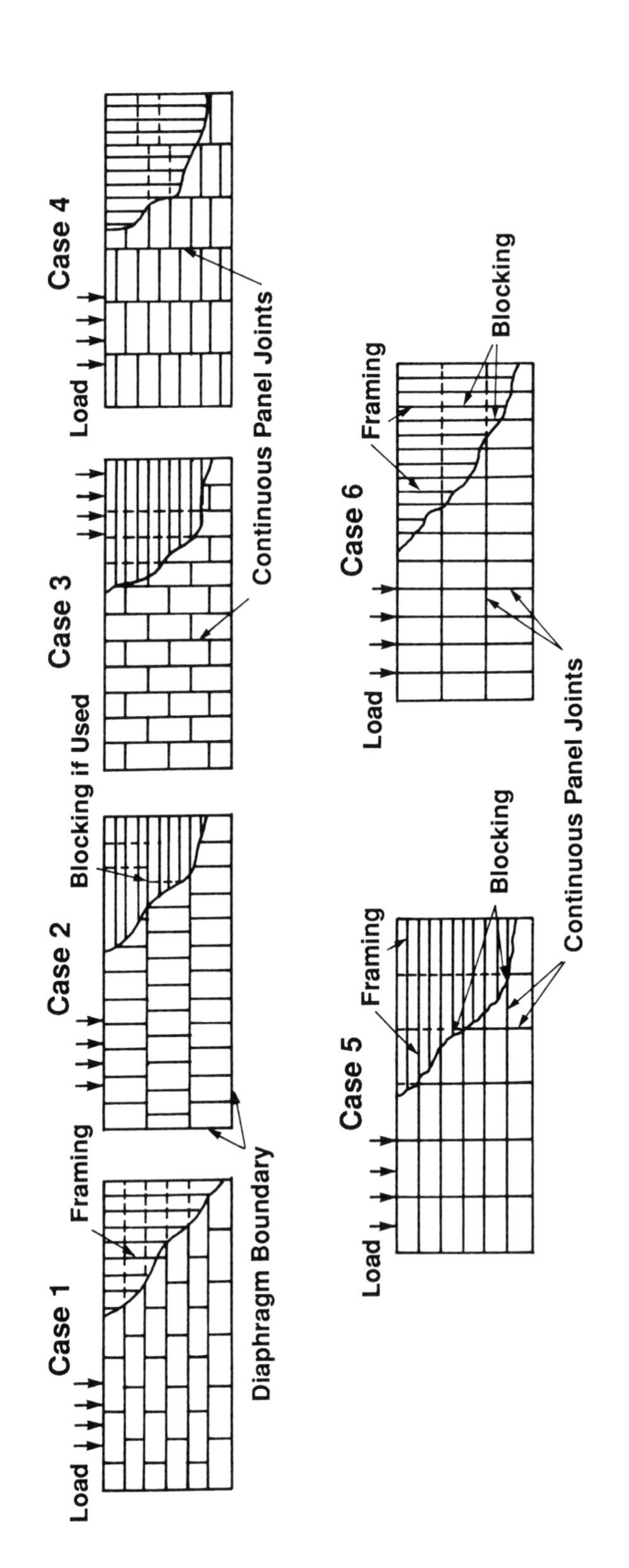

Figure 5.10.1 Shear panel diaphragms

der. For wind forces the basis for selection of degree of hazard is the wind contour map from ANSI A58/ASCE Standard 7, as noted in Chapter 3, or variations thereof in the model codes. For seismic forces the basis is more divergent. Since this handbook is intended to acquaint the user with the latest developments on ground motions and bridge the gap between these developments and the design codes that must, as a minimum, be met for approval of a project it was decided that the most recent contour mapping of probable ground motions would be incorporated. It should be noted that both wind and earthquake mapping is based on a 0.02 annual probability of occurrence of the design contour values. We will determine the required design base shears to which each example building would be subject in order to conform to the two criteria.

The earthquake force base shear is determined utilizing the mapping shown in Chapter 2 and the procedure in Chapter 4 or the mapping and procedure in the UBC 1991 edition. The selection of "v" or "Z" will be from the same map vicinity.

The base shear formula, from the Appendix to Chapter 1 of the 1988 NEHRP document, employed with the new contour mapping is: $V = C_sW$, as compared to the Uniform Building Code Formula which is: $V = ZICW/R_w$. The Formula for C_s, $C_s = 0.014vS/RT^{2/3}$ has most of the same parameters as ZC/R_w in the UBC Formula where $C = 1.25S/T^{2/3}$. The importance factor, I, as a multiplier on forces is not used in NEHRP but a more stringent story drift control is used to recognize the need for damage control and improved performance of critical facilities. The NEHRP Formula uses the velocity, v, from the velocity contour map while the UBC Formula uses the zone factor, Z, which is the contour of a combined EPA/EPV value approximating a design acceleration. You will note that both formulae have the soil factor, S, and the period, T, raised to the $^{2/3}$ power. They both have a response modification factor, R or R_w, but the values are different since NEHRP uses a strength design basis while UBC uses an allowable working stress basis. The base shear quantity from the NEHRP procedure will, on the average, be 30% higher than that obtained from the UBC procedure.

Both procedures have a limiting value which C_s or C need not exceed. For most low-rise buildings the limiting value will be the design value and the building period need not be calculated.

DESIGN EXAMPLE 5.11.1 One story steel frame building

A one-story steel frame industrial building will be designed for a site in the mid-west. An isometric view of the building with plan and elevation dimensions is given in Figure 5.11.1-1. The basic design information is given below:

DESIGN LOADS AND FORCES

Live Load	Roof		20 PSF
Lateral Forces	Wind	ASCE 7 1988:	70 mph
	Seismic	NEHRP 1988: A = 0.15	v = 15
		UBC 1991:	ZONE 2A
Material Properties	Steel	ASTM A36	36,000PSI

GRAVITY LOADS

		ROOF PSF		WALLS PSF
DEAD	Metal deck (Insulated)	3	Metal panel (Insulated)	3
LOAD	Purlins	2	Girts	2
	Beams	4	Columns	5
	Lighting & mechanical	2		
TOTAL DEAD LOAD		11		10
LIVE LOAD		20		
TOTAL LOAD		31		

WIND FORCES

For the mid-west region selected the Basic Wind Speed from Figure 1 is 70 mph. For a building with an Importance Factor of 1 located in an area with an Exposure Category C determine the wind pressures and forces at the various levels [ASCE 7, 1988]:

$$p = q * [(G_h * C_{pw}) - (G_h * C_{pl})]$$
$$q = 0.0256 * K_z * (I * 70)^2 = 12.544 * K_z$$

From Table 6 [ASCE 7, 1988], K_z, for 0' - 15', average upper wall, eave, mid roof and ridge heights are 0.800, 0.835, 0.870, 0.917 and 0.963.

From Table 8, G_z, for 0' - 15', eave, mid roof and ridge heights are 1.32, 1.29, 1.273 and 1.263. For the main building frame $G_h = G_z$ at eave = 1.29 since roof slopes less than 10°.

From Figure 2 the value of C_p is: for windward wall, 0.8; for leeward wall, -0.5 for E-W and -0.45 for N-S; for side walls, 0.7; for windward roof, -0.27 or -0.8 and for leeward roof, -0.7 and for wind parallel to the ridge, -0.7.

The wind forces on the building, $p = q * K_z * G_h * C_p$:

p_{w1} = 12.544 * 0.800 * 1.29 * 0.8 = 10.36 psf windward
p_{w2} = 12.544 * 0.835 * 1.29 * 0.8 = 10.81 psf windward
p_{w3} = 12.544 * 0.917 * 1.29 * 0.8 = 11.87 psf windward
p_{wl} = 12.544 * 0.870 * 1.29 * 0.5 = -7.04 psf E-W leeward
p_{wl} = 12.544 * 0.870 * 1.29 * 0.45 = -6.34 psf N-S leeward
p_r = 12.544 * 0.870 * 1.29 * [-0.27 or -0.8 or -0.7] = -3.80 or -11.26 E-W roof windward and -9.85 psf leeward and side walls.
p_i = 12.544 * 0.870 * 0.25 = 2.73 psf internal (internal forces cancel for total force on building)

The total force on the building at the roof equals the area tributary to the roof times the wind pressure on the tributary area, see Table 5.11.1-A.

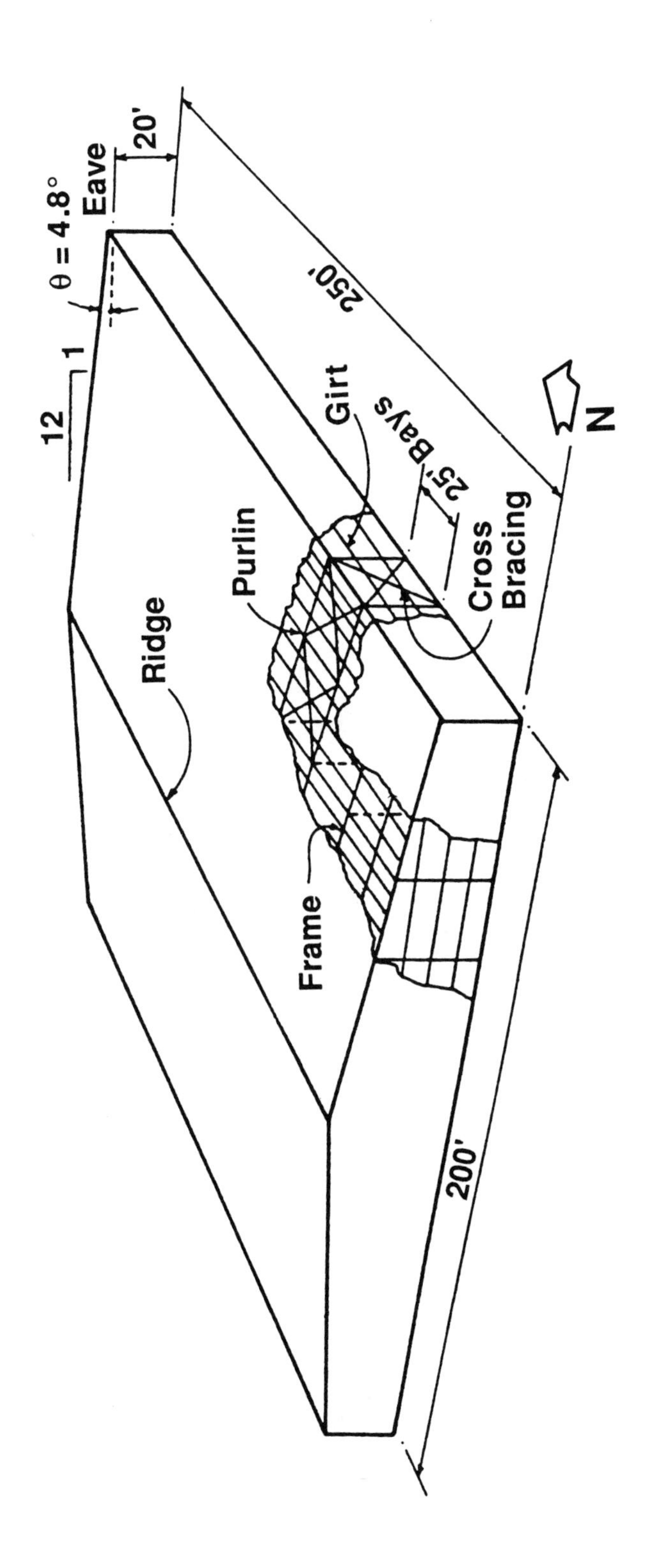

Figure 5.11.1-1 One story frame building: dimensions and structural framing system

SEISMIC FORCES

Calculate total base shear.

NEHRP	**UBC**
$V = C_s * W$: $v = 15$	$V = Z * I * C/R_w * W$: $Z = 0.15$
$C_s = 0.014 * v * S/T^{2/3}$	$I = 1$: $C = 1.25 * S/T^{2/3}$
S, Soil factor = 1.2	S, Soil factor = 1.2
R is 4.5 E-W and 5 N-S	R_w is 6 E-W and 8 N-S

W = weight tributary to roof

$T_a = C_t * h_n^{3/4}$ E-W, $C_t = 0.035$	$T = C_t * h_n^{3/4}$
$T_a = 0.05 * h_n/L^{1/2}$ N-S	$C_t = 0.035$ E-W and 0.020 N-S

$h_n = 24.2'$ (mean roof height)

$T_a = 0.035 * 24.2^{3/4} = 0.382$	$T = 0.035 * 24.2^{3/4} = 0.382$ E-W
$T_a = 0.05 * 24.2/250^{1/2} = 0.077$	$T = 0.020 * 24.2^{3/4} = 0.218$ N-S
$C_s = 0.014 * 15 * 1.2/R * Ta^{2/3}$	$C = 1.25 * 1.2/T^{2/3}$
$C_s = 0.252/4.5 * 0.382^{2/3} = 0.1064$	$C = 1.5/0.382^{2/3} = 2.849$ E-W
$C_s = 0.252/5 * 0.077^{2/3} = 0.2796$	$C = 1.5/0.218^{2/3} = 4.138$ N-S
C_s (limit) = $2.5 * A/R$: $A = 0.15$	C (limit) = 2.75: Base Shear Coefficient, $C_b = Z * I * C/Rw$
$C_s = 2.5 * 0.15/4.5 = 0.0833$ E-W	$C_b = 0.15 * 1 * 2.75/6 = 0.0688$
$C_s = 2.5 * 0.15/5 = 0.0750$ N-S	$C_b = 0.15 * 1 * 2.75/8 = 0.0516$

For a low building the limiting value of C_s or C will govern and the period need not have been calculated. See Table 5.11.1 (A) for W and V for both directions.

It would appear that the NEHRP seismic forces would govern the design rather than wind while is clear that wind forces govern over UBC seismic forces for this example. The difference between strength design and allowable stress design will show wind to govern over NEHRP seismic in the E-W direction and be approximately the same for the N-S direction. It should be remembered, however, that there are detailing requirements associated with seismic design which must be met and it is recommended that seismic detailing be followed since the NEHRP seismic forces are higher than 50 percent, and the UBC seismic forces are higher than 40 percent, of the wind forces.

Design For Lateral Forces

The design will follow the UBC allowable stress procedure. For the N-S direction the lateral force resistance will be provided by diaphragm action (bracing) in the plane of the roof and by bracing in the side walls. For the E-W direction the moment frames will provide the resistance.

The force to each bracing element in the N-S direction with two bays braced is:

$$F = 42.61/(2 * 2) = 10.65 \text{ kips}$$

The maximum load on a roof brace assuming tension only bracing is:

$$P = 10.65/\cos 45 = 15.06 \text{ kips}$$

The maximum load on a wall brace assuming tension only bracing is:

$$P = 10.65 \cos 38.66 = 13.64 \text{ kips}$$

The uplift forces on the roof can contribute to the overturning moment on the braced bays depending on their relative location. The columns must be designed for the most critical combination of axial loads.

For the moment frames the critical loading may be one of the possible combinations of external and internal pressures. The forces on the frames are the 25 foot tributary width times the pressures. The values and their location on the frame are shown on Figure 5.11.1-2. There is the primary loading condition for gravity dead and live loads which is not shown on this figure.

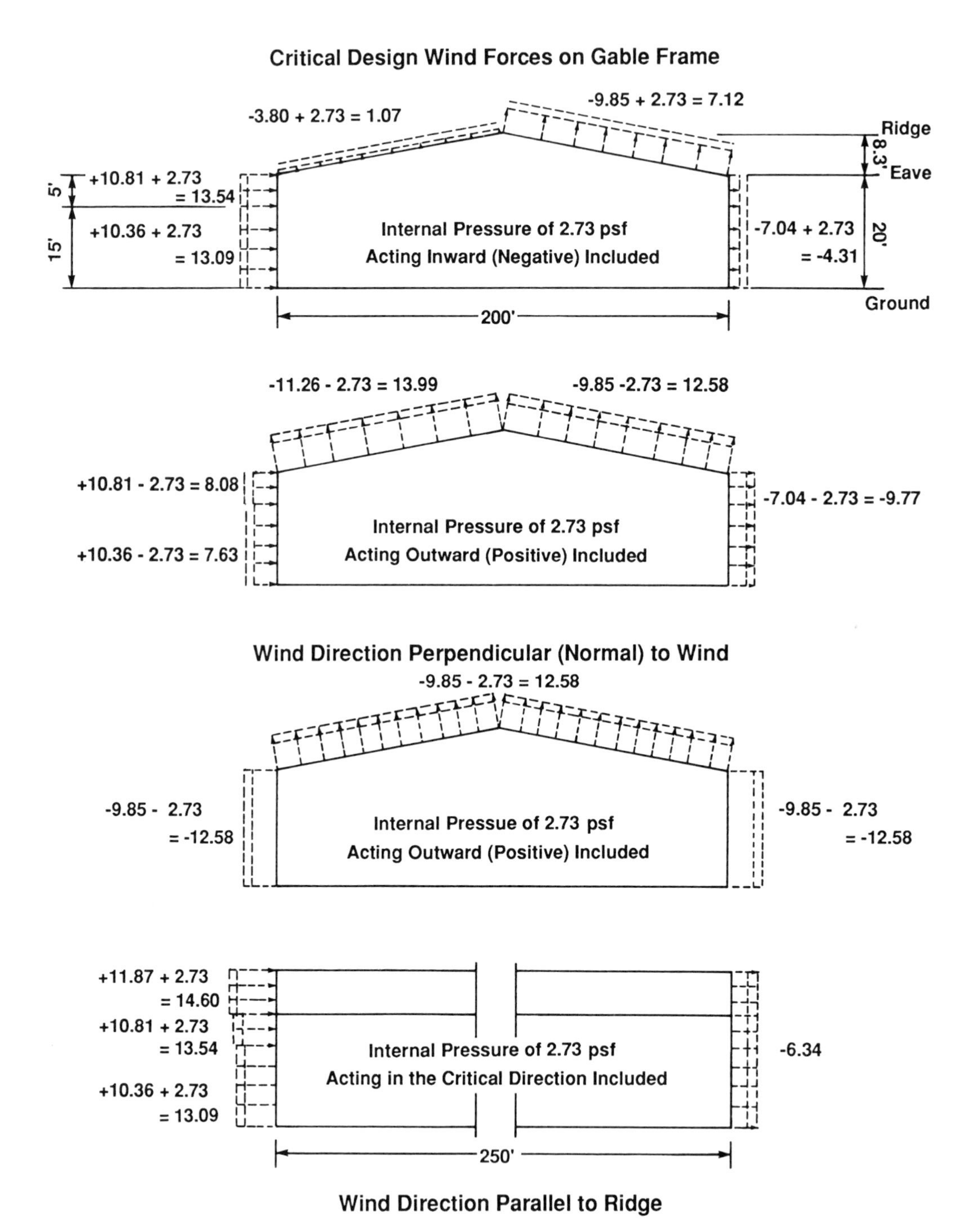

Figure 5.11.1-2 Critical design wind forces on gable frame

TABLE 5.11.1 (A): NORTH - SOUTH DIRECTION

		WIND				SEISMIC					
LEVEL	Height from base h_x (feet)	Area Tributary to level (1000 sq. ft.)	Wind Pressure (pounds per sq. ft.)	Story Shear [F] Story Shear V (kips)	Overturning Moment [Story] Total (foot kips)	Weight Tributary to Level (kips)	Distribution Factor*	N E H R P		U B C	
								Story Shear V (kips)	Over-turning Moment (ft.kips)	Story Shear V (kips)	Over-turning Moment (ft.kips)
ROOF	0 to 15 15 to 20 0 to 20 20 to 28.3	3.00 x 0.3103 1.00 x 0.7256 0.83 x 0.9924 4.83 x 0.5050	10.36 10.81 11.87 6.34	[9.65] [7.84] [9.78] [15.46]	[1032.88]	(Rf 1 x 50.0 x 0.011 = 550.0) (Wl .5 x 9.67 x 0.010 = 48.3)	1.000				
				42.74	1032.88	598.3		44.87	1084.4	30.87	746.1

*For NEHRP and UBC

TABLE 5.11.1 (B): EAST - WEST DIRECTION

		WIND				SEISMIC					
LEVEL	Height from base h_x (feet)	Area Tributary to level (1000 sq. ft.)	Wind Pressure (pounds per sq. ft.)	Story Shear [F] Story Shear V (kips)	Overturning Moment [Story] Total (foot kips)	Weight Tributary to Level (kips)	Distribution Factor*	N E H R P		U B C	
								Story Shear V (kips)	Overturning Moment (ft.kips)	Story Shear V (kips)	Overturning Moment (ft.kips)
ROOF	0 to 15 15 to 20 0 to 20 20 to 28.3	3.75 x 0.3103 1.25 x 0.7586 5.00 x 0.4138 2.083 x 1.0000	10.36 10.81 7.04 6.05	[12.06] [10.25] [14.57] [12.60]	[1195.77]	(Rf 1 x 50.00 x 0.011 = 550.0) (Wl .5 x 10.00 x 0.010 = 50.0)	1.0000				
				49.48	1195.77	600.0		49.48	1207.9	41.25	996.9

* For NEHRP and UBC

DESIGN EXAMPLE 5.11.2 One story industrial building with mezzanine office

A one-story industrial building will be designed for a site in the mid-west. The steel framing supports metal roof decking. The mezzanine construction consists of concrete fill on metal decking on open web steel joists. The mezzanine floor is not connected to the walls and is, in effect, a structure within a structure with concentric braces providing lateral force resistance. Plan views of the roof and mezzanine are shown in Figure 5.11.2 -1. Perimeter walls are reinforced masonry and vertical steel bracing is provided where shown on the mezzanine plan.

The basic design information is given below:

DESIGN LOADS AND FORCES

LIVE LOAD	ROOF		20 PSF
	FLOOR		100 PSF
SNOW LOAD			16 PSF
LATERAL FORCES	WIND	ASCE 7 1988	70 MPH
	SEISMIC	NEHRP 1988: A = 0.30,	v = 30
		UBC 1988	ZONE 3

MATERIAL PROPERTIES

STEEL	ASTM A36		36,000PSI
CONCRETE	ASTM C94	f'_c	3,000PSI
REINFORCING	ASTM A615,GRADE 60		60,000PSI
MASONRY	ASTM C90	f'_m	1,500PSI

GRAVITY LOADS

		ROOF PSF		FLOOR PSF
DEAD	Roofing	3	Concrete fill	37
LOAD	Metal deck	2	Metal deck	3
	Insulation	5		
	Mech. & Elect.	4	Joists	3
	Sprinkler	5	Mech. & Elect.	2
	Framing	3	Sprinkler	3
		5	Framing	2
	TOTAL DEAD	25		50
	LOAD	20		100
	LIVE LOAD	45		150
	TOTAL LOAD			
	WALL	(10 inch reinforced hollow concrete block)		75

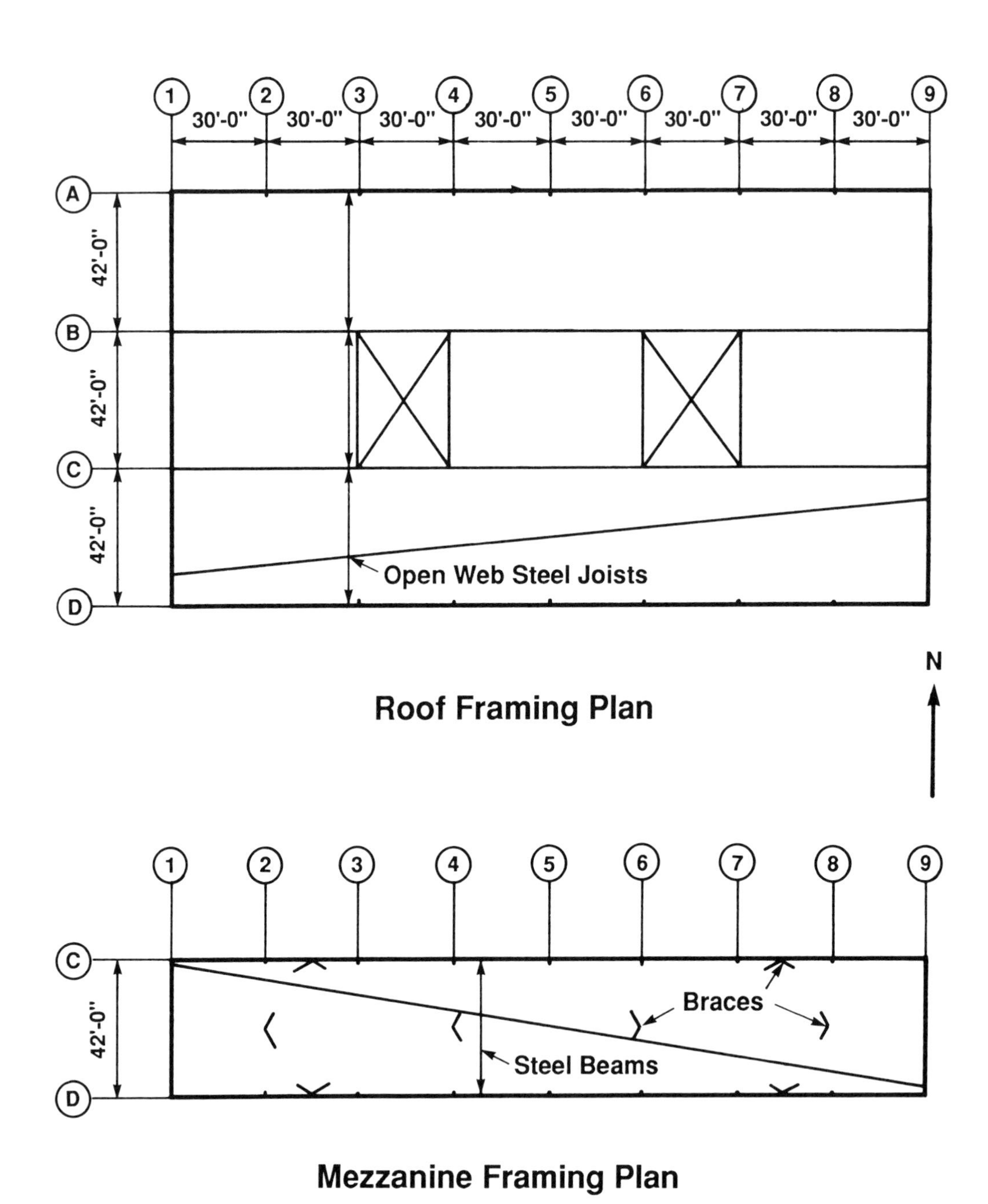

Figure 5.11.2-1 Framing plans for a one story industrial building

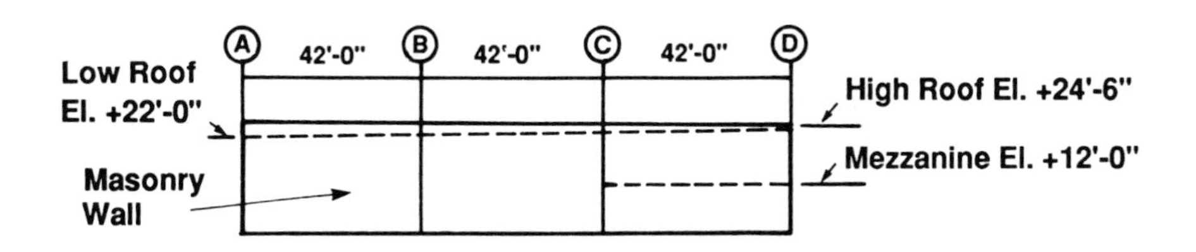

East Wall Elevation

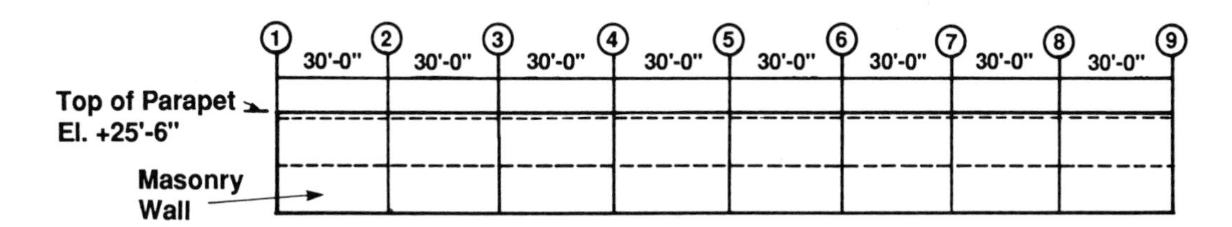

South Wall Elevation

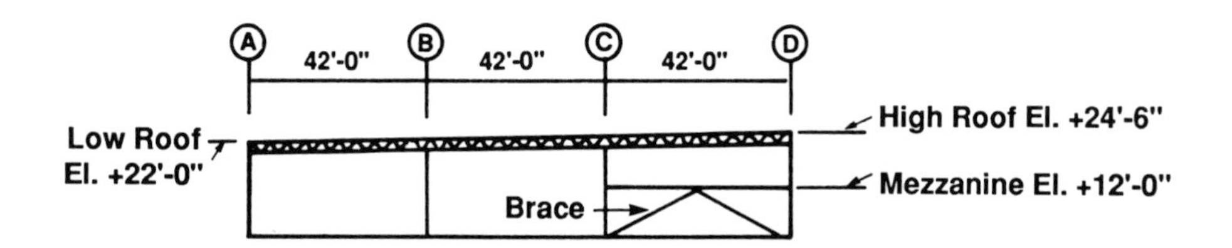

Line 2, 4, 6, & 8 Elevation

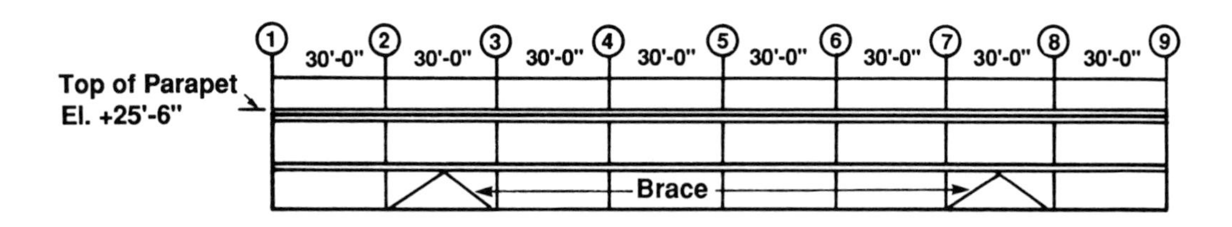

Line C Elevation

Figure 5.11.2-2 Wall elevations

WIND FORCES

For the mid-west region selected the Basic Wind Speed from Figure 1 is 70 mph. For a building with an Importance Factor of 1 located in an Exposure Category C area determine the wind pressures and forces at the various levels:

$p = q * [(G_h * C_{pw}) - (G_h * C_{pl})]$
$q = 0.0256 * K_z * (I * 70)2 = 12.544 * K_z$

From Equation A3, K_z, at 15 feet, at mean roof level (23.25 feet) and at the top of the wall (25.5 feet) are 0.801, 0.908 and 0.932.
Average K_z in the upper portion of the wall is 0.867.
From Equation A5, G_z, at 15 feet and at mean roof level are 1.30 and 1.271.
For the main building frame $G_h = G_z$ at mean roof level = 1.271.
From Figure 2 the value of C_p is: for windward wall, 0.8; for leeward wall, -0.5 for N-S and -0.32 for E-W.
Wind pressures on the building are determined as follows:

$F = [p - (-p)] * A$
$p_{w1} = 12.544 * 0.801 * 1.271 * 0.8 = 10.22$ psf windward
$p_{w2} = 12.544 * 0.867 * 1.271 * 0.8 = 11.05$ psf windward
$p_{l1} = 12.544 * 0.908 * 1.271 * 0.5 = -7.24$ psf N-S leeward
$p_{l2} = 12.544 * 0.908 * 1.271 * 0.32 = -4.63$ psf E-W leeward
$p_i = 12.544 * 0.908 * 0.25 = 2.85$ psf internal
(internal pressures cancel for total force on building)

The wind forces per foot at the roof line in the north-south, critical, direction are:
$F_h = 15 * 0.0102 * 0.323 + 10 * 0.0111 * 0.860 + 20 * 0.0072 * 0.5 = 0.217$ k/ft.
For total wind forces on the building see Table 5.11.2 - A & B

SEISMIC FORCES

Calculate total base shear:

NEHRP	**UBC**
$V = C_s * W$	$V = Z * I * c/R_w * W$
$C_s = 0.014 * v * S/(R * T^{2/3})$	$I - 1.0$: $C = 1.25 * S/T^{2/3}$
$V = 30$	$Z = .30$

S = Soil factor = 1.2

$R = 4.5$ for main structure;	$R_w = 8$ for main structure
Main Structure	
$T = 0.05 * h_n/L^{1/2}$	$T = C_t * h_n^{3/4}$: $C_t = 0.020$

$h_n = 23.25'$ (mean roof height)

$T_a = 0.05 * 23.25/126^{1/2} = 0.1036$	$T = 0.020 * 23.25^{3/4} = 0.212$
$T_a = 0.05 * 23.25/240^{1/2} = 0.0750$	

$C_s = 0.014 * 30 * 1.2/(4.5 * 0.136^{2/3}) = 0.2359$	$C = 1.25 * 1.2/0.212^{2/3} = 4.218$
C_s (limit) = 2.5 * A/R	C (limit) = 2.75 Base ShearCoefficient, Cb = Z * I * C/Rw
$C_s = 2.5 * 0.30/4.5 = 0.1667$	$C_b = 0.30 * 1 * 2.75/8 = 0.1031$
R = 5 for mezzanine structure	R_w = 6 for mezzanine structure
C_s limit = 2.5 * 0.30/5 = 0.1500	$C_b = 0.30 * 1 * 2.75/8 = 0.1375$

For a low building the limiting value of C_s or C will govern and the period need not have been, and the mezzanine structure period will not be, calculated. See Tables 5.11.2 (A), (B) and (C) for W and V.

The UBC seismic force per foot at the roof of the main structure in the north-south direction is:

$F_h = [0.075*\{22* 0.473 + 25.5 * 0.548\} + 0.025 * 126] * 0.1031 = .513$ k/ft

It is clear that seismic forces govern for this example.

Design for Lateral Forces

For both directions the lateral force resistance will be provided by the metal deck roof diaphragm and the exterior reinforced masonry walls for the main structure. The metal deck and concrete floor diaphragm and the interior steel bracing at lines C, D, 2, 4, 6 and 8 will provide the lateral force resistance for the mezzanine structure.

Determine the rigidities of the vertical resisting elements and the ratio of diaphragm deflection to the story drift in order to establish if diaphragm is flexible.

The metal roof diaphragm critical shear is at the end walls with the forces acting in the north-south direction. UBC forces and capacities will be used since they are readily available and do not require conversion.

v = 122.98/(2 * 126) = 0.488 k/ft.

Roof diaphragm deflection contributed by shear in the decking is

$= v * L * F = 488 * 240 * 16.4/10^6 = 1.92$ inches.

Where v if the shear in pounds per foot, L is the span of the diaphragm and F is the flexibility from the manufacturers approved tables for 20 gauge deck having supports at 6 foot centers, 5 puddle welds per sheet at supports and 1 1/2 inch seam welds at 12 inch centers.

The wall deflection with all cells filled and considering shear deformation contribution (major) would be

$= 1.2*v*L*h/(A*E_m) = 1.2*0.488*306/(9.63*2250*12) = 0.0007$ inches

where v is the shear per foot, L is the length of the wall, h is the height of the wall, A is the area of the wall which is the length times the thickness so the lengths cancel and E_m is the modulus of the masonry.

The diaphragm is flexible as defined by Section 2312(e)6. The design diaphragm shear, incorporating the uncertainties required by Section 2312(e)5 is determined to be

v = 122.98 * 0.55/126 = 0.537 k/foot

Since the diaphragm is flexible, no torsion can be transmitted by diaphragm action. There is the primary loading condition for gravity dead and live loads which is not shown on this figure.

The critical unit stress in the shear walls is determined as follows:

v = (122.98 * 0.55)/(127.5 * 12 * 9.63) = 0.005 ksi.

Check the diaphragm shears and determine the "chord" forces and the forces to be transferred at the corners of the openings in the diaphragm.

At end walls V = 122.98 * 0.55 = 67.64 kips
v = 67.64/126 = 0.537 kips/foot

At ends of openings V = 67.54 - 60 * 0.513 = 37.76 kips and
37.76 - 30 * 0.513 = 21.36 kips
v = 37.76/ 84 = 0.438 k/foot: within allowable capacity.

Critical F_h to be transferred at corners of openings is:
F_h = [37.76/126] * 21 = 6.29 kips.

Chord forces:
At centerline: 0.513 * 240^2/(124 * 8) = 29.31 kips

At ends of opening: [67.64 * 60 - 0.513 * 60^2/2]/126 + 6.29
= 31.17 kips
[67.64 * 90 - 0.513 * 902/2]/126 + 3.56
= 35.39 kips

For the mezzanine structure the forces will be distributed to the braced frames at lines C, D, 2, 4, 6 and 8. With the length width ratio in the north-south direction under 1 to 1.5 and less than 1 for the east-west direction the concrete filled metal deck diaphragm is rigid. The accidental torsion will not be critical for the braced frames due to the configuration. Check the forces in the braces along lines C and D ignoring the minor contribution from torsion.

P = 74.25/(8 * cos 38.66°) = 11.89 kips

Design force for chevron brace, P = 1.5 * 11.89 = 17.83 kips, per UBC Section 2722(g)4. Note that the beams and girders at the chevron braces are to be designed for the full gravity loads ignoring any contribution to support from the bracing.

TABLE 5.11.2 (A): NORTH - SOUTH DIRECTION

		WIND				SEISMIC					
								NEHRP		UBC	
LEVEL	Height from base h_x (feet)	Area Tributary to level (1000 sq. ft.)	Wind Pressure (pounds per sq. ft.)	Story Shear [F] Story Shear V (kips)	Overturning Moment [Story] Total (foot kips)	Weight Tributary to Level (kips)	Distribution Factor*	Story Shear V (kips)	Overturning Moment (ft.kips)	Story Shear V (kips)	Overturning Moment (ft.kips)
ROOF	0 to 15 15 to 25.5 0 to 25.5	3.50 x 0.3061 2.52 x 0.8265 6.12 x 0.5483	10.36 11.05 7.24	[11.10] [23.02] [24.30]	[1358.27]	(Rf 1 x 30.24 x 0.025) (Wl 0.5135x 11.40 x 0.075)	1.0000				
				58.42	1358.27	1192.9		198.85	4623.4	122.98	2859.3

* For NEHRP and UBC

TABLE 5.11.2 (B): EAST - WEST DIRECTION

		WIND				SEISMIC					
								N E H R P		U B C	
LEVEL	Height from base h_x (feet)	Area Tributary to level (1000 sq. ft.)	Wind Pressure (pounds per sq. ft.)	Story Shear V (kips)	Overturning Moment [Story] Total (foot kips)	Weight Tributary to Level (kips)	Distribution Factor*	Story Shear V (kips)	Overturning Moment (ft.kips)	Story Shear V (kips)	Overturning Moment (ft.kips)
ROOF	0 to 15 15 to 25.5 0 to 25.5	1.89 x 0.3226 1.32 x 0.8710 3.21 x 0.5484	10.36 11.05 4.63	6.32 12.73 8.16		(Rf 1 x 30.24 x 0.025) (Wl 0.5484x 6.43 x 0.075)	1.000				
				27.21	632.56	1020.3		170.08	3954.5	105.19	2445.7

* For NEHRP and UBC

TABLE 5.11.2 (C): MEZZANINE - BOTH DIRECTIONS

LEVEL	SEISMIC ONLY, NO WIND	SEISMIC					
		Total Weight (kips)	Distribution Factor*	N E H R P		U B C	
				Story Shear V (kips)	Overturning Moment (ft.kips)	Story Shear V (kips)	Overturning Moment (ft.kips)
MEZZ		1.0 x 10.80 x 0.050	1.000				
		540.0		81.00	972.0	74.25	891.0

* For NEHRP and UBC

DESIGN EXAMPLE 5.11.3 Three story residential building

A three-story residential building will be designed for a site near the east coast. The building is L shaped in plan with separation joints dividing the structure into three separate units, each of which is approximately symmetrical (see Figures 5.11.3-1 and -2). Masonry bearing walls support hollow-core concrete floors with concrete topping. The wood frame roof is plywood sheathed. The vertical resisting elements are the masonry bearing walls and longitudinal masonry corridor walls.

The basic design information is given below:

DESIGN LOADS AND FORCES

LIVE LOAD	ROOF		20 PSF
	FLOOR		40 PSF
	ATTIC		10 PSF
	CORRIDOR		80 PSF
LATERAL FORCES	WIND	ASCE 7 1988	90 MPH
	SEISMIC	NEHRP 1988: A = 0.10	v = 10
	UBC 1991		ZONE 1

MATERIAL PROPERTIES	STEEL	A36		36,000PSI
	CONCRETE	C94	f'_c	4,000PSI
	REINFORCEMENT		A615, GRADE 60	60,000PSI
	MASONRY	C90	f'_m	1,500PSI
	WOOD	SOUTHERN YELLOW PINE, NO. 2	f_b	1,150PSI

GRAVITY LOADS

		ROOF PSF		ATTIC PSF	FLOOR PSF
DEAD LOAD	Shingles	3	Concrete fill	---	25
	Plywood, 1/2 inch	2	Hollow core slab	45	45
	Truss & bracing	3	Mechanical	5	---
	Miscellaneous	2	Ceiling finish	5	5
	TOTAL DEAD LOAD	10		55	75
	LIVE LOAD	16		10	40
	TOTAL LOAD	26		65	115
	WALL (10 inch reinforced hollow concrete block)			55	

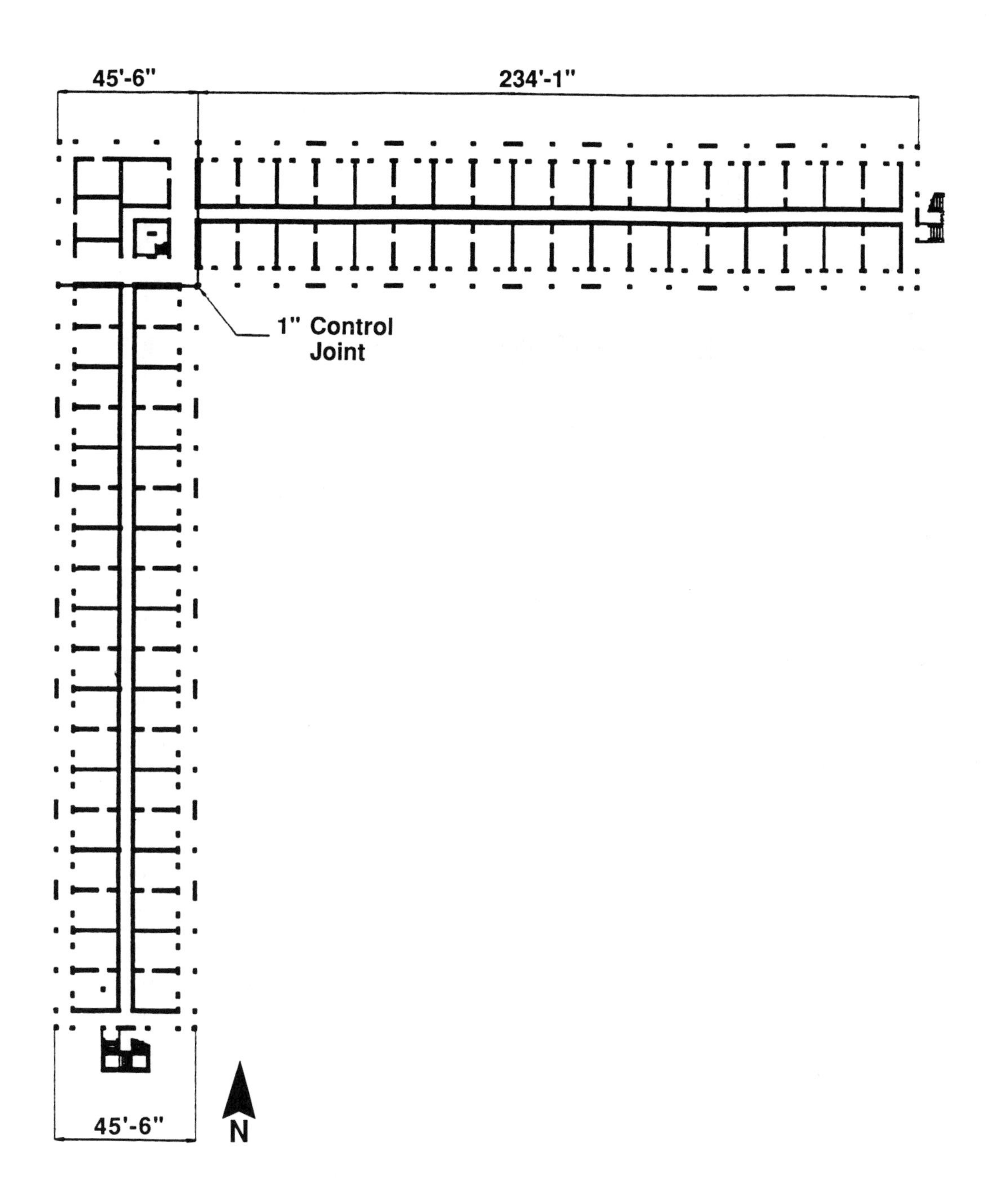

Figure 5.11.3-1 Floor plan for three story residential building

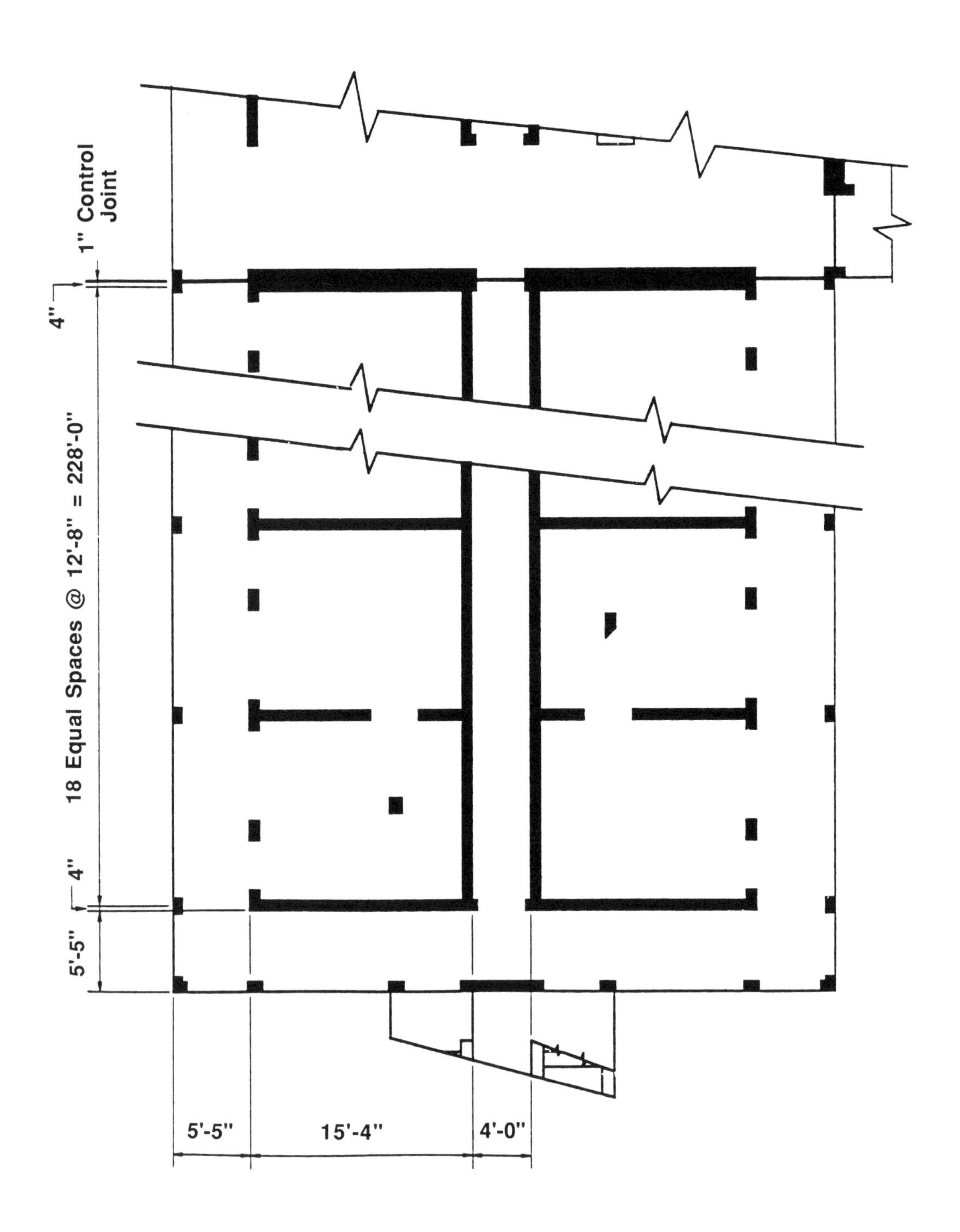

Figure 5.11.3-2 Detailed floor plan

WIND FORCES

For the east coast region selected the Basic Wind Speed from Figure 1 is 90 mph. For a building with an Importance Factor of 1.05 located in an Exposure Category C area determine the wind pressures and forces at the various levels:

$$p = q *[(G_z * C_{pw}) - (G_h * C_{pl})]$$

$$q = 0.00256 * K_z * (I * 90)2 = 22.861 * K_z$$

From Equation A3, K_z at 15', 17.33', 25', at mean roof height and at roof ridge are 0.801, 0.835, 0.927, 957 and 0.985.

From Equation A5, G_z at 15', 17.33' 25' area and mean roof height are 1.32, 1.306, 1.273 and 1.263. For the main building frame $G_h = G_z$ at mean roof level = 1.263.

From Figure 2 the value of C_p is: for windward wall, 0.8; for leeward wall, -0.5 for transverse and -0.2 for longitudinal; for windward roof, 0.2; for leeward roof and parallel to ridge, -0.7.

The wind pressures on the building at the various levels are:

p_{w1} = 22.861 * 0.801 * 1.263 * 0.8 = 18.50 psf windward wall
p_{w2} = 22.861 * 0.835 * 1.263 * 0.8 = 19.29 psf windward wall
p_{w3} = 22.861 * 0.927 * 1.263 * 0.8 = 21.41 psf windward wall
p_{wl} = 22.861 * 0.957 * 1.263 * 0.5 = 13.82 psf leeward trans.
p_{wl} = 22.861 * 0.957 * 1.263 * 0.2 = 5.53 psf leeward long
p_{rw} = 22.861 * 0.957 * 1.263 * 0.2 = 5.53 psf windward roof
p_{rl} = 22.861 * 0.957 * 1.263 * 0.7 = 19.34 psf leeward roof
p_i = 22.861 * 0.957 * 0.25 = 5.47 psf internal
(internal pressures cancel for total force on building)

Total wind forces on building are given in Table 5.11.3 - A and B as determined from the following equation:

$$F = [p - (-p)] * A$$

SEISMIC FORCES

Calculate total base shear.

NEHRP		UBC
$V = C_s * W$: v = 7.5		$V = Z * I * C/R_w * W$: Z = 0.075
$C_s = 0.014 * v * S/T^{2/3}$		I = 1: $C = 1.25 * S/T^{2/3}$
	S = Soil factor = 1.2	
R = 3.5		$R_w = 6$
	W = sum of weights tributary to each level.	
$T_a = 0.05 * h_n^{1/2}$		$T = C_t * h_n^{3/4}$: $C_t = 0.020$
	h_n = 25.11' (height to top level)	
$T_a = 0.05 * 25.11/45.5^{1/2} = 0.186$		$T = 0.020 * 25.11^{3/4} = 0.224$
C_s (limit) = 2.5 * A/R: A = 0.075		C (limit) = 2.75 Base Shear Coefficient, $C_b = Z * I * C/Rw$
$C_s = 2.5 * 0.075/3.5 = 0.0536$		$C_b = 0.075 * 1 * 2.75/6 = 0.0344$

For a low building the limiting value of C_s or C will govern. The period need not have been calculated for the narrow dimension of the building for NEHRP. From the period calculated for UBC it is clear that the calculated value of C will be greater than the limit. See Tables 5.11.3 (A) and (B) for W and V.

It is clear that wind forces govern for the transverse direction and seismic forces govern for the longitudinal direction for this example.

Design for lateral forces

For both directions the lateral force resistance will be provided by the plywood roof diaphragm or the untopped concrete hollow core attic floor diaphragm, the concrete topped concrete hollow core floor diaphragms and the exterior and interior masonry walls.

The wind forces are applied at the level at which they occur while the seismic force distribution is shown in Tables 5.11.3 (A) and (B).

The symmetry of the vertical resisting elements and the small length-width ratios of the diaphragms indicate that the distribution of forces can be made by tributary areas and that torsional effects can be neglected. The values and their location on the building are shown on Figure 5.11.3-2. There is the primary loading condition for gravity dead and live loads which is not shown on this figure.

TABLE 5.11.3 (A): NORTH - SOUTH DIRECTION

		WIND				SEISMIC					
LEVEL	Height from base h_x Story height h_s (feet)	Area Tributary to level (1000 sq. ft.)	Wind Pressure (pounds per sq. ft.)	Story Shear [F] Story Shear V (kips)	Overturning Moment [Story] Total (foot kips)	Weight Tributary to Level (kips)	Weight times height (ft.kips) Distribution Factor*	N E H R P		U B C	
								Story Shear V (kips)	Overturning Moment (ft.kips)	Story Shear V (kips)	Overturning Moment (ft.kips)
3 [R]	25.11 [7.78]	2.170	24.87	[41.27] 41.27	[320.94] 320.94	795.2	19967.5 0.4138	[65.72] 65.72	[511.3]	[42.18] 42.18	[328.0] 328.0
2	17.33 [8.67]	1.820	33.12	[60.27 101.54	[522.34] 843.28	1096.0	18997.3 0.3937	[62.53] 128.25	[1111.9] 1623.2	[40.13] 82.31	[713.4] 1041.4
1	8.67 [8.67]	1.820	32.32	[58.82] 160.36	[509.78] 1353.06	1071.8	9289.0 0.1925	[30.57] 158.82	[1377.0] 3000.2	[19.62] 101.93	[883.4] 1924.8
						2963.0	48253.8	158.82		101.93	

* For NEHRP and UBC

TABLE 5.11.3 (B): EAST - WEST DIRECTION

		WIND				SEISMIC					
								N E H R P		U B C	
LEVEL	Height from base h_x Story height h_s (feet)	Area Tributary to level (1000 sq. ft.)	Wind Pressure (pounds per sq. ft.)	Story Shear [F] Story Shear V (kips)	Overturning Moment [Story] Total (foot kips)	Weight Tributary to Level (kips)	Weight times height (ft.kips) Distribution factor*	Story Shear V (kips)	Overturning Moment (ft.kips)	Story Shear V (kips)	Overturning Moment (ft.kips)
3 [R]	25.1 [7.78]	0.335	24.90	[8.34] 8.34	[64.86] 64.86	810.5	2035.7 0.4224	[66.79] 66.79	[519.6] 519.6	[42.86] 42.86	[333.3] 333.3
2	17.33	0.396	24.81	[9.93] 18.27	[158.34] 223.20	1071.8	18577.8 0.3856	[60.97] 127.75	[1107.2] 1626.8	[39.13] 81.99	[710.6] 1043.9
1	8.67	0.396	24.03	[9.52] 27.79	[240.85] 464.05	1067.6	9252.6 0.1920	[30.36] 158.11	[1370.8] 2997.6	[19.49] 101.48	[879.5] 1923.4
						2949.9	48182.1	158.11		101.48	

* For NEHRP and UBC

DESIGN EXAMPLE 5.11.4 Five Story Steel Frame Building

A five-story steel frame office building will be designed for a site located in California near a coastal region. A plan view of the building is shown in Figure 5.11.4-1. The building will be designed with the lateral forces in the E-W direction resisted by special moment resisting space frames (SMRSF) and in the N-S direction by concentric chevron braced frames. A parapet five feet high is provided at the roof.

To simplify the example, it will be assumed that the nominal weight of the stairs and the mechanical equipment in the service core is equal to the design dead loads used for each floor. This means that the center of mass is located at the geometric center of the building. Normally this will not be true and the center of mass will need to be computed.

The basic design information is given below:

DESIGN LOADS AND FORCES

LIVE LOADS:	ROOF		20 PSF
	FLOOR		50 PSF
LATERAL FORCES	WIND	ASCE 7 1988	70 MPH
	SEISMIC	NEHRP 1988: A = 0.40	v = 40
	UBC 1991		ZONE 4
MATERIAL PROPERTIES	STEEL	A36	36,000PSI
	CONCRETE		4,000PSI
	REINFORCEMENT	A615,GRADE 60	60,000PSI

GRAVITY LOADS

		ROOF PSF		FLOOR PSF
	Insulation plus membrane	11	Finish	1
DEAD	Deck plus concrete	47		47
LOAD	Mechanical plus electrical	5		5
	Ceiling and lighting	5		5
	Steel framing & fireproofing	16		20
	Partitions	--		20
	TOTAL DEAD LOAD	84		98
	LIVE LOAD	20		50
	TOTAL LOAD	104		148

WALL DEAD LOAD (Prefabricated panels) 18

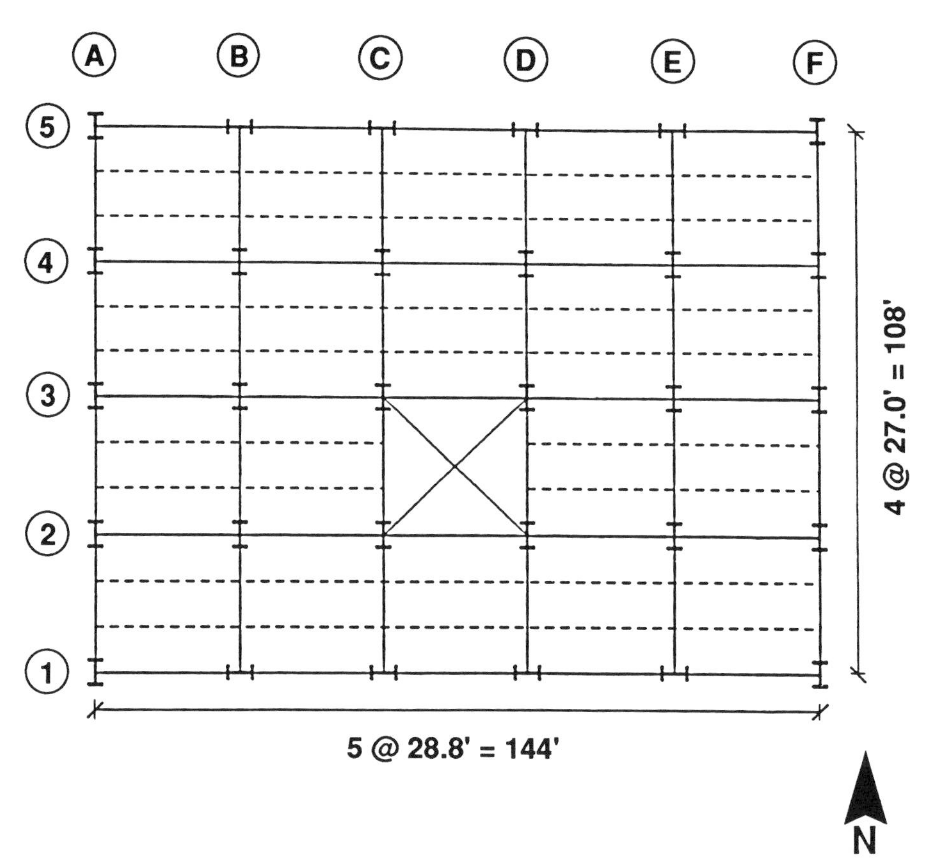

Figure 5.11.4-1 Floor plan of five story frame building

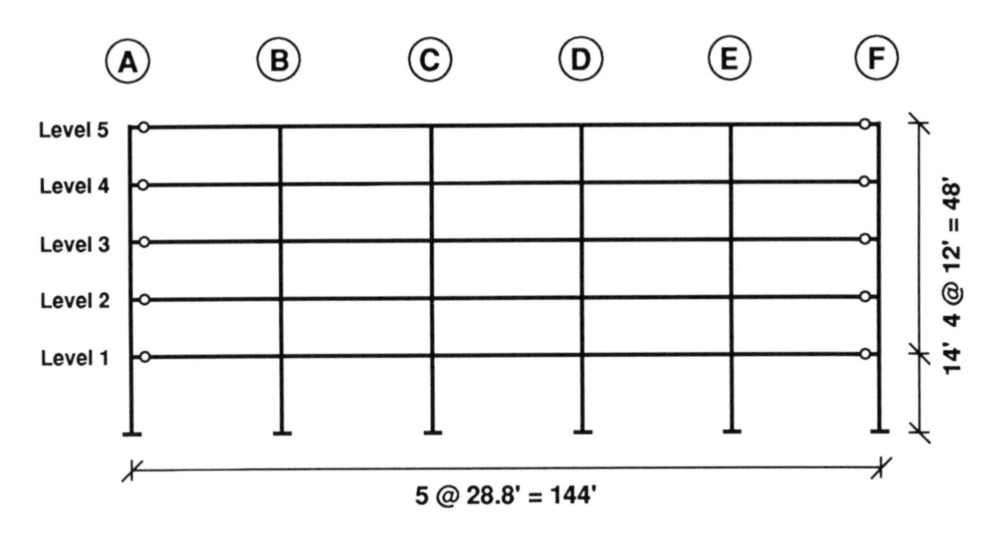

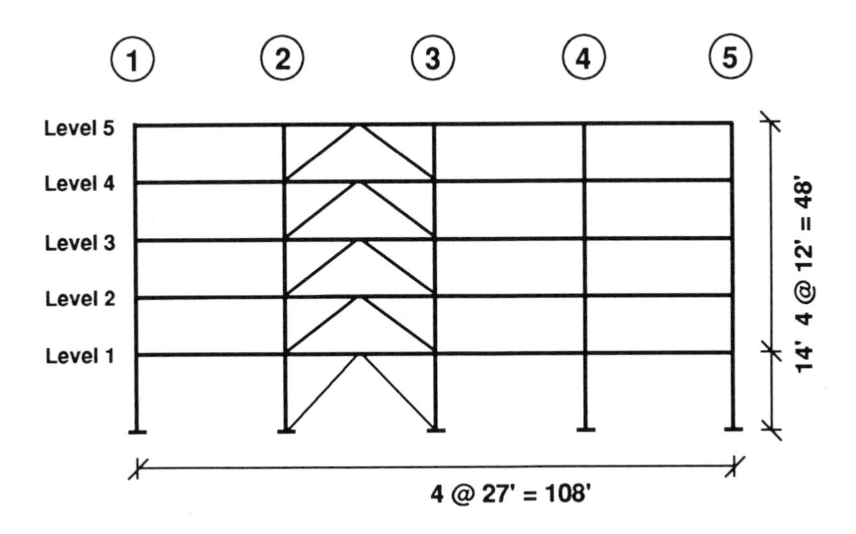

Figure 5.11.4-2 Elevations of frames on lines 1 and D

Note: The full partition allowance is used in calculation of seismic forces in lieu of using some part of the design live load.

There are a number of engineers who design steel frame buildings so that only the perimeter frames are moment frames that carry lateral forces. All interior frames are designed with simple (pinned) beam-column connections. If that is the case for this building, frames at lines 1 and 5 would each be designed to carry half of the design forces in the E-W direction. Braced frames at lines C and D would each carry half of the design forces in the N-S direction. Elevation views of Frame 1 and Frame D are shown in Figure 5.11.4-2.

WIND FORCES

For the California coastal region, the Basic Wind Speed from Figure 1 is 70 mph. For a building with an importance factor of 1 located in an exposure C determine wind pressures and forces at the various levels:

$p = q * [(G * C) - (G * C)]$

$q = 0.00256 * K * (1 * 70) = 12.544 * K$

From Equation A3, K_z, for average pressure at mid-point of tributary area and at roof = 0.814, 0.932, 1.045, 1.130, 1.199 and 1.201.

From Equation A5, G_z, interpolated for mid-point of tributary area and at roof = 1.314, 1.272, 1.238, 1.213, 1.199 and 1.198. For the main building frame G_h = G_z at roof = 1.198.

From Figure 2 C_p = 0.8 for windward and -0.5 or -0.44 for leeward.

$F = p * A$

$p_r = 12.544 * 1.196 * 1.197 * 0.8 = 14.40$ psf windward

$p_5 = 12.544 * 1.130 * 1.197 * 0.8 = 13.57$ psf windward

$p_4 = 12.544 * 1.045 * 1.197 * 0.8 = 12.55$ psf windward

$p_3 = 12.544 * 0.932 * 1.197 * 0.8 = 11.19$ psf windward

$p_2 = 12.544 * 0.814 * 1.197 * 0.8 = 9.78$ psf windward

$p_1 = 12.544 * 1.201 * 1.198 * 0.5 = -9.02$ psf leeward N-S

$p_1 = 12.544 * 1.201 * 1.198 * 0.44 = -7.94$ psf leeward E-W

Tributary area for each floor level is assumed to be the width times the height to mid-story of adjacent stories or to top of parapet. Total width of exposure:

N-S 144 + 2 * 2 = 148 ft.

E-W 108 + 2 * 2 = 112 ft

The force calculations can be more easily accomplished, and comparison made with seismic forces, by tabularizing as illustrated in Tables 5.11.4 (A) and (B).

SEISMIC FORCES

Calculate total base shear.

NEHRP

$V = C_s * W$: $v = 40$

$C_s = 0.014 * v * S/T^{2/3}$

S = Soil factor = 1.2

$R = 5$ N-S or 8 E-W

$T_a = 0.05 * h_n^{1/2}$ N-S

UBC

$V = Z * I * C/R_w * W$: $Z = 0.4$

$I = 1.0$: $C = 1.25 * S/T^{2/3}$

$R_w = 8$ N-S or 12 E-W

$T = C_t * h_n^{3/4}$: $C_t = 0.020$ N-S

$h_n = 62.0$

$T_a = 0.05 * 62/110^{1/2} = 0.296$

$T_a = C_t * h_n^{3/4}$: $C_t = 0.035$ E-W

$T_a = 0.035 * 62^{3/4} = 0.773$

$C_s = 0.014 * 40 * 1.2/(R * T_{2/3})$
$= 0.672/(5 * 0.296^{2/3})$
$= 0.3026$ N-S: $C_s = 0.672/(8 * 0.772^{2/3})$
$= 0.0997$ E-W

C_s (limit) $= 2.5 * A/R$: $A = 0.40$

$C_s = 1/5 = 0.2000$ N-S: $1/8$
$= 0.1250$ E-W

$T = 0.020 * 62^{3/4} = 0.442$

$T = C_t * h_n^{3/4}$: $C_t = 0.035$ E-W

$T = 0.0350 * 62^{3/4} = 0.773$

$C = 1.25 * 1.2/0.442 = 2.585$ N-S
$C = 1.25 * 1.2/0.733 = 1.780$ E-W
C limit = 2.75
Base Shear Coefficient,

$C_b = Z * I * C/R_w = 0.40 * 1 * 2.585/8 = 0.1293$ E-W:

$C_b = 0.40 * 1 * 1.780/12 = 0.0596$ N-S

Determine weights tributary to each level and sum the weights.

$W_5 = 108 * 144 * 0.084 * 2 * (108 + 144) * (6 + 5) * 0.018$
$= 1306.4$ kips + 99.9 kips = 1406.2 kips

$W_f = 108 * 144 * 0.098 + 2 * (108 + 144) * 12 * 0.018$
$= 1524.1$ kips + 108.9 kips = 1633 kips

$W_1 = 108 * 144 * 0.098 + 2 * (108 + 144) * (7 + 6) * 0.018$
$= 1524.1$ kips + 117.9 kips = 1642 kips

See Tables 5.11.4 (A) and (B) for W and V.
It is clear that earthquake forces govern for this example.

Design for lateral forces

Calculate vertical distribution of forces. The distribution of forces in the N-S direction is the same for both NEHRP and UBC procedures. For the E-W direction, however, the vertical distribution of forces at the various levels varies. Since the period is between 0.5 and 2.5 seconds the exponent, k, is 1.1165 from Section 4.3 of NEHRP. For UBC the force, F_t, from formula 34-7 plus the force at that level determined by multiplying the base shear by the distribution factor from the following formula: (Modified formula 34-8 from UBC)

$$DF = W_i * h_i / \Sigma(W_i * h_i)$$

Note that the total force at the roof level for UBC is F_t plus F_5, 25.5 + 131.3 = 156.8 kips, Table 5.11.4 (B), Seismic Column 5.

See Table 5.11.4 (A), Seismic Column 2 for distribution factors for both NEHRP and UBC. See Table 5.11.4 (B), Seismic Columns 1 and 2 for distribution factors for NEHRP and UBC respectively.

Since the building is symmetric in both mass and stiffness, the floor-level forces would be distributed equally to each frame in the direction under consideration. The design forces applied to each frame and their preliminary designs are shown in Figure 5.11.4 - 3 for the braced frame and Figure 5.11.4 - 4 for the moment frame. For the braced frame in the N-S direction, C1 = W14x176, C2 = W14x90, G1 = W14x22, B1 = 2L8x8x7/8, and B2 = 2L8x8x5/8. For the SMRSF in the E-W direction, C3= W24x146, C4 = W24x84, G2 = W27x102 and G3 = W24x68.

The only step remaining to be completed is to include the effects of accidental torsion, recognizing the symmetry of mass and stiffness. The torsional moment at each level is calculated by applying the seismic story shear in a particular direction at an eccentricity equal to 5 percent of the building dimension normal to the direction of the force. Since the moment will cause the building to rotate in plan, the frames in both directions will be required to carry additional shear in accordance with their contribution to the torsional rigidity. The calculation of the additional force at each story of each frame that results from the torsion is straightforward and is similar to calculating the shear force on bolts in an eccentrically loaded connection. The stiffness, K_i , of each frame at each story must be computed. Next, the polar moment of inertia (or polar moment of stiffness) must be calculated for each story. Call this J_i. Then, the force carried by a frame is simply $F_i = M_i * r_i * K_i / J_i$ where M_i is the moment in the story and r_i is the distance from the center of rigidity to the frame under consideration. These relationships are valid only if it is assumed that the floor slab is rigid in its plane. This assumption should be checked as discussed earlier.

To calculate the braced frame stiffness, K_b, at each story, apply a unit load at the top level and calculate the deflection at each floor. Since the shear in each story is one kip, the story stiffness is simply one divided by the relative story deflection, or the story drift. Calculate the moment frame stiffness, K_m, at each level in the same manner. Since there are two identical frames in each direction located the same distance from the center of rigidity, the polar moment of stiffness at each story is simply $J = 2(K_b x^2 + K_m y^2)$, where x and y are the distances from the center of rigidity to the braced frames and the moment frames, respectively. The distribution of the shear due to torsional moment to each of the resisting elements is the proportion of the torsional stiffness of that element to the total torsional stiffness: $DF = K_b * x^2/J_i$ or $K_m * y^2/J_i$.

The story stiffnesses for each frame and the story torsional rigidities are given in Table 5.11.4 (C). Even though the braced frame is six times stiffer than the moment frame, it provides less than half of the torsional rigidity of the moment frame. This is because the moment frame is much farther from the center of rigidity. This illustrates the importance of placing as many of the vertical resisting elements of the lateral force resisting system at the perimeter of the building as possible.

Calculate Frame Forces Due to Torsion

1. Seismic Design Moment from Forces Acting in NS Direction

M_i $= V_i * e_i$ Where e = eccentricity = 0.05 * 144 $M_i = F_i * 7.2$

M_5 = 302.3 * 7.2 = 2176.6 k-ft

M_4 = 585.4 * 7.2 = 4212.7 k-ft

M_3 = 800.5 * 7.2 = 5763.6 k-ft

M_2 = 947.7 * 7.2 = 6823.4 k-ft

M_1 = 1027.4 * 7.2 = 7397.3 k-ft

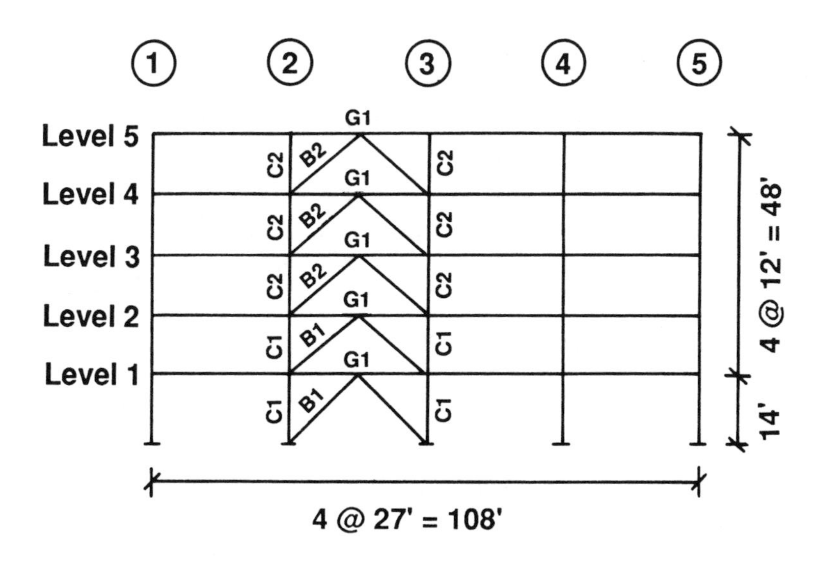

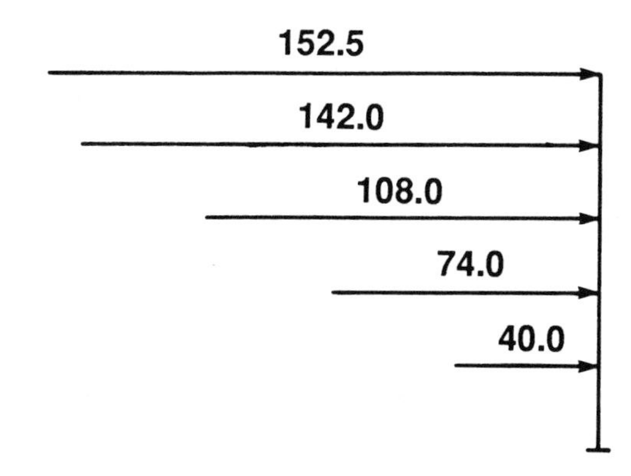

Figure 5.11.4-3 Frame D and the equivalent lateral forces acting on it

Elevation of Frame 1

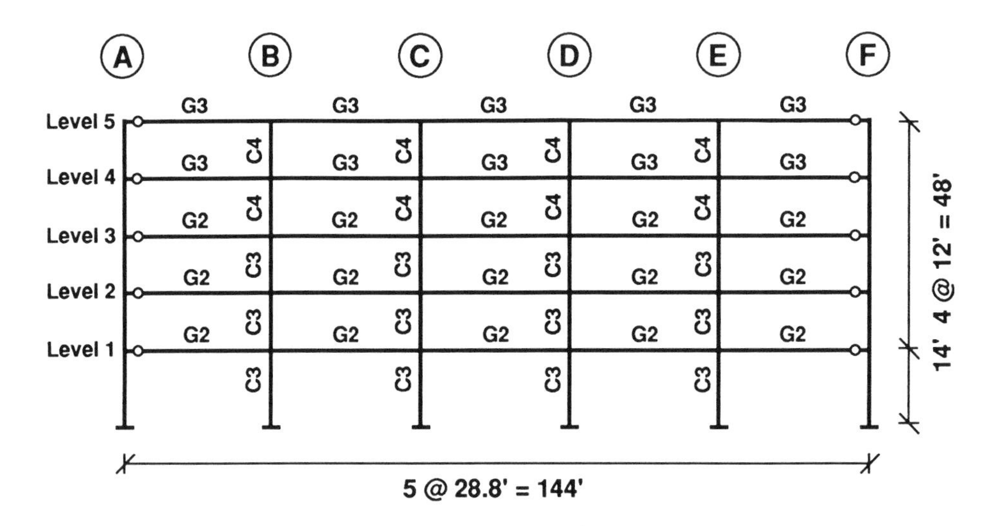

Equivalent Lateral Forces

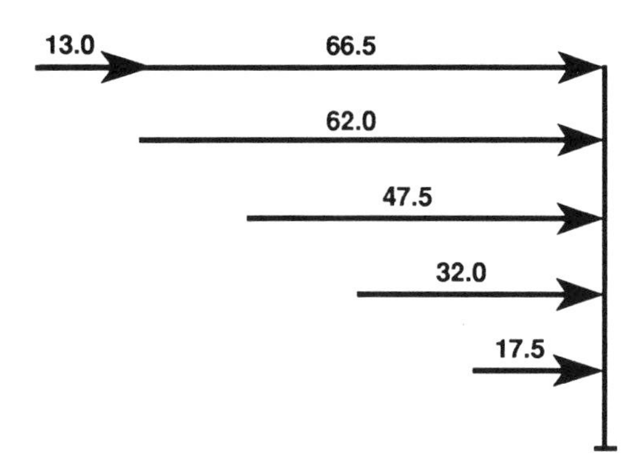

Figure 5.11.4-4 Frame 1 and the equivalent lateral forces acting on it

Elevation of Frame 1
Equivalent Lateral Forces

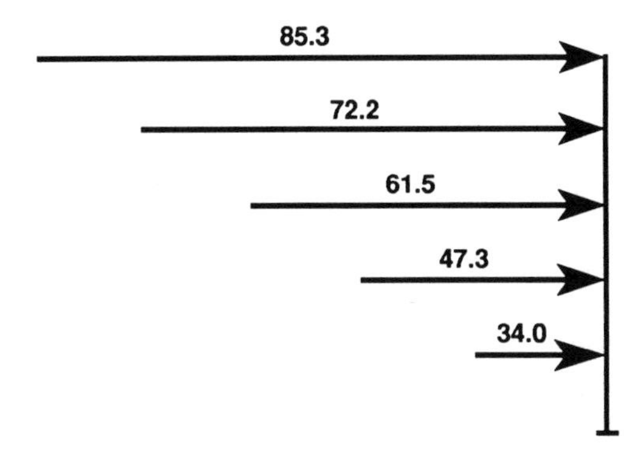

Elevation of Frame D
Equivalent Lateral Forces

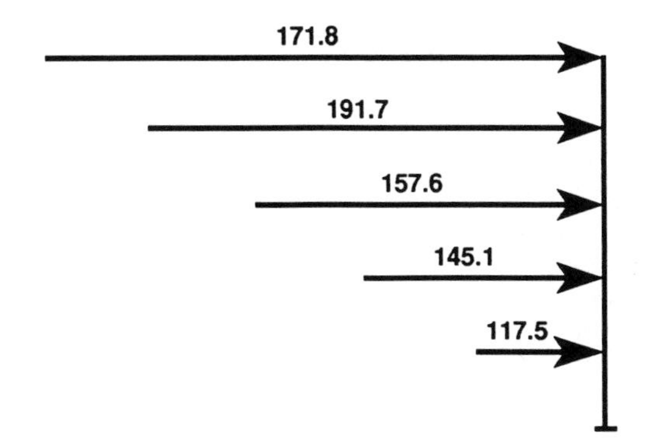

Figure 5.11.4-5 Final design forces for frames on lines 1 and D

2. Distribution of shear from torsional moments:

Shear in Braced Frame Due to NS Force

$F_i = M_i * DF/x$ where x = 14.4 ft for all stories
$F_5 = 2176.6 * 0.1311/14.4 = 19.82$ kips
$F_4 = 4212.7 * 0.1357/14.4 = 39.70$ kips
$F_3 = 5763.6 * 0.1238/14.4 = 49.55$ kips
$F_2 = 6823.4 * 0.1500/14.4 = 71.08$ kips
$F_1 = 7397.3 * 0.1508/14.4 = 77.47$ kips

The additional shear in Moment Frames due to torsion from N-S forces is, by observation, not critical compared with direct forces plus additional shear due to torsion from E-W forces.

3. Seismic Design Moment from Forces Acting in the E-W Direction
$M_i = V_i * e$ where e = 0.05 * 108 = 5.4 $M_i = F_i * 5.4$
$M_5 = 156.8 * 5.4 = 846.7$ k-ft
$M_4 = 279.7 * 5.4 = 1510.4$ k-ft
$M_3 = 373.1 * 5.4 = 2014.7$ k-ft
$M_2 = 437.0 * 5.4 = 2359.8$ k-ft
$M_1 = 471.6 * 5.4 = 2546.6$ k-ft

The additional shear in Braced frames due to torsion from E-W forces is not critical.

4. Distribution of shear from torsional moment:

Shear in Moment Frames due to torsion from E-W forces.

$F_i = M_i * DF/x$ where x = 54 ft. for all stories
$F_5 = 846.7 * 0.3689/54 = 5.78$ kips
$F_4 = 1510.4 * 0.3643/54 = 10.19$ kips
$F_3 = 2014.7 * 0.3762/54 = 14.04$ kips
$F_2 = 2359.8 * 0.3500/54 = 15.30$ kips
$F_1 = 2546.6 * 0.3492/54 = 16.47$ kips

There is a very low probability that the maximum forces in the N-S direction and the maximum forces in the E-W direction will occur simultaneously. This is because the actual earthquake motions will be different in each direction and the structure has different periods of vibration in each direction. It would be overly conservative to take the maximum forces induced in a particular frame by the E-W forces and add them to those from the N-S forces. Generally one would take the maximum from the direction under consideration plus 0.3 of the maximum from the other direction.

The final design forces at each level are shown in Figure 5.11.4 - 5 for both frames. The members in each frame must be designed to withstand the stresses produced by these earthquake forces. The calculated story drift ratios for these earthquake forces must satisfy these allowable drifts:

Braced Frame

$$\delta_i = 0.04 * h_i/R_w = 0.005 * h_i$$

$$\delta_{5\text{-}2} = 0.005 * 144 = 0.72 \text{ inches} \quad \delta_1 = 0.005 * 168 = 0.84 \text{ inches}$$

Moment Frame

$$\delta_i = 0.04 * h_i/R_w = 0.0033 * h_i$$

$$\delta_{5\text{-}2} = 0.0033 * 144 = 0.48 \text{ inches} \quad \delta_1 = 0.0033 * 168 = 0.56 \text{ inches}$$

It is most often the case that a steel braced frame will satisfy the drift requirements but a SMRSF will not. Therefore, the member sizes in a moment frame will usually need to be increased beyond those required to resist the forces within allowable stresses.

Although earthquake forces govern in this example for the main lateral force resisting system, this may not be true for the parapet or the cladding. These elements are designed using different procedures that are described separately in Chapter 6.

Even if wind forces govern the design, it is quite possible that small earthquakes could cause forces in the building that greatly exceed those arising from the design wind. This is because when we design for earthquakes, we design for forces that are significantly lower than those that the building could experience during the design earthquake. Thus, even when wind forces govern, it is essential to use good detailing and inspection practices as required by UBC.

Finally, it should be noted that the purpose of this example was to demonstrate how one would calculate and distribute the lateral forces for design for wind and earthquake forces. Many engineers would find the structural scheme used here to be undesirable in highly seismic areas. Only two load resisting frames are provided in each direction, so the redundancy is not high. Also, the use of only chevron braces in the N-S direction is risky since these are not particularly ductile elements and because only four members at each floor level carry all of the earthquake forces. A dual system composed of two braced frames and two SMRSF in each direction would be far superior. The braces would provide great stiffness under moderate earthquakes and the SMRSF would serve as a backup in case some of the braces are lost during a major earthquake.

TABLE 5.11.4 (A): - NORTH - SOUTH DIRECTION

		WIND				SEISMIC					
								NEHRP		UBC	
LEVEL	Height from base h_x Story height h_s (feet)	Area Tributary to level (1000 sq. ft.)	Wind Pressure (pounds per sq. ft.)	Story Shear [F] Story Shear V (kips)	Overturning Moment [Story] Total (foot kips)	Weight Tributary to Level (kips)	Weight times height (ft.kips) Distribution Factor*	Story Force [F] Story Shear V (kips)	Overturning Moment [story] Total (ft.kips)	Story Force [F] Story Shear V (kips)	Overturning Moment [story] Total (ft.kips)
5	62.0 [12.0]	1.628	14.40 - [-9.02]	[38.13] 38.13	[457.56] 457.56	1406.2	87184.4 0.2942	[467.61] 467.61	[5611.4] 5611.4	[302.25] 302.25	[3627.0] 3627.0
4	50.0 [12.0]	1.776	13.57 - [-9.02]	[40.12] 78.25	[939.00] 1396.56	1633.0	81650.0 0.2755	[437.89] 905.50	[10866.0] 16477.4	[283.04] 585.29	[7023.5] 10650.5
3	38.0 [12.0]	1.776	12.55 - [-9.02]	[38.31] 116.56	[1398.72] 2795.28	1633.0	63054.0 0.2094	[322.83] 1238.33	[14860.0] 31337.4	[215.13] 800.42	[9605.1] 20255.6
2	26.0 [12.0]	1.776	11.19 - [-9.02]	[35.89] 152.45	[1829.40] 4624.68	1633.0	42458.0 0.1433	[227.77] 1466.10	[17593.2] 38930.6	[147.22] 947.64	[11371.7] 31627.3
1	14.0 [14.0]	1.924	9.78 - [-9.02]	[36.17] 188.62	[2640.68] 7265.36	1642.0	22988.0 0.0776	[123.34] 1589.44	[22252.2] 71182.8	[79.72] 1027.37	[14383.2] 46010.5
						7947.2	296334.4	1589.44		1027.37	

* For NEHRP and UBC

TABLE 5.11.4 (B): - EAST - WEST DIRECTION

		WIND				SEISMIC					
LEVEL	Height from base h_x Story height h_s (feet)	Area Tributary to level (1000 sq. ft.)	Wind Pressure (pounds per sq. ft.)	Story Shear [F] Story Shear V (kips)	Overturning Moment [Story] Total (foot kips)	Weight Tributary to Level (kips) Distribution Factor^	Weight times height (ft.kips) Distribution Factor+	NEHRP		UBC	
								Story Force [F] Shear V (kips)	Overturning Moment [story] Total (ft.kips)	Story Force [F] Shear V (kips)	Overturning Moment [story] Total (ft.kips)
5	62.0 [12.0]	1.232	14.40 - [-7.94]	[27.52] 27.52	[330.27] 2330.27	1406.2 0.3103	87184.4 0.2942	[245.87 245.87	[2950.4] 2950.4	F_t24.20] [131.64] 155.84	[1870.0] 1870.0
4	50.0 [12.0]	1.344	13.57 - [-7.94]	[28.91] 56.43	[667.19] 1007.46	1633.0 0.2822	81650.0 0.2755	[223.60] 469.47	[5633.6] 8584.0	[123.27] 279.11	[3349.3] 5219.3
3	38.0 [12.0]	1.344	12.55 - [-7.94]	[27.54] 83.97	[1007.65] 2015.11	1633.0 0.2066	63054.0 0.2094	[163.70] 633.17	[7598.0] 16182.0	[93.69] 372.80	[4473.6] 9692.9
2	26.0 [12.0]	1.344	11.19 - [-7.94]	[25.71] 109.68	[1316.18] 3331.29	1633.0 0.1342	42458.0 0.1433	[106.33] 739.50	[8874.0] 25056.0	[64.12] 436.92	[5243.6] 14936.0
1	14.0 [14.0]	1.456	9.78 - [-7.94]	[25.80] 135.48	[1896.75] 5228.04	1642.0 0.0668	22988.0 0.0776	[52.93] 792.35	[11092.9] 36148.9	[34.72] 471.64	[6603.0] 21538.9
						7947.2	296334.4	792.35		471.64	

* For NEHRP and UBC   ^ For NEHRP   + For UBC

TABLE 5.11.4 (C): CALCULATION OF TORSIONAL RIGIDITY

Story	Braced Frame		SMRSF		J
	K_y (Kip/in)	$\bar{x}$	K_x (Kip/in)	$\bar{y}$	(kip.ft²/in)
1 st	3280.	14.4	540.	54.0	4.51×10^6
2 nd	2410.	14.4	400.	54.0	3.33×10^6
3 rd	1620.	14.4	350.	54.0	2.71×10^6
4 th	1310.	14.4	250.	54.0	2.00×10^6
5 th	1200.	14.4	240.	54.0	1.90×10^6

Note: $J = 2(K_y \bar{x}^2 + K_x \bar{y}^2)$

DESIGN EXAMPLE 5.11.5 Five Story Concrete Frame-Shear Wall Building

A five-story concrete frame and shear wall office building will be designed for a site located in Nevada. A plan view of the building is shown in Figure 5.11.5-1. The building will be designed with the lateral forces in the E-W direction resisted by special moment resisting space frames (SMRSF) and in the N-S direction by concrete shear walls. A parapet five feet high is provided at the roof.

To simplify the example, it will be assumed that the nominal weight of the stairs and the mechanical equipment in the service core is equal to the design dead loads used for each floor. This means that the center of mass is located at the geometric center of the building. Normally this will not be true and the center of mass will need to be computed.

The basic design information is given below:

DESIGN LOADS AND FORCES

LIVE LOADS:	ROOF		20 PSF
	FLOOR		50 PSF
LATERAL FORCES	WIND	ASCE 7 1988	70 MPH
	SEISMIC	NEHRP 1988: A = 0.15	v = 40
		UBC 1988:	ZONE 4
MATERIAL PROPERTIES	CONCRETE	ASTM C94	5,000PSI
	REINFORCEMENT	A615, GRADE 60	60,000PSI

GRAVITY LOADS

		ROOF PSF		FLOOR PSF
DEAD LOAD	Insulation plus membrane	11	Finish	1
	Concrete slab	60		60
	Mechanical and electrical	5		5
	Concrete framing	50		60
	Ceiling and lighting	5		5
	Partitions	--		20
	TOTAL DEAD LOAD	131		151
	LIVE LOAD	20		50
	TOTAL LOAD	151		201

WALL DEAD LOAD (GFRC and glass) 18

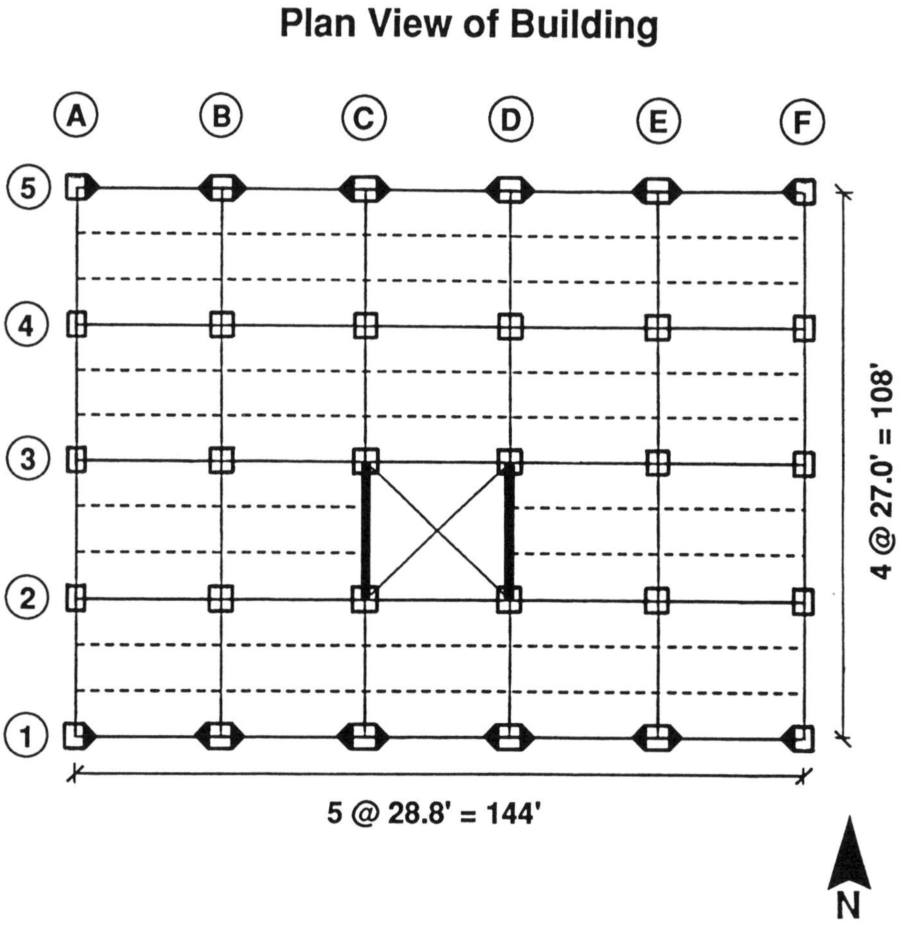

Plan View of Building

Figure 5.11.5-1 Plan view of five story concrete frame-shear wall building

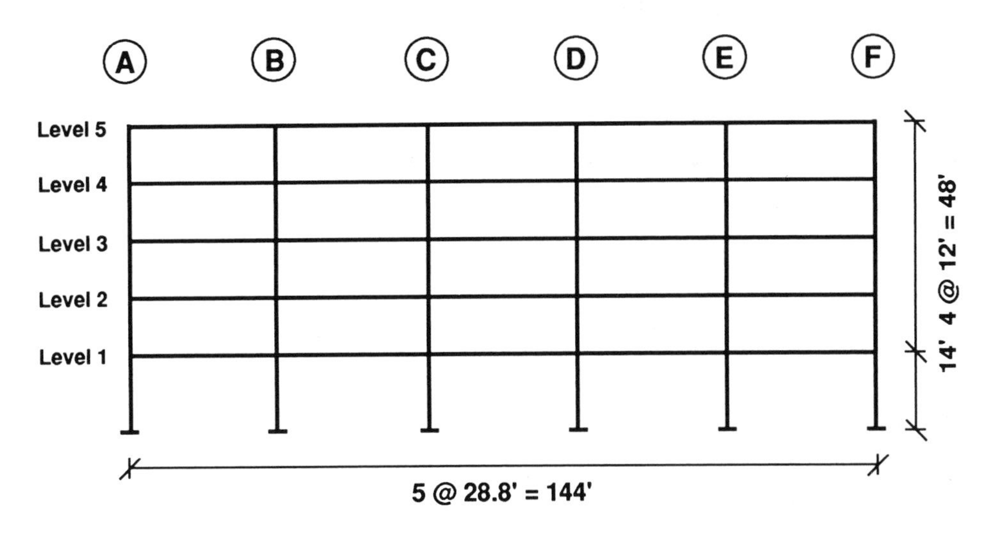

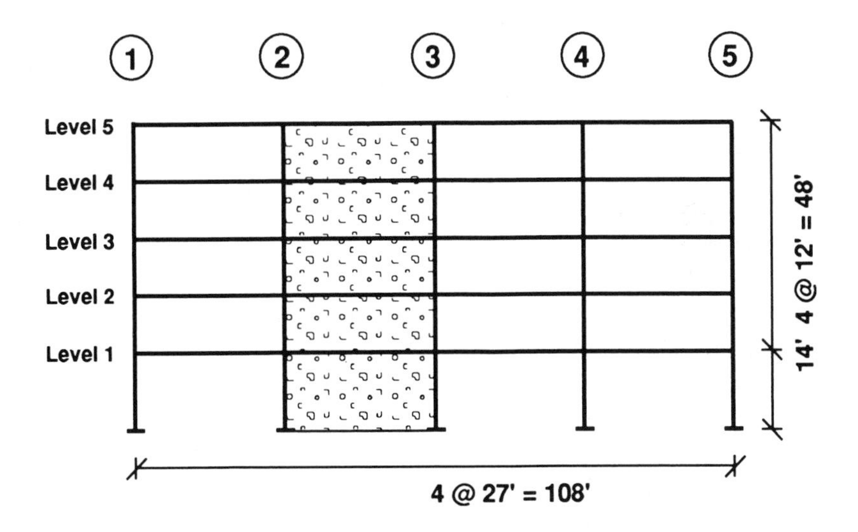

Figure 5.11.5-2 Elevations of frames on lines 1 and D

Note: The full partition allowance is used in calculation of seismic forces in lieu of using some part of the design live load.

Reinforced concrete frame buildings do not readily adapt to the use of perimeter frames only resisting the lateral forces. It will be assumed that all lines of framing in the east west direction will be designed and detailed as special moment resisting frames. Shear walls at lines C and D would each carry half of the design forces in the N-S direction. Elevation views of Frame 1 and Frame D are shown in Figure 5.11.5-2.

WIND FORCES

For the California coastal region, the Basic Wind Speed from Figure 1 is 70 mph. For a building with an importance factor of 1 located in an exposure C determine wind pressures and forces at the various levels:

$p = q * [(G * C) - (G * C)]$

$q = 0.00256 * K * (1 * 70) = 12.544 * K$

From Equation A3, K_z, for average pressure at mid-point of tributary area and at roof = 0.814, 0.932, 1.045, 1.130, 1.199 and 1.201.

From Equation A5, G_z, interpolated for mid-point of tributary area and at roof = 1.314, 1.272, 1.238, 1.213, 1.199 and 1.198. For the main building frame $G_h = G_z$ at roof = 1.198.

From Figure 2 C_p = 0.8 for windward and -0.5 or -0.44 for leeward.

$F = p * A$

$p_r = 12.544 * 1.196 * 1.197 * 0.8 = 14.40$ psf windward

$p_5 = 12.544 * 1.130 * 1.197 * 0.8 = 13.57$ psf windward

$p_4 = 12.544 * 1.045 * 1.197 * 0.8 = 12.55$ psf windward

$p_3 = 12.544 * 0.932 * 1.197 * 0.8 = 11.19$ psf windward

$p_2 = 12.544 * 0.814 * 1.197 * 0.8 = 9.78$ psf windward

$p_l = 12.544 * 1.201 * 1.198 * 0.5 = -9.02$ psf leeward N-S

$p_l = 12.544 * 1.201 * 1.198 * 0.44 = -7.94$ psf leeward E-W

Tributary area for each floor level is assumed to be the width times the height to mid-story of adjacent stories or to top of parapet. Total width of exposure:

N-S 144 + 2 * 2 = 148 ft.

E-W 108 + 2 * 2 = 112 ft

The force calculations can be more easily accomplished, and comparison made with seismic forces, by tabularizing as illustrated in Tables 5.11.5 (A) and (B).

SEISMIC FORCES

Calculate Total Base Shear

NEHRP	**UBC**
$V = C_s * W$: v = 40	$V = Z * I * C/R_w * W$: Z = 0.4
$C_s = 0.014 * v * S/T^{2/3}$	I = 1.0: $C = 1.25 * S/T^{2/3}$

S = Soil factor = 1.2

R = 5.5 N-S or 8 E-W	R_w = 8 N-S or 12 E-W
$T_a = 0.05 * h_n/L^{1/2}$ N-S	$T = C_t * h_n^{3/4}$: C_t = 0.020 N-S

$h_n = 62.0$

$T_a = 0.05 * 62/110^{1/2} = 0.296$

$T_a = C_t * h_n^{3/4}$: $C_t = 0.030$ E-W

$T_a = 0.030 * 62^{3/4} = 0.663$

$C_s = 0.014 * 40 * 1.2/(R * T_a^{2/3})$

$C_s = 0.672/(5.5 * 0.296^{2/3})$
$= 0.2751$ N-S: $C_s = 0.672/(8 * 0.663^{2/3})$
$= 0.1105$ E-W

C_s limit $= 2.5 * A/R$: $A = 0.40$

$C_s = 1/5.5 = 0.1818$ N-S

$C_s = 1/8 = 0.1250$ E-W

$T = 0.020 * 62^{3/4} = 0.442$

$T = C_t * h_n^{3/4}$: $C_t = 0.035$ E-W

$T = 0.030 * 62^{3/4} = 0.663$

$C = 1.25 * 1.2/T^{2/3}$

$C = 1.5/0.442^{2/3} = 2.585$ N-S
$C = 1.5/0.663^{2/3} = 1.973$
C limit = 2.75

Base Shear Coefficient

$C_b = Z * I * C/R_w = 0.40 * 1 * 2.585/8$
$= 0.1293$ E-W: $C_b = 0.40 * 1 * 1.973/12$
$= 0.0658$ N-S

Determine weights tributary to each level and sum the weights.

$W_5 = 108 * 144 * 0.131 * 2 * (108 + 144) * (6 + 5) * 0.018$
$= 2037.3$ kips $+ 99.8$ kips $= 2137.1$ kips

$W_f = 108 * 144 * 0.151 + 2 * (108 + 144) * 12 * 0.018$
$= 2348.3$ kips $+ 108.9$ kips $= 2457.2$ kips

$W_1 = 108 * 144 * 0.098 + 2 * (108 + 144) * (7 + 6) * 0.018$
$= 2348.3$ kips $+ 117.9$ kips $= 2466.3$ kips

See Tables 5.11.5 (A) and (B) for W and V.
It is clear that earthquake forces govern for this example.

DESIGN FOR LATERAL FORCES

Calculate vertical distribution of forces. The distribution of forces in the N-S direction is the same for both NEHRP and UBC procedures. For the E-W direction, however, the vertical distribution of forces at the various levels varies. Since the period is between 0.5 and 2.5 seconds the exponent, k, is 1.0814 from Section 4.3 of NEHRP. For UBC there is no force, F_t, since the period is less than 0.7 seconds. The force at each level is determined by multiplying the base shear by the distribution factor from the following formula: (Modified formula 34-8 from UBC)

$$DF = W_i * h_i / \Sigma(W_i * h_i)$$

See Table 5.11.5 (A), Seismic Column 2 for distribution factors for both NEHRP and UBC. See Table 5.11.5 (B), Seismic Columns 1 and 2 for distribution factors for UBC and NEHRP respectively

Since the building is symmetric in both mass and stiffness, the floor-level forces would be distributed equally to each frame in the direction under consideration. The design forces applied to each resisting element and their preliminary designs are shown in Figure 5.11.5-3 for the shear wall and Figure 5.11.5-4 for the moment frame. Use an 8" shear wall in the N-S direction. For the SMRSF in the E-W direction, columns are 18" square and girders are 14" x 30".

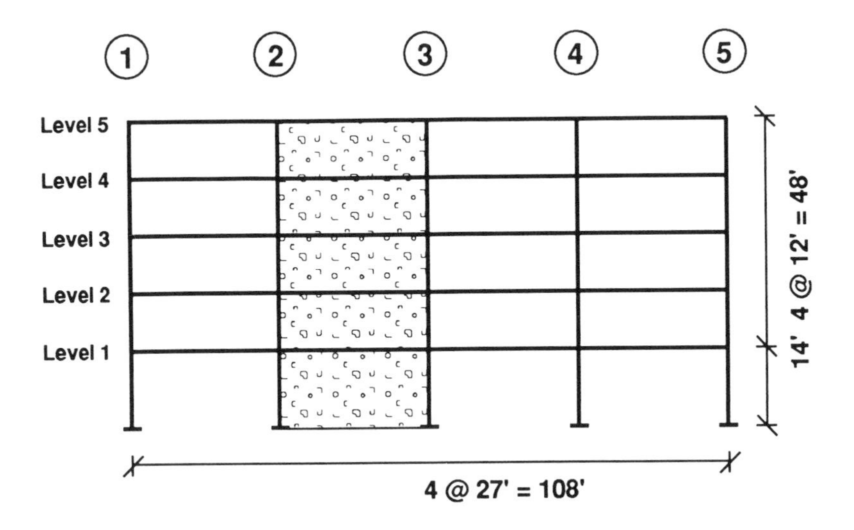

Columns 18" square; girders 14" x 30";shear walls 8" thick.

Equivalent Lateral Forces

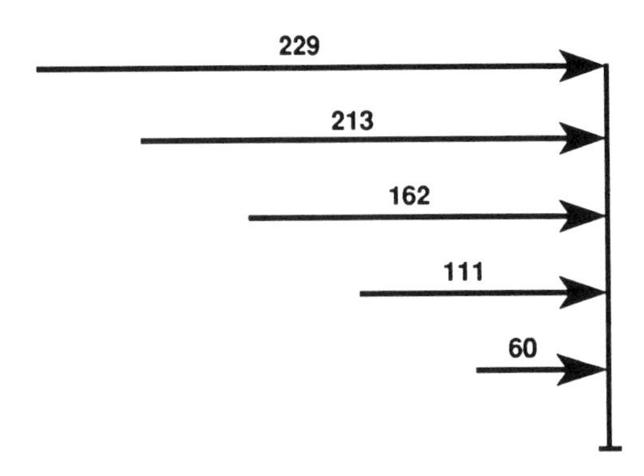

Figure 5.11.5-3 Frame D and the equivalent lateral forces acting on it

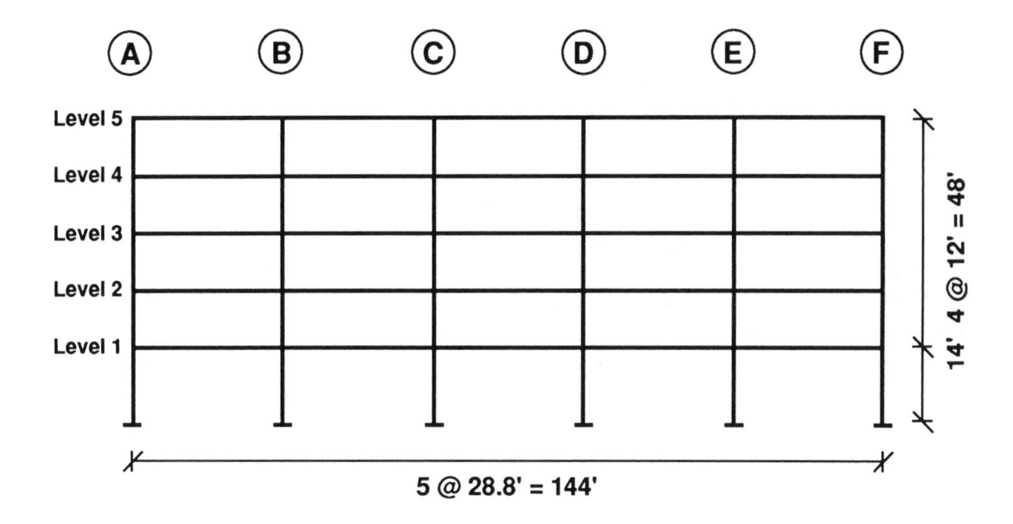

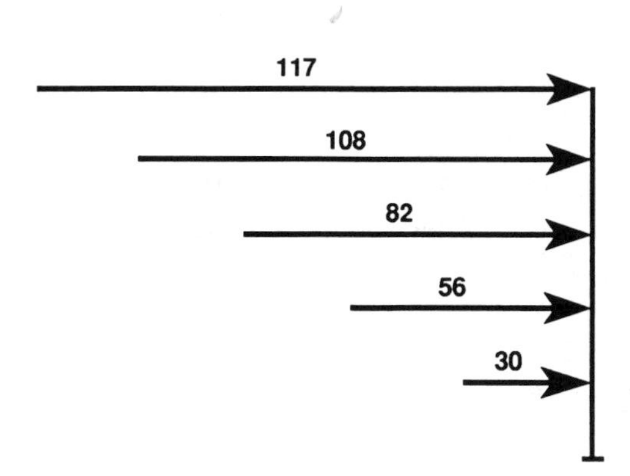

Figure 5.11.5-4 Frame 1 and the equivalent lateral forces acting on it

The only step remaining to be completed is to include the effects of accidental torsion, recognizing the symmetry of mass and stiffness. The torsional moment, at each level is calculated by applying the seismic story shear in a particular direction at an eccentricity equal to 5 percent of the building dimension normal to the direction of the force. Since the moment will cause the building to rotate in plan, the vertical resisting elements in both directions will be required to carry additional shear in accordance with their contribution to the torsional rigidity. The calculation of the additional force at each story of each frame that results from the torsion is straightforward, and is similar to calculating the shear force on bolts in an eccentrically loaded connection. The stiffness, K_i, of each resisting element at each story must be computed. Next, the polar moment of inertia (or polar moment of stiffness) must be calculated for each story. Call this J_i. Then, the force carried by a resisting element is simply $F_i = M_i * r_i * K_i/J_i$ where M_i is the moment in the story and r_i is the distance from the center of rigidity to the element under consideration. These relationships are valid only if it is assumed that the floor slab is rigid in its plane. This assumption should be checked as discussed earlier.

To calculate the shear wall stiffness, K_w, at each story, apply a unit load at the top level and calculate the deflection at each floor. Since the shear in each story is one kip, the story stiffness is simply one divided by the relative story deflection, or the story drift. Calculate the moment frame stiffness, K_m, at each level in the same manner. Since there are two identical resisting elements in each direction located the same distance from the center of rigidity, the polar moment of stiffness at each story is simply $J = 2(K_w x^2 + K_m y^2)$, where x and y are the distances from the center of rigidity to the braced frames and the moment frames, respectively. The distribution of the shear due to torsional moment to each of the resisting elements is the proportion of the torsional stiffness of that element to the total torsional stiffness: DF $= K_w * x^2/J_i$ or $K_m * y^2/J_i$.

The story stiffnesses for each frame and the story torsional rigidities are given in Table 5.11.5 (C). Even though the shear wall is 25 times stiffer than the moment frame, it provides a torsional rigidity of about one and a half times that of the moment frame. This is because the moment frame is much farther from the center of rigidity. This illustrates the importance of placing as many of the vertical resisting elements of the lateral force resisting system at the perimeter of the building as possible.

Calculate Resisting Element Forces Due to Torsion

1. Seismic Design Moment from Forces Acting in NS Direction

$M_i = V_i * e_i$ Where e = eccentricity = 0.05 * 144 $M_i = F_i * 7.2$

M_5 = 458.7 * 7.2 = 3302.6 k-ft
M_4 = 884.1 * 7.2 = 6365.5 k-ft
M_3 = 1207.3 * 7.2 = 8692.8 k-ft
M_2 = 1428.6 * 7.2 = 10285.6 k-ft
M_1 = 1548.1 * 7.2 = 11146.0 k-ft

2. Distribution of shear from torsional moments:

Shear in Shear Wall Due to NS Force

$F_i = M_i * DF/x$ Where x = 14.4 ft for all stories

$F_5 = 3302.6 * 0.3850 / 14.4 = 88.3$ kips
$F_4 = 6365.5 * 0.3160 / 14.4 = 139.7$ kips
$F_3 = 8692.8 * 0.2902 / 14.4 = 175.2$ kips
$F_2 = 10285.6 * 0.2777 / 14.4 = 198.4$ kips
$F_1 = 11146.0 * 0.2661 / 14.4 = 206.0$ kips

The additional shear in Moment Frames due to torsion from N-S forces is, by observation, not critical compared with direct forces plus additional shear due to torsion from E-W forces.

3. Seismic Design Moment from Forces Acting in the E-W Direction

$M_i = V_i * e$ Where e = 0.05 * 108 = 5.4 $M_i = F_i * 5.4$
$M_5 = 233.4 * 5.4 = 1260.1$ k-ft
$M_4 = 444.7 * 5.4 = 2428.5$ k-ft
$M_3 = 614.2 * 5.4 = 3316.5$ k-ft
$M_2 = 726.7 * 5.4 = 3924.0$ k-ft
$M_1 = 587.5 * 5.4 = 4252.4$ k-ft

The additional shear in shear walls due to torsion from E-W forces is not critical.

4. Distribution of shear from torsional moment:
Shear in Moment Frames due to torsion from E-W forces.

$F_i = M_i * DF/x$ Where x = 54 ft. for all stories

$F_5 = 1260.1 * 0.1152 / 54 = 2.7$ kips
$F_4 = 2428.5 * 0.1840 / 54 = 8.3$ kips
$F_3 = 3316.5 * 0.2098 / 54 = 12.9$ kips
$F_2 = 3924.0 * 0.2224 / 54 = 16.2$ kips
$F_1 = 4252.4 * 0.2338 / 54 = 18.4$ kips

The lateral force resisting elements do not have any commonality and orthogonal effects are not considered.

The final design forces at each level are shown in Figure 5.11.5-5 for both the shear walls and the frames. The wall and the members in each frame must be designed to withstand the stresses produced by these earthquake forces. The calculated story drift ratios for these earthquake forces must satisfy these allowable drifts:

Shear Wall

$\delta_i = 0.04 * h_i/R_w = 0.005 * h_i$
$\delta_{5\text{-}2} = 0.005 * 144 = 0.72$ inches $\delta_1 = 0.005 * 168 = 0.84$ inches

Moment Frame

$\delta_i = 0.04 * h_i/R_w = 0.0033 * h_i$

$\delta_{5-2} = 0.0033 * 144 = 0.48$ inches $\delta_1 = 0.0033 * 168 = 0.56$ inches

It is most often the case that a shear wall will satisfy the drift requirements but a concrete SMRSF may not. Therefore, the member sizes in a moment frame may need to be increased beyond those required to resist the forces within allowable stresses.

Although earthquake forces govern in this example for the main lateral force resisting system, this may not be true for the parapet or the cladding. These elements are designed using different procedures that are described separately in Chapter 6.

As noted in design example 5.11.4 the design wind forces may exceed the design earthquake forces but may not exceed the real earthquake forces. Thus, even when wind forces govern, it is essential to use good detailing and inspection practices as required by UBC.

Finally, it should be noted that the purpose of this example was to demonstrate how one would calculate and distribute the lateral forces for design for wind and earthquake forces. Many engineers would find the structural scheme used here to be undesirable in highly seismic areas. Only two vertical lateral force resisting elements are provided in each direction, so the redundancy is not high. A dual system composed of two shear walls and two SMRSF in each direction would be far superior. The shear walls would provide great stiffness under moderate earthquakes and the SMRSF would serve as a backup in case some of the shear wall capacity is lost during a major earthquake.

Elevation of Frame 1
Equivalent Lateral Forces

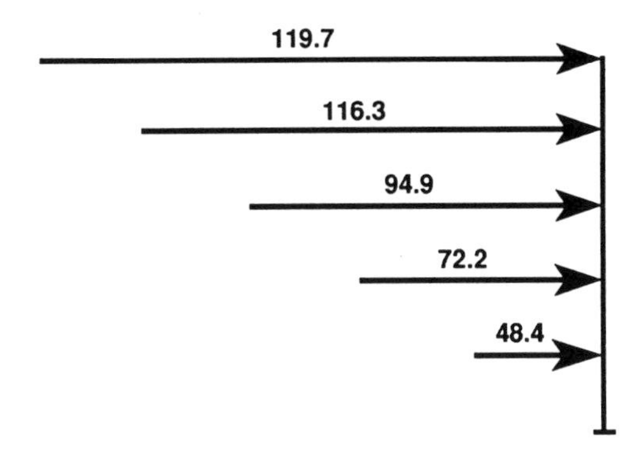

Elevation of Frame D
Equivalent Lateral Forces

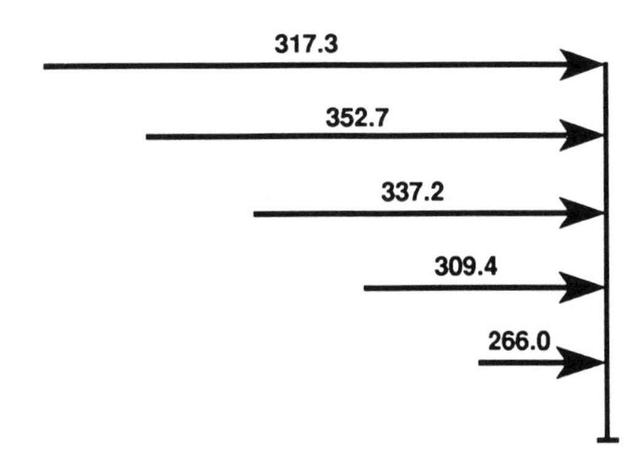

Figure 5.11.5-5 Final design forces for frames on lines 1 and D

TABLE 5.11.5 (A): - NORTH - SOUTH DIRECTION

		WIND				SEISMIC					
LEVEL	Height from base h_x Story height h_s (feet)	Area Tributary to level (1000 sq. ft.)	Wind Pressure (pounds per sq. ft.)	Story Shear [F] Story Shear V (kips)	Overturning Moment [Story] Total (foot kips)	Weight Tributary to Level (kips)	Weight times height (ft.kips) Distribution Factor*	NEHRP		UBC	
								Story Force [F] Story Shear V (kips)	Overturning Moment [story] Total (ft.kips)	Story Force [F] Story Shear V (kips)	Overturning Moment [story] Total (ft.kips)
5	62.0 [12.0]	1.628	14.40 - [-9.02]	[38.13] 38.13	[457.56] 457.56	2137.1	132500.2 0.2963	[645.06] 645.06	[7740.7] 7740.7	[458.69] 458.69	[5504.3] 5504.3
4	50.0 [12.0]	1.776	13.57 - [-9.02]	[40.12] 78.25	[939.00] 1396.56	2457.2	122860.0 0.2748	[598.25] 1243.31	[14919.8] 22660.5	[425.41] 884.10	[10609.2] 16113.5
3	38.0 [12.0]	1.776	12.55 - [-9.02]	[38.31] 116.56	[1398.72] 2795.28	2457.2	93373.6 0.2088	[454.57] 1697.88	[20374.5] 43035.0	[323.23] 1207.33	[14488.0] 30601.5
2	26.0 [12.0]	1.776	11.19 - [-9.02]	[35.89] 152.45	[1829.40] 4624.68	2457.2	63887.2 0.1429	[311.10] 2008.98	[24107.8] 66142.8	[221.22] 1428.55	[18571.1] 49172.6
1	14.0 [14.0]	1.924	9.78 - [-9.02]	[36.17] 188.62	[2640.68] 7265.36	2466.3	34528.8 0.0772	[168.07] 2177.05	[30478.7] 96621.5	[119.51] 1548.06	[21672.8] 70845.5
					11975.0	447149.2	2177.05		1548.06		

* For NEHRP and UBC

TABLE 5.11.5 (B): - EAST - WEST DIRECTION

		WIND				SEISMIC					
								NEHRP		UBC	
LEVEL	Height from base h_x Story height h_s (feet)	Area Tributary to level (1000 sq. ft.)	Wind Pressure (pounds per sq. ft.)	Story Shear [F] Story Shear V (kips)	Overturning Moment [Story] Total (foot kips)	Weight Tributary to Level (kips) Distribution Factor^	Weight times height (ft.kips) Distribution Factor+	Story Force [F] Shear V (kips)	Overturning Moment [story] Total (ft.kips)	Story Force [F] Shear V (kips)	Overturning Moment [story] Total (ft.kips)
5	62.0 [12.0]	1.232	14.40 - [-7.94]	[27.52] 27.52	[330.27] 2330.27	2137.1 0.2963	185403.2 0.3062	[392.10] 392.10	[4705.3] 4705.3	[223.35] 233.35	[2800.2] 2800.2
4	50.0 [12.0]	1.344	13.57 - [-7.94]	[28.91] 56.43	[667.19] 1007.46	2457.2 0.2748	168517.5 0.2783	[363.58] 755.68	[9068.1] 13773.4	[216.37] 449.72	[5396.6] 8196.8
3	38.0 [12.0]	1.344	12.55 - [-7.94]	[27.54] 83.97	[1007.65] 2015.11	2457.2 0.2088	125550.5 0.2073	[276.32] 1032.00	[6878.8] 21452.2	[164.44] 614.16	[7369.9] 15566.7
2	26.0 [12.0]	1.344	11.19 - [-7.94]	[25.71] 109.68	[1316.18] 3331.29	2457.2 0.1429	83290.0 0.1375	[189.06] 1221.06	[9947.5] 31399.7	[112.51] 726.67	[8720.1] 24286.8
1	14.0 [14.0]	1.456	9.78 - [-7.94]	[25.80] 135.48	[1896.75] 5228.04	2466.3 0.0772	42802.5 0.0707	[102.18] 1323.24	[18525.4] 49925.1	[60.81] 787.48	[11024.7] 35311.5
						11975.0	605563.7	1323.24		787.48	

* For NEHRP and UBC ^ For NEHRP + For UBC

TABLE 5.11.5 (C): CALCULATION OF TORSIONAL RIGIDITY

Story	Concrete Shear Wall		Concrete SMRSF		J
	K_y (Kip/in)	$\bar{x}$	K_x (Kip/in)	$\bar{y}$	(kip.ft²/in)
1 st	2.679×10^7	14.4	5.698×10^5	54.0	1.443×10^{10}
2 nd	1.216×10^7	14.4	5.033×10^5	54.0	7.979×10^9
3 rd	8.598×10^6	14.4	4.419×10^5	54.0	6.144×10^9
4 th	7.194×10^6	14.4	4.097×10^5	54.0	5.373×10^9
5 th	6.653×10^6	14.4	4.160×10^5	54.0	5.186×10^9

Note: $J = 2(K_y \bar{x}^2 + K_x \bar{y}^2)$

DESIGN EXAMPLE 5.11.6 One story wood frame building

A one-story wood frame commercial building will be designed for a site on the west coast. The plan and elevations of the building are shown on Figure 5.11.6-1. The basic design information is given below:

DESIGN LOADS AND FORCES

LIVE LOADS:	ROOF		20 PSF
	SNOW		30 PSF
LATERAL FORCES	WIND	ASCE 7 1988:	70 MPH
	SEISMIC	NEHRP 1988: A = 0.40	v = 40
		UBC 1991:	ZONE 4
MATERIAL PROPERTIES	WOOD	DF-L #2	

GRAVITY LOADS

		ROOF PSF		WALL PSF
DEAD	Roofing, 3 ply	2.2	Stucco	10.0
LOAD	Sheathing, 1/2"	1.5	Sheathing, 1/2"	1.5
	Sub-purlins & purlins	2.0	Framing	2.5
	Girders	4.1	Gypsum board	2.5
	Insulation	1.5	Insulation	1.5
	Lights, mechanical, etc.	2.7		
	TOTAL DEAD LOAD	14.0		18.0

WIND FORCES

For the west coast region selected the Basic Wind Speed from Figure 1 is 70 MPH. For a building with an importance factor of 1 located in an area with an exposure category B determine the wind pressures and forces at various levels:

$p = q * [(G_h * C_{pw}) - (G_h * C_{pl})]$

$q = 0.0256 * K_z * (I * 70)2 = 12.544 * K_z$

From Table 6, K_z, for 0' to 15' and 15' to 20' are 0.37 and 0.42.

From Table 8, G_z, for 0' to 15' and 15' to 20' are 1.65 and 1.59.

From Figure 2 the value of Cp is: for windward wall, 0.8; for leeward wall, longitudinal direction 0.5 and transverse direction 0.3.

The wind forces on the building, $p = q * K_z * G_h * C_p$ are:

$p_{w1} = 12.544 * 0.37 * 1.65 * 0.8 = 6.13$ psf windward

$p_{w2} = 12.544 * 0.42 * 1.59 * 0.8 = 6.70$ psf windward

$p_{w1} = 12.544 * 0.42 * 1.59 * 0.5 = 4.19$ psf leeward long.

or $= 12.544 * 0.42 * 1.59 * 0.3 = 2.52$ psf leeward trans.

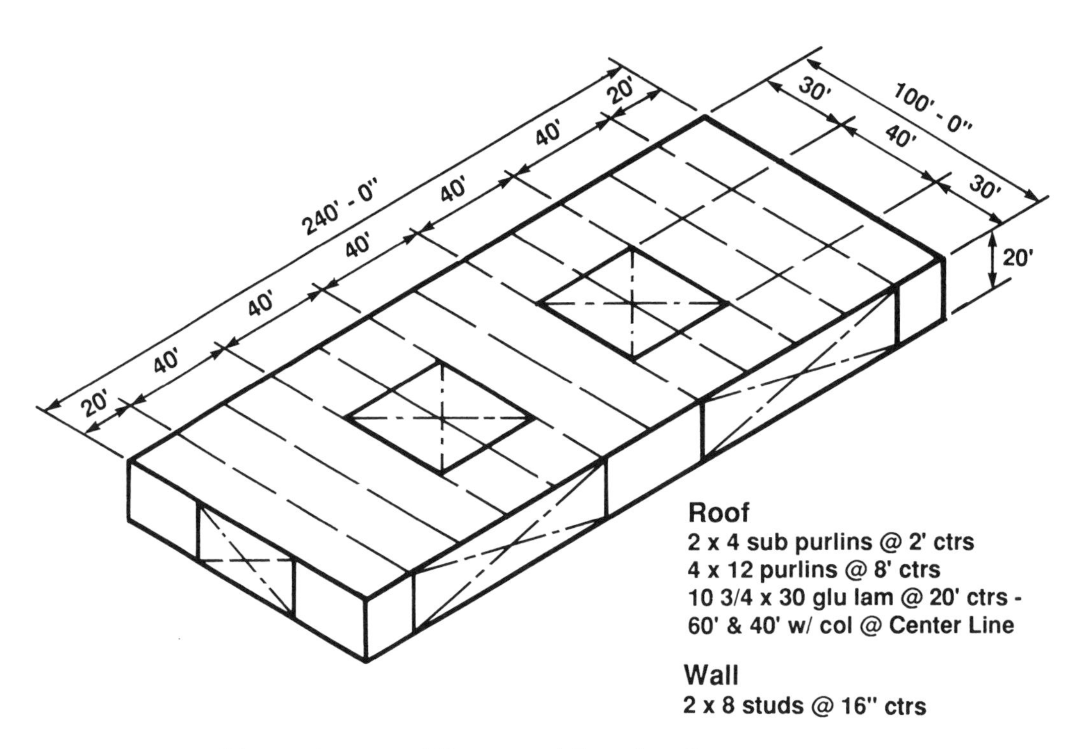

Dimensions and Structural Framing System

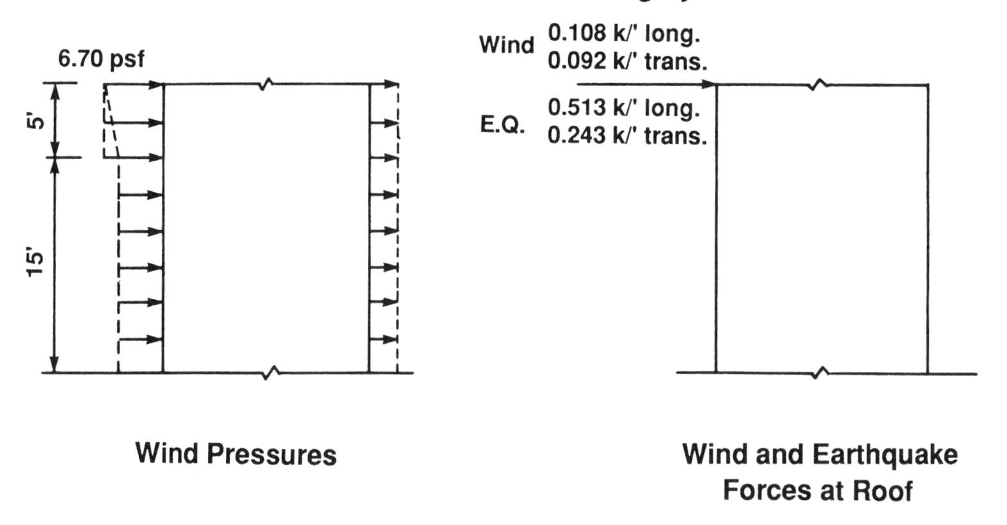

Figure 5.11.6-1 One story wood frame commercial building

Force at the roof line is 15 * 0.00613 * 15 * 0.375 + 5 * 0.00670 * 0.875 + 20 * (0.00419 or 0.00252) * 0.5 = 0.108 K/ft longitudinal or 0.092 K/ft transverse.
See Table 5.11.6 for total wind forces.

SEISMIC FORCES

Calculate Total Base Shear

N E H R P	**U B C**
$V = C_s * W$: $v = 40$	$V = Z * I * C/R_w * W$: $Z = 0.40$
$C_s = 0.014 * v * S/T^{2/3}$ or	$I = 1.0$: $C = 1.25 * S/T^{2/3}$
C_s limit = 2.5 * A/R: A = 0.40	C limit = 2.75

S, Soil factor = 1.2

R = 6.5 N-S or 8 E-W	R_w = 8 N-S and E-W

For h_n = 20' the period will not be computed and the limiting values of C_s and $C_b = Z * I * C/R_w$ will be used.

$C_s = 2.5 * 0.40/6.5 = 0.154$	$C_b = 0.40 * 1 * 2.75/8 = 0.138$

Force at roof line is [20 * 0.018 * 2 * .5 + 0.014 * {100 or 240}] * C_s (or Cb) = 0.271 K/ft or 0.573 K/ft for NEHRP and 0.243 K/ft or 0.513 K/ft for UBC.

DESIGN FOR SEISMIC FORCES

For both directions the seismic forces will be resisted by the plywood sheathed diaphragm and shear walls. Critical direction for seismic forces is east-west. The maximum diaphragm shears occur at the end walls and are
58.29 * 0.55/100 = 0.321 k/ft for UBC forces.

Select the nailing required for 15/32" Structural 1 Rated Sheathing fastened with 10^d nails from Table 25-J-1:

With 2x framing and 6-inch nail spacing at panel edges the allowable shear is 0.320 k/ft.

Determine the shear forces at the ends of the skylight openings to be transmitted into the body of the diaphragm and the secondary bending in the diaphragm elements at the openings.

V = 32.06 - [60 * 0.243] = 17.47 kips and 17.47 - [40 * 0.243] = 7.76 kips.
v = 17.47/60 = 0.291 k/ft therefore 10d @ 6" centers O. K.
V to be transmitted by connections at critical corner of openings is [17.47/100] * 20 = 3.49 kips.

The "chord" forces, ignoring the reduction due to the action of the parallel shear walls, would be as follows:

$T = C = 0.243 * 240^2/[8 * 100] = 17.50$ kips @ center line
$T = C = \{32.07 * 60 - 0.243 * 60^2 * 0.5\}/100 + 3.49$
$= 18.36$ kips

and $T = C = \{32.07 * 100 - 0.243 * 100^2 * 0.5\}/100 + 1.55$
$= 21.47$ kips, maximum, at the ends of the openings.

Shear walls at ends have 40 foot display windows located at center of the length. Critical wall shears are:

$$v = 58.29/[2 * 60] = 0.486 \text{ k/ft.}$$

From Table 25-K-1 the allowable shear for 15/32" Structural 1 Rated Sheathing with 10d nails at 4 inch centers at panel edges is 0.510 k/ft.

Overturning on the wall segments is [58.29/4] * 20 = 291.5 ft.kips.
Righting moment = 0.85[{20 * 0.018 + 10 * 0.014} * 30 * 15 + {20 * 0.018 * 10} * 30] = 283.0 ft kips. Provide hold downs.

TABLE 5.11.6 (A): NORTH - SOUTH DIRECTION

		WIND				SEISMIC					
LEVEL	Height from base h_x (feet)	Area Tributary to level (1000 sq. ft.)	Wind Pressure (pounds per sq. ft.)	Story Shear [F] Story Shear V (kips)	Overturning Moment [Story] Total (foot kips)	Weight Tribu-tary to Level (kips)	Distribution Factor*	N E H R P		U B C	
								Story Shear V (kips)	Overturning Moment (ft.kips)	Story Shear V (kips)	Overturning Moment (ft.kips)
ROOF	0 to 15 15 to 20	3.00 x 0.375 1.00 x 0.875	10.32 10.89	11.61 9.53		(Rf 100 x 240 x 0.014 = 336.0) (Wl 20 x 100 x 0.018 x 2 x 0.5 = 32.0)	1.00				
				21.14	1029.74	372.0		57.29	1145.8	51.34	1026.7

* For NEHRP and UBC

TABLE 5.11.6 (B): EAST - WEST DIRECTION

		WIND				SEISMIC					
LEVEL	Height from base h_x (feet)	Area Tributary to level (1000 sq. ft.)	Wind Pressure (pounds per sq. ft.)	Story Shear [F] Story Shear V (kips)	Overturning Moment [Story] Total (foot kips)	Weight Tributary to Level (kips)	Distribution Factor*	N E H R P		U B C	
								Story Shear V (kips)	Overturning Moment (ft.kips)	Story Shear V (kips)	Overturning Moment (ft.kips)
ROOF	0 to 15 15 to 20	7.20 x 0.375 2.40 x 0.875	8.65 9.22	23.36 19.36		(Rf 336.0) (Wl 20 x 240 x 0.018 x 2 x 0.5 = 86.4)	1.000				
				42.72	854.44	422.4		65.05	1300.1	58.29	1165.8

* For NEHRP and UBC

6 Connections

6.1 INTRODUCTION

The quality of connectivity within the structural system is probably the most critical factor in the performance of a building subjected to high wind and seismic forces. This concept is clearly emphasized in both the SEAOC and NEHRP seismic provisions which specify design requirements under the category of *Ties and Continuity*, and in their commentaries, discuss the importance of tying the structure together to act as a unit. Furthermore, in low-rise construction, which commonly utilizes combinations of materials with differing properties, the design and detailing of connections takes on even greater importance.

The fundamental differences between wind and seismic forces have been discussed in Chapters 1 and 5. From the standpoint of connection design, two basic differences between these forces should be noted: first, the high ductility demands on a system under seismic loading require special emphasis on the strength and ductility of connections in general; and second, wind forces on light weight building portions, create special detailing requirements to address issues such as uplift effects on roof systems and negative pressures at discontinuities.

Because the actual seismic forces on a building may greatly exceed the design forces, there must be a strong emphasis on detailing of the system to achieve maximum ductility. The inherent ductility demands require that the structural system as a whole, not just individual members, perform inelastically without collapse. Thus, the fundamental seismic design requirement of providing a continuous load path must be enforced. This is accomplished by attention to connection design.

In order to assure adequate ductility whenever significant seismic forces may occur, it is often appropriate to use special seismic detailing even where wind forces appear to "govern" the design. Recommended guidelines for determining when seismic versus wind forces govern the design of the lateral force resisting system are contained in Chapter 5. Both the SEAOC and NEHRP provisions contain requirements which are intended to insure adequate ductility under seismic forces through design and detailing requirements which vary by seismic zone. The responsibility lies with the designer, however, to recognize special conditions which dictate that greater redundancy or ductility be incorporated into the structural system by adding members or improving certain connections.

The SEAOC Provisions require that connections intended to resist seismic forces be designed and detailed on the drawings, rather than by the contractor or fabricator. The intent is to eliminate connection failures which commonly occur in earthquakes. This approach is highly recommended as is the practice of careful checking of shop drawing details and field review of connection elements during construction.

Definitions

A **joint** is the entire assemblage at the intersection of the members.
A **connection** is the group of elements that attach a member at the joint, e.g. welds, plates, angles and bolts.

6.1.1 General Considerations

Static loads due to *gravity* place a particular system of forces on the connection, which can be determined once the design dead and live loads are known. These forces act in a specific direction and hence the elements (welds, bolts, nails, etc.) of the connection can be placed and sized to optimize the resistance to these forces. In gravity load applications, invariant tensile and compression zones exist on the connection, and the design can be tailored accordingly. Lateral forces due to wind or seismic effects can occur in either sense and hence the capacity of the connection must be investigated for both possible loading directions. In these applications, any portion of the connection can act in either tension or compression at different times.

The demands placed on a connection by wind and earthquake forces are somewhat different in nature from each other. Hence, the design limit states will not be the same. Connections for wind forces differ from gravity connections mainly because the direction of wind forces can change throughout time. Structures are designed for both gravity and wind loads with the expectation that the structure will respond elastically. In general, the earthquake environment will be more demanding than the wind environment since force levels experienced by a structure during an earthquake can far exceed the design forces. A connection which is well designed for seismic applications will be good for wind applications as well.

Seismic excitation places demands on the structure that require a ductile response. Generally, the forces on the connections for seismic excitation are severe but few in number. The forces that the connections are subjected to are determined not only by the applied loads but also by the response of the structure. In both seismic and wind environments, low-cycle fatigue is likely to be the final cause of failure of the connections (if the connection fails) regardless of the method of connection.

Many low-rise buildings are subject to special diaphragm design and detailing concerns. These are related to common plan irregularities and to diaphragm flexibility problems due to the stiffness of the vertical lateral force resisting elements. Distress associated with such problems is best overcome by providing adequate continuity to take advantage of any redundancy in the lateral force resisting system and to permit it to dissipate energy inelastically as is assumed in a code level design.

Both the SEAOC and NEHRP provisions emphasize the importance of continuity within a structural system. Both documents require that all parts of a structure be interconnected and contain specific requirements for ties and continuity related to portions of buildings and to individual members. NEHRP requires that any smaller portion of a building be tied to the remainder of the building with elements having the strength to transfer at least $A_v/3$ times the weight of the smaller portion, but not less than 5 percent of the portion's weight. In addition, it requires that the connection of each beam, girder or truss be capable of transferring a force parallel to the member of at least 5 percent of the dead and live load reaction. SEAOC provisions for portions of buildings and individual members are similar, as are the UBC-91 requirements except that the velocity-related acceleration coefficient, A_v, is replaced by the zone factor, Z.

The transfer of forces originating in one portion of a building to the lateral force resisting elements designed to resist those forces is accomplished through the use of collectors. The SEAOC and NEHRP provisions specifically require that collector elements be provided to make this transfer.

Buildings with plan irregularities require particular attention to chord and collector design. The SEAOC provisions define plan irregularities as types A,B,C,D or E, as described in Table 6-1 and shown in Figure 6-1 [SEAOC, 1988]. In seismic zones 3 and 4, collector connections for buildings with irregularity types A,B,C or D are required to be designed without the allowable one-third stress increase usually permitted in allowable stress design. Also in seismic zones 3 and 4, for buildings with irregularity type B, it is required that chords and collectors be designed considering independent movement of the projecting wings. The diaphragm elements are required to be designed for the more severe of the following two assumptions:

(a) Motion of the projecting wings in the same direction.
(b) Motion of the projecting wings in opposing directions.

This requirement is deemed satisfied if a three-dimensional dynamic analysis is performed.

6.2 CONNECTIONS IN WOOD FRAME BUILDINGS

6.2.1 Introduction

Wood frame construction is prevalent in the construction of low-rise buildings. Over the years, methods of construction have been developed by engineers and builders for the connections between the various structural elements. Additionally, significant effort has been made in research of connections for wood frame construction and many prefabricated connection devices are currently available to aid designers in their goal of tying the structural system together. Technical information about these devices is readily available and their use can be quite beneficial.

The loading criteria for wood frame, low-rise construction have been discussed in previous chapters. These buildings are relatively light weight, and the generated seismic loads for these buildings is often be less than the wind design loads. In general, the construction details for both types of loading are quite similar. Whether for wind or earthquake, the connections between the structural elements must serve to tie the various structural elements together so that the building can act as a unit. The consideration of continuity through the connection, attention to eccentricities, and a continuous load path are most vital design issues. Prevention of collapse depends on the energy absorption characteristics of the structural elements, so the connections should be well detailed and have enough strength to develop the useful strength of the structural elements.

Two types of wood frame construction are common in this country, namely platform framing and balloon framing. The more common platform framing is shown schematically in Figure 6.2.1. In balloon framing, the vertical members are continuous. The lateral load paths for these two systems can differ, but properly detailed, either system can be considered effective under lateral force conditions.

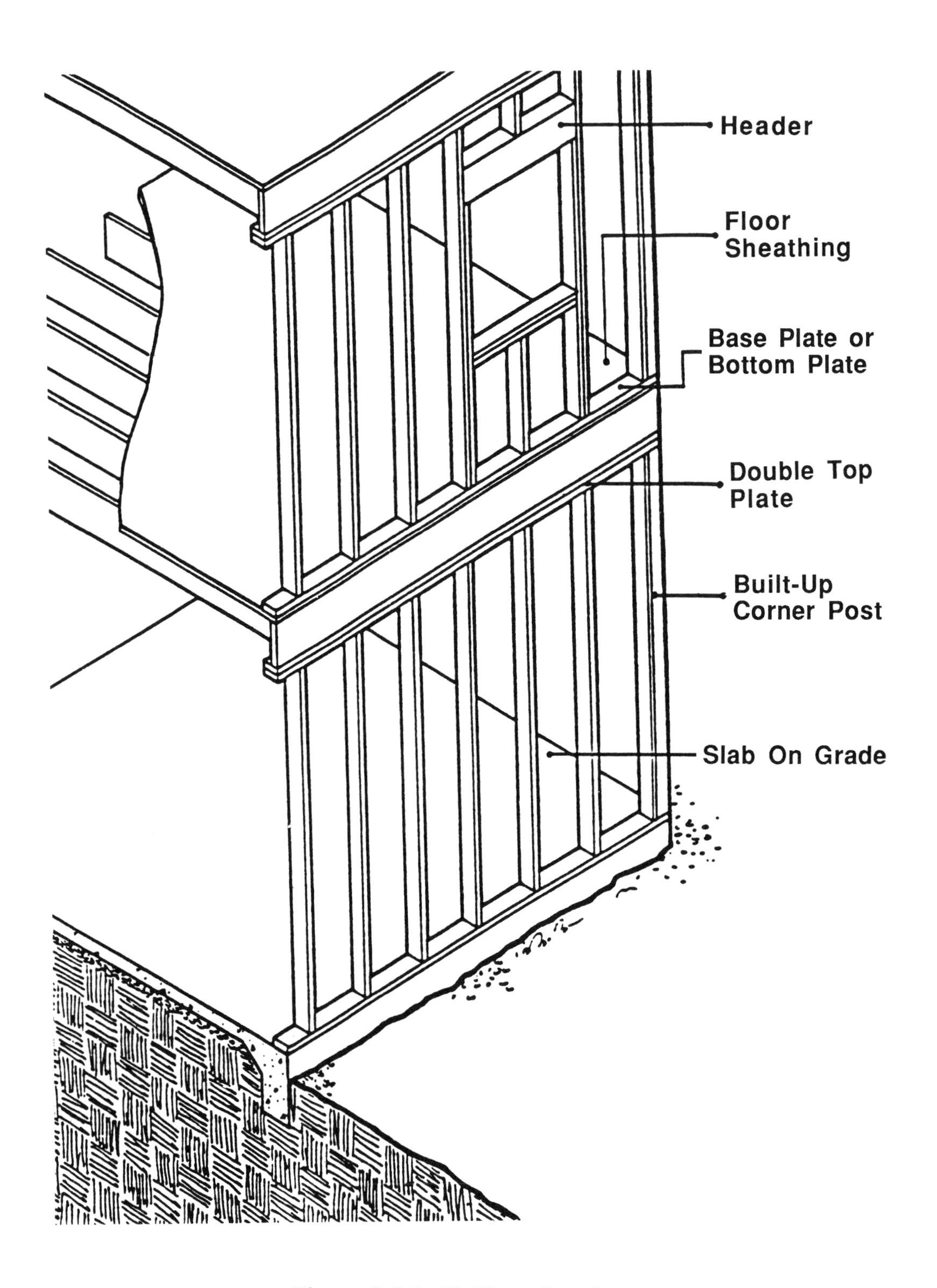

Figure 6.2.1 Platform framing

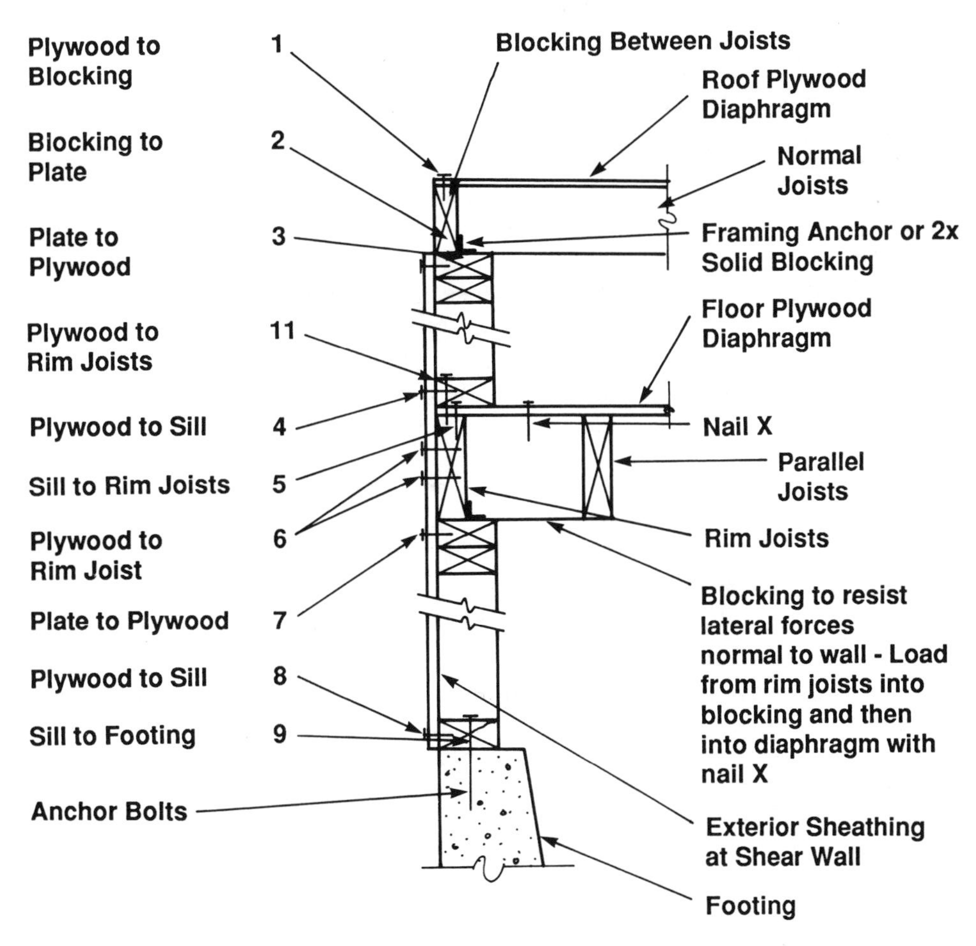

Moisture and Termite Protection Not Shown

1 - 9 Path of Forces From Roof to Foundation

11 , 6 - 9 Path for Forces from Floor Diaphragm

Details above are schematic. The purpose is to show the path of forces in a particular arrangement of framing elements.

Figure 6.2.2 Wood diaphragm and shear wall nailing

6.2.2 Detailing the Load Path from Roof to Foundation

In detailing a lateral force resistance system, each element and connection in the load path must be considered. In wood construction, the lateral force resisting elements typically consist of floor or roof diaphragms and framed shear panel walls. In Figure 6.2.2, the load path can be traced from the roof to the foundation through a typical platform framed wall. The following connections must be considered in order to assure the completeness of the lateral force resisting system:

1. Connection of structural and nonstructural elements to the diaphragms.
2. Interconnection of the key structural elements within the diaphragm itself.
3. Connections to transfer lateral forces from the diaphragm to the shear walls.
4. Connections within shear walls to tie the walls together and to transfer forces down the walls from upper to lower stories.
5. Connections between the shear walls and the foundations.

6.2.3 Connection of Structural and Nonstructural Elements to the Diaphragms

To transmit the lateral loads through the diaphragm to the vertical load resisting elements, the loads must first be transferred from the other elements into the diaphragm sheathing. These loads include those generated by the nonstructural elements, and the in-plane wall loads (shear) and those normal to the walls. The connection of nonstructural elements such as partitions, window walls, ceilings and mechanical equipment will be considered in Chapter 7 on nonstructural elements. These forces can make a major contribution to the total loads in the diaphragm, but the connections between the structural walls (shear walls and/or bearing walls) are of primary consideration since the stability of the structure may depend on it.

Figure 6.2.3 shows a typical platform framing connection between the exterior walls and roof diaphragm where it is necessary to tie the wall to the diaphragm to resist wind loads. In this condition, the roof rafters (or floor joists) are running parallel to the wall. The wind load normal to the surface of the wall is transferred into the diaphragm, which will span between other perpendicular walls. The wind load is transferred from he studs to the double top plate through end nailing in the top of the studs, through the sheet metal framing clip to the bottom of the blocking and into the diaphragm sheathing. The blocking, located between the rim joist and adjacent joist, is spaced as required to carry the load. Where the joists or rafters are perpendicular to the wall, the load can be transferred directly into the joists without the blocking.

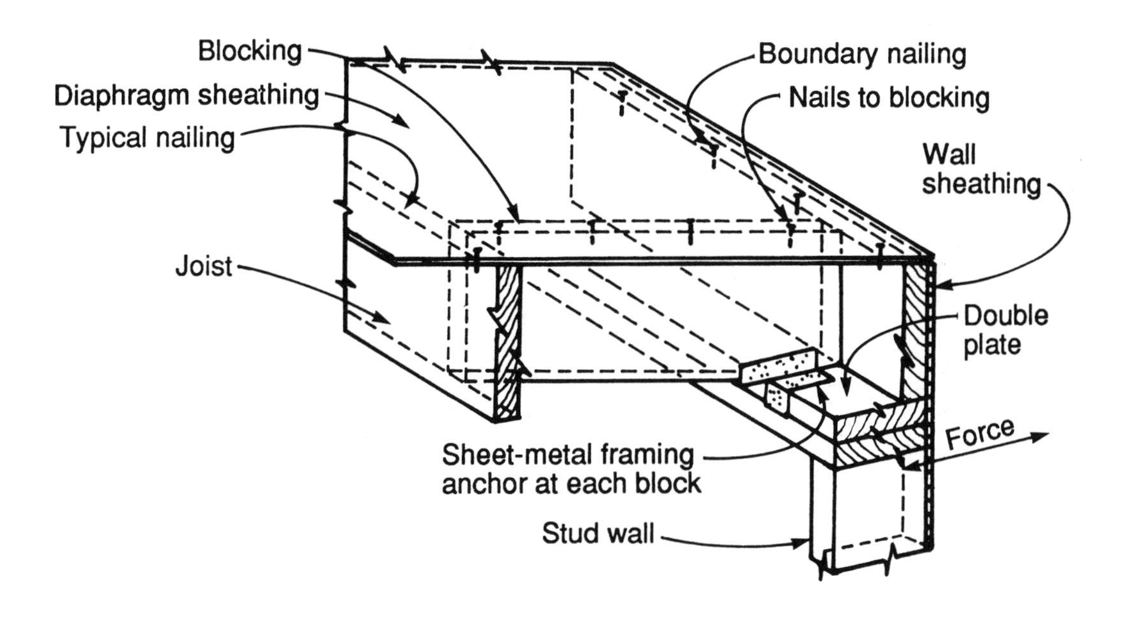

Figure 6.2.3 Diaphragm to wall nailing

6.2.4 Interconnection of the Key Structural Elements Within the Diaphragm Itself

The critical diaphragm elements can be identified by using a girder analogy. As shown in Figure 6.2.4, the plywood sheathing can be compared with the girder web, and the double top plate at the perimeter walls can be compared to the girder flanges. Interconnection within the diaphragm field is accomplished between the plywood sheets through the nailing to joists and blocking at the sheet edges. Shear transfer to the "flanges" and to the end wall reactions occurs through the nailing of sheathing to blocking, then to the top plates. Continuous rim joists at the flanges are often used in place of the blocking along the diaphragm edges for better continuity. In regular shaped diaphragms, one must consider the load transfers between the sheathing and the edge plates (the web-to-flange connections), and the splicing together of the continuous elements in the diaphragm acting as chords and/or collectors.

The connections between sheathing and edge framing can be governed by lateral forces from either direction. First, forces perpendicular to the direction of the edge members cause web-to-flange shear transfer (VQ/I), and second, the shear caused by the reaction force for loads parallel to the edge members must transfer to the vertical element. These forces vary within the diaphragm, and the governing forces are provided in the boundary nailing of the sheathing to the edge blocking or rim joists.

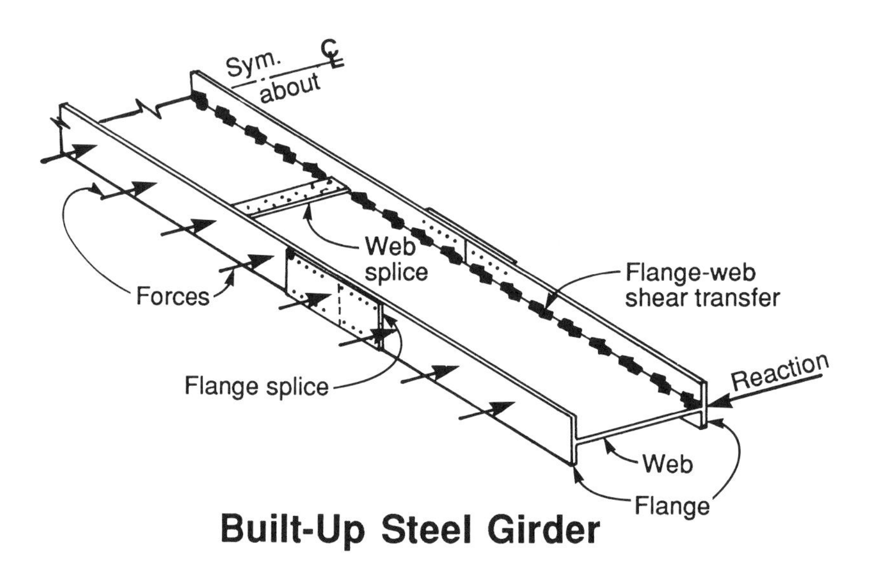

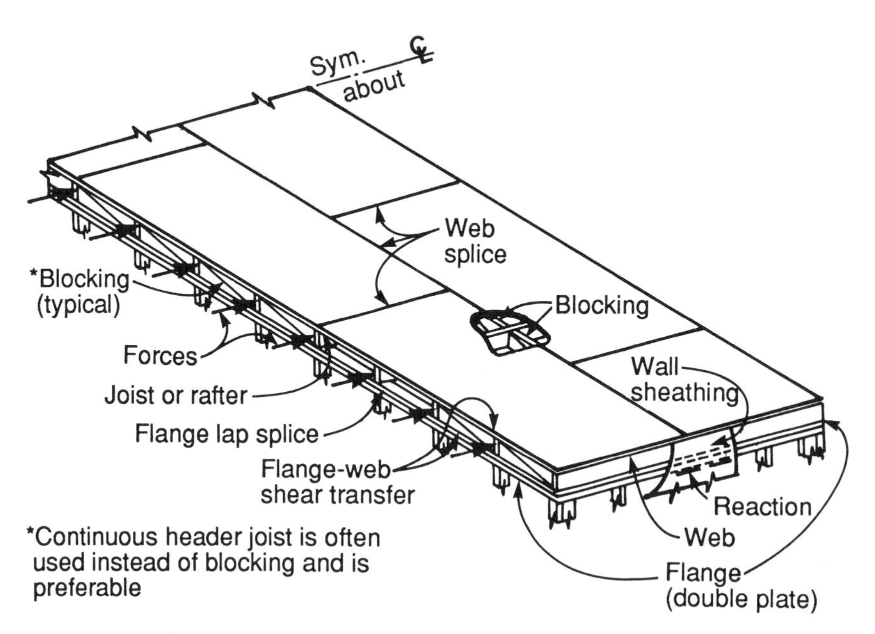

Figure 6.2.4 Girder analogy

These forces are then transferred to the double top plate acting as the continuous flange or acting as the reaction point on the edges. Once the governing forces are calculated, the boundary nailing requirements can be taken directly from the Code tables for allowable diaphragm shears. For a discussion of diaphragm amplifications, see Chapter 4.

See Example 6.2.1 for the calculations for the nailing of a regular-shaped diaphragm. The edge members acting as chords or collectors must be interconnected. For most diaphragms, it is not possible to provide a single piece of lumber to run the full length of the diaphragm. Continuity of the double top plate can be accomplished by lapping of the plates with adequate length for proper nailing to transfer the load from one plate to the other, or by the use of metal straps. All splices should be designed to transfer the full design chord or collector force at the splice location. Chord splices should be located as far as possible from the point of maximum moment in the diaphragm. Because the actual splice location cannot always be controlled, the it is generally advisable to design the connection for the maximum chord force. It may be practical to simply specify a typical minimum distance between the plate splices and the number of nails required between the top and bottom plates at the splice. For higher demands, metal straps and/or bolts may be required, such as at a collector splice at the end of a shear panel. Also, metal straps will be needed at angle connections where continuity of the chord is still required. In addition, where the plates might be interrupted by conduits or pipes, metal straps are required for continuity. It is necessary to check these straps for compression and provide nailing to prevent buckling. Some typical plate splices are shown in Figure 6.2.5.

The span-width ratio is limited in the UBC to 4:1 for plywood diaphragms to prevent excessive diaphragm deflections. The ratios for other horizontal diaphragm materials, and for vertical diaphragms (shear walls) are also provided in the Code.

6.2.5 Connections Between Diaphragm and Shear Walls

Forces must be considered for the in-plane direction of the shear panel walls, and for loads that are out-of-plane to the panel walls. In the case of in-plane loads, the connections must transmit the loads from the roof or floor diaphragms into the wall panels. Conversely, for the out-of-plane direction, the connections should be adequate to prevent the wall panels from pulling away from the floor diaphragms.

Figure 6.2.6 shows several examples of the connections between structural walls and wood diaphragms. In detailing this type of connection, the loads normal to the wall must be transferred from the wall to the diaphragm, and the in-plane loads must be transferred from the diaphragm to the shear wall below. Section A and Section B show connections for the condition where the floor joist are perpendicular to the wall, in platform and ballroom framing, respectively. In these connections, the loads normal to the wall are transferred from the wall to the diaphragm along the entire length of the joist itself. The use of sheet metal framing anchors, as shown in Section B, is quite common and can be useful in situations where significant loads must be transferred, in both platform and balloon framing. The similar connections, with parallel joists, are shown in Section C and Section D. In these conditions, the normal loads are transferred into the blocking between the joists, as described in Section 6.2.3.

The connection between diaphragm and walls for in-plane shears is of critical importance. Referring to Section A, the in-plane wall shear loads are transferred through nails in the 2x plate, or in Section B, through the nails into the horizontal

Unless Shown Otherwise In Special Details Lap And Nail Top Plate At Intersection As Shown Below.

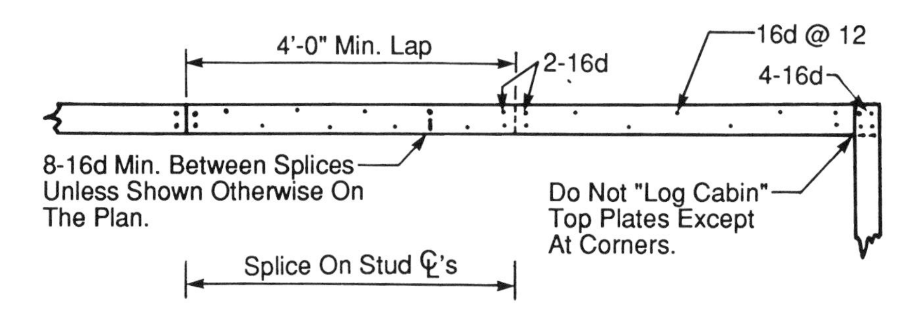

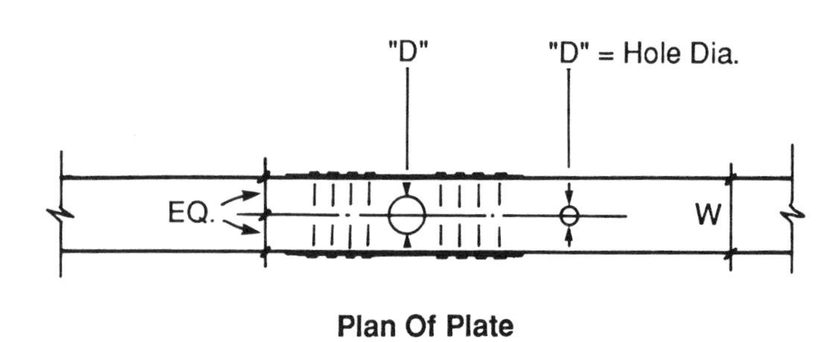

Plan Of Plate

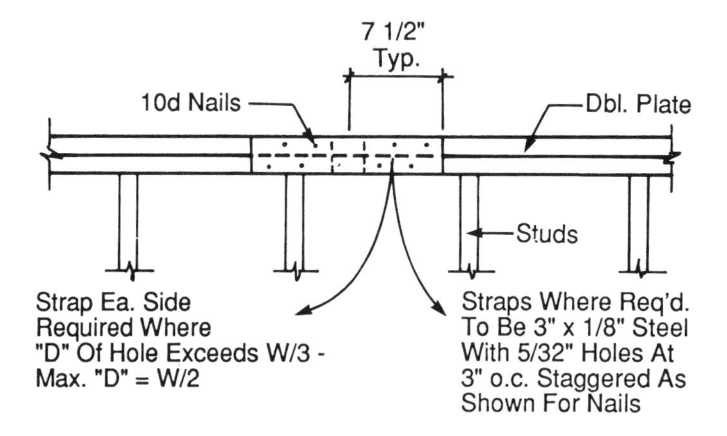

Elevation Of Plate

Figure 6.2.5(Top) Typical top plate splice (Bottom) Typical detail for holes in plate of non-bearing partition

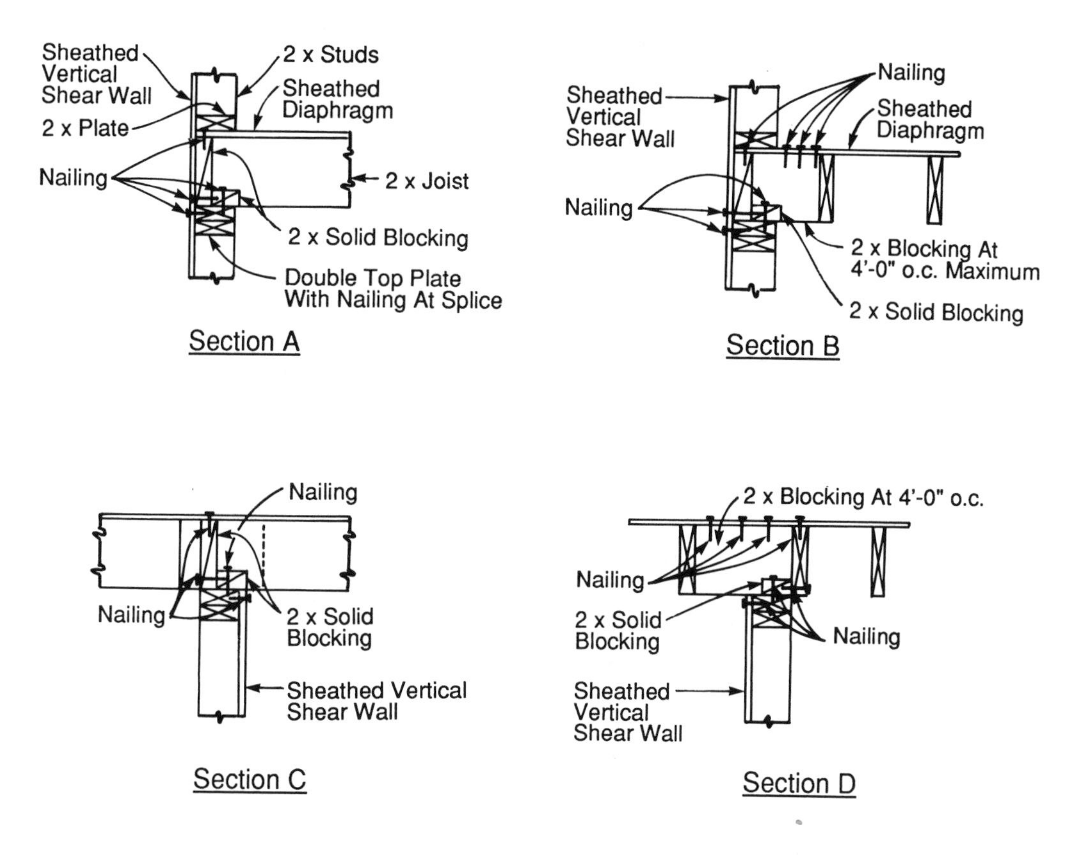

Figure 6.2.6 Wood diaphragms - typical connection details

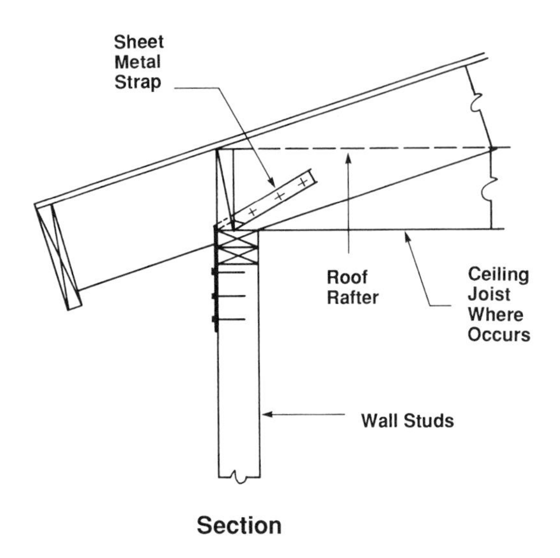

Figure 6.2.7 Rafter to wall connection

blocking between studs and into the continuous ledger nailed to the face of the studs. The nailing requirement through the plate in Section A is the same as that calculated for the edge nailing of the floor diaphragm, as discussed in Section 6.2.4. The shear loads are transferred into the wall sheathing below through nailing into the blocking or plates as shown in Section A, or through nailing into the horizontal blocking as shown in Section B.

Special hold-down detailing may be required due to high wind loads acting upward on roof overhangs or on open portions of structures. A typical case involves either sloped or flat rafters which cantilever past an exterior wall as shown in Figure 6.2.7. The net uplift (after deducting the roof dead load) should be taken through a steel strap fastened to rafters or blocking, and to studs or sheathing below. Depending on nail withdrawal or toenails for resistance to these loads is not recommended. In all cases, the flow of forces must be continued down to the point where sufficient building dead load has been engaged to resist the uplift.

6.2.6 Connections to Transfer Lateral Forces Within the Shear Walls

Lateral loads must be transferred from the top of the shear walls to the foundations. In multi-story buildings, a critical connection is located at the intermediate floor levels. Shear, and possibly overturning forces, must be considered. In Figure 6.2.6a, shear is transferred through this juncture by the nailing of the plywood above the sole plate, then into the rim joist, through the double plates and back into the lower plywood sheathing. Loads from the floor diaphragm are transferred into the blocking or rim joist, then into the double plate, as described earlier, and then transmitted also into the lower plywood sheathing. Loads from the floor diaphragm are transferred into the blocking or rim joist, then into the double plate, as described earlier, and then transmitted also into the lower plywood sheathing.

If overturning forces must be transferred to achieve equilibrium of the shear wall, it can be accomplished with hold-downs. Net uplift forces should be resisted

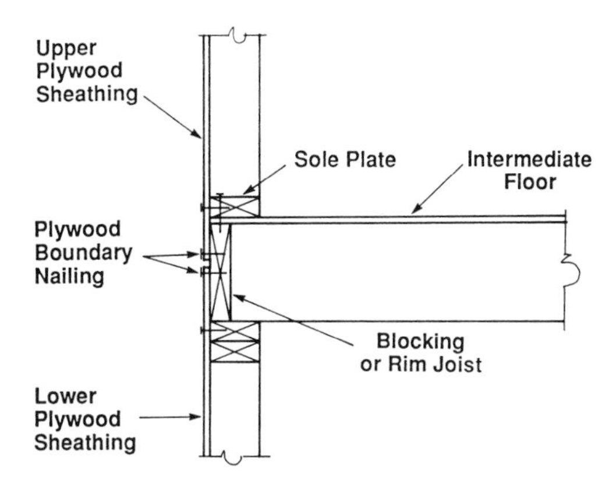

Figure 6.2.8 Floor to frame connection detail in platform framing

through a steel strap nailed to the face of the framing, or through the use of some other hold-down device. If bolts through the studs are used for fastening of the hold-down device, the proper end distance for the bolts from the end of the studs must be provided. Code requirements for end distance with load parallel to grain is 7 bolt diameters, and many manufactured hold-down devices are designed to automatically provide this distance. For large overturning moments, the use of a hold-down on both sides of a double stud is recommended for improved behavior.

The plywood sheets used on the shear wall panels must be nailed in accordance with Code requirements. Similar to the calculations for diaphragm nailing, the shear demand along the panel edges should be compared to the values in the tables for allowable shears in the Code. Special attention should be paid to the footnotes regarding framing size requirements for closer nail spacing, and for allowable load increases under certain conditions.

6.2.7 Connections to Transfer Lateral Forces from Shear Walls to Foundations

The final connection in the lateral load path has often proved to be the most critical to the overall performance of the building when subjected to earthquake loads. For wood frame buildings constructed without attention to lateral force detailing, failures have most commonly occurred at the foundation level.

Of primary concern at this connection, are the transfer of shear and overturning forces between the walls and the foundation. The shear transfer is accomplished through nailing of the plywood sheathing to the mud sill, and then through the anchor bolts holding the mud sill to the top of the foundation wall. Figure 6.2.8 shows the foundation conditions for a building with a basement or crawl space. The short stud wall is a cripple wall. In past construction, it was common to omit the bracing of the cripple wall, thus leaving a weak plane for shear transfer. It is necessary to transfer the shear, and overturning, through the cripple wall stud framing down to the foun-

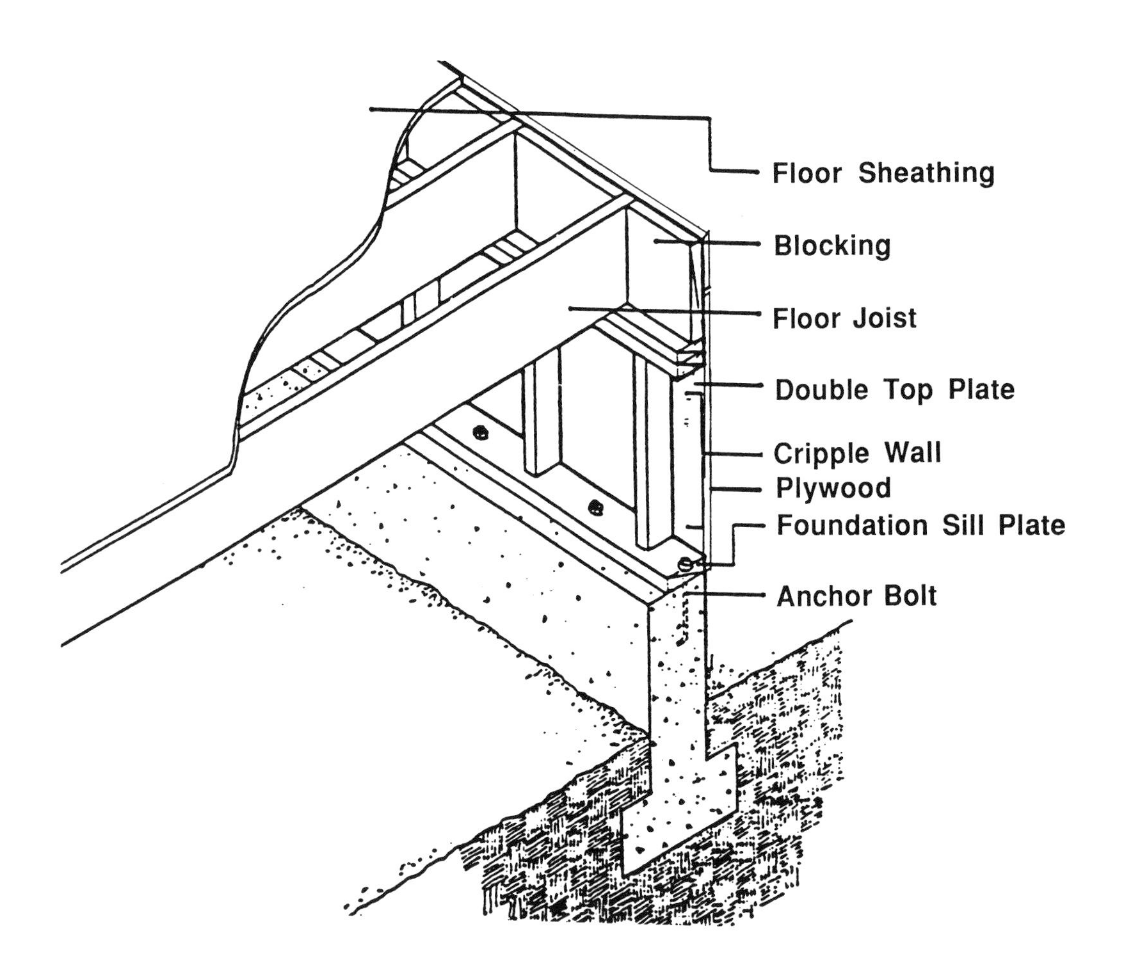

Figure 6.2.9 Foundation details for wood framing

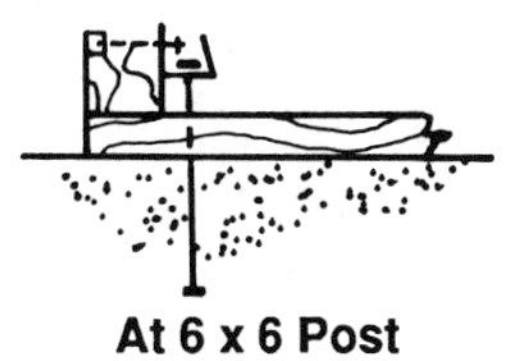

At 6 x 6 Post

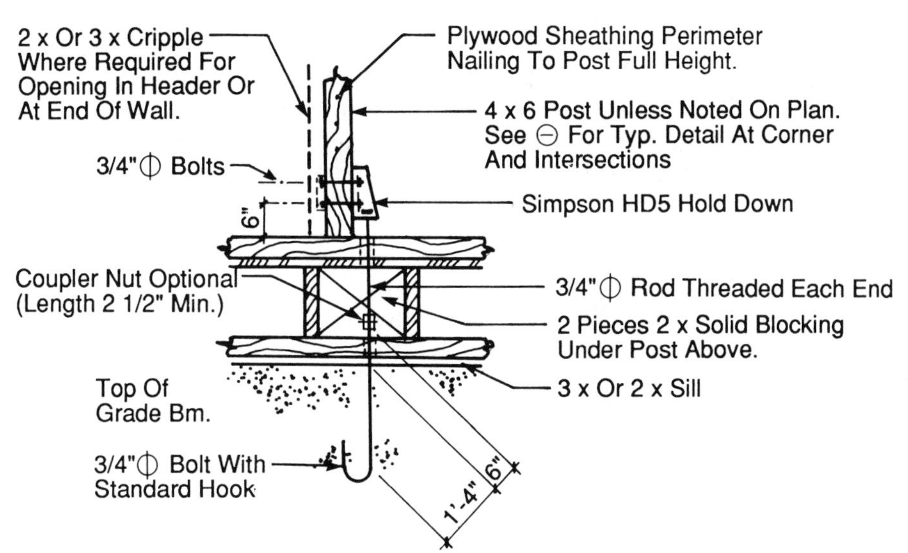

Floor Bearing On Grade Beam

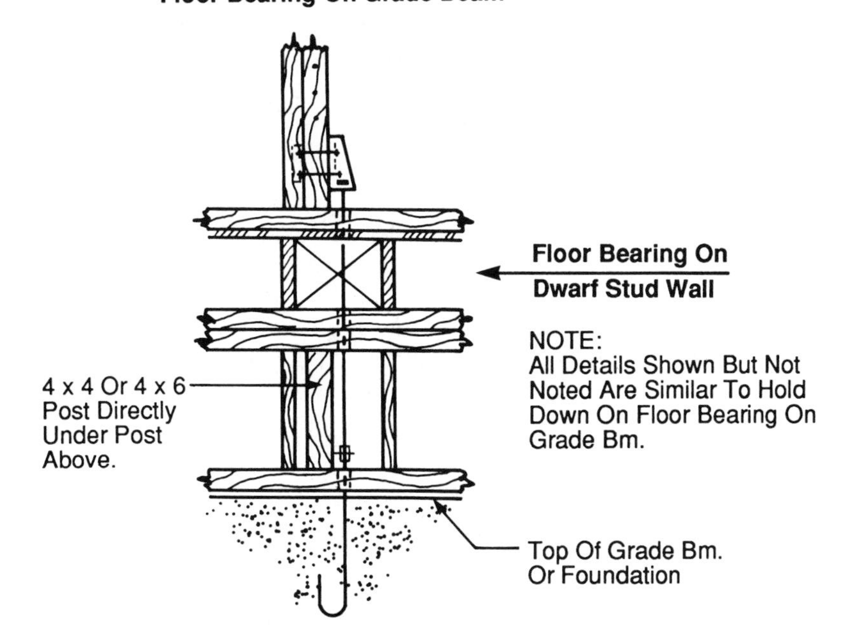

Figure 6.2.10 Typical holding down details

dation. The details for the sheathing are similar to those used in the first story above. Failure to provide for this load path has been the cause for numerous collapses in past earthquakes.

Overturning must be resolved at this level. Hold-downs to resist net uplift forces are required, and are similar to those used between floor levels as discussed previously. The anchor bolts must have proper embedment into the foundation, and the hold-down devices should be properly fastened to the stud framing to prevent a premature failure. A simple hold-down is shown in Figure 6.2.9.

Finally, at the foundation level, one must consider the continuity of the mud sill plate when the shear wall panels are not continuous along the entire length of the building. The shear load transferred from the shear wall panel must be developed in the shear bolts in the foundation. If the length of mud sill directly beneath the shear wall is not adequate for the required number of anchor bolts, the mud sill must be continuous until the load can be delivered through the bolts into the foundation. Either a continuous mud sill is required, or the sill plate should be spliced with a metal strap.

6.2.8 Other Considerations

All of the details described in this Section for lateral force resistance can be installed at a very minor cost relative to the over-all wood-frame construction cost. However, it has been proven to be important that the engineer verify the proper installation of special lateral force framing elements such as straps, angles, hold-downs and anchor bolts, and that the integrity of timber chords and collectors be inspected after utilities have been installed by cut-in. Even the most conscientious of carpenters often do not appreciate the function of an element that provides resistance to only lateral forces.

The simple hold-down detail shown in Figure 6.2.10 has often caused problems during construction. Since the detail is called for at ends of the shear wall panels for overturning, it frequently interferes with door jambs and is eliminated in the field. This is particularly the case where hold-down anchors are called for on both sides of the double studs, as for high seismic loading. An alternative approach is to use a second pair of double studs just 6 inches inside the jamb to use for the hold-down attachment.

Design example 6.2.1

Refer to design example 5.11.6 in Chapter 5, One-story wood frame building in seismic zone 4, 100' x 240' in plan. Consider diaphragm and shear wall connections for forces in transverse direction. Use *National Design Specification* for timber connection values.

A. Diaphragm Boundary Element Connections:

- *Chords at longitudinal walls:*

$$M = \frac{WL}{8} = \frac{58.29^k \times 240'}{8} = 1749'k$$

$$Max\ T\ or\ C\ in\ chord = \frac{M}{100'} = 17.5\ k$$

Assume Douglas Fir - Larch No. 1 Grade Top Plates

F_a = 1000 psi

$$Area\ reqd. = \frac{17{,}500}{1000 \times 1.33} = 13.1\ in^2$$

Assume 2 - 2x6 plates spliced with steel straps.
Net area, subtracting for 3/4" dia. bolts:
A = 2 x 1.5 x 5.5 - 0.75 x 1.5 x 2 = 14.25 in^2
∴ 2 - 2x6 s adequate even if splice is made at point of maximum moment.
Size plates and bolts for M_{max}:
3/4" dia. bolts: Increase allowable stress for duration of load (1.33) and metal plates (1.62). Decrease for no. bolts in row.
For 4 bolts: P_{allow} = 4 x 1.33 x 1.62 x 0.88 x 2630#
= 20.07 k > 17.5 k
Steel Plates: Use A36 steel

$$A\ reqd. = \frac{17.5\ k}{1.33 \times 0.6 \times 36\ ksi} = 0.61\ in^2$$

Use 2 - 1/4" x 2 1/2" straps: A = 2 x 0.25 (2.5 - 0.75) = 0.875 in^2

- *Collectors at End Walls:*
  Max Collector Force = 292#/ft x 20' = 5840#
  Use 2 - 2x6 top plates spliced with thru bolts (no steel plates)

$$Area\ of\ wood\ plate\ reqd. = \frac{5840\ lb.}{1.33 \times 1000} = 4.38\ in^2$$

(if splice were made at point of max collector force)
A = 1.5 (5.5 - 1.0) = 7.12 in^2
Try 7 - 1" bolts in single shear:
Allowable P = 7 x 1.33 x 0.71 x 945# = 6260#

B. <u>Connections at End Walls:</u>
v = 292#/ft. along 100 ft. diaphragm
v = 486#/ft. in transverse shear walls

Roof Diaphragm to End Walls:
- *Plywood to Blocking:* v = 292 #/ft.
  Use 10^d nails 6"cc - Diaphragm Boundary Nailing per UBC 25-J-1
  Allowable = 320 #/ft.
- *Blocking to blocking and blocking to double top plate collector:*
  v = 292 #/ft., using double top plate as collector.
  Use 3 - 16^d nails per block:
  Allowable = 1.33 x 3 x 107 / 1.33 = 321 #/ft
  Note: Could also use metal framing anchors.
- *Plywood wall sheathing to top and sill plates.*
  v = 486 #/ft; use 10^d nails @ 4"cc per UBC Table 25-K-1

- *Sill plate to foundation.*
  v = 486 #/ft; use 7/8" dia. bolts @ 24 "cc
  Allowable bolt load 7/8" dia. bolt, single shear parallel to grain = 830# (UBC Table 25-F)
  Allowable = 1.33 x 830 / 2 = 553 #/ft
- *Anchor for uplift at ends of walls.*
  Nett overturning from Example 5.11.6 is very small
  291.5 k.ft overturning moment versus 283.0 k.ft resisting moment
  Provide anchor bolt with hold-down similar to that shown in Figure 6.2.10.

6.3 MASONRY AND TILT-UP CONNECTIONS

6.3.1 Introduction

In masonry and precast building construction, a continuous load path must be provided between the floor/roof diaphragms and the walls. These diaphragm to wall connections must resist forces both in the plane of and perpendicular to the walls and also often serve as the gravity load connection between the roof/floor and the wall. The following discussion will concentrate on connections which transfer lateral forces between horizontal diaphragms and masonry or precast walls as well as connections within horizontal diaphragms and between precast wall elements.

6.3.2 Loading

Diaphragm to wall connections may be required to resist forces in three mutually perpendicular directions as shown in Figure 6.3.1. Forces parallel to the plane of the wall (F1) are resisted by shear in the connection. The design shear force is based upon the over-all lateral analysis of the building and is equal to the shear force along the boundary of the diaphragm. The connection also must resist forces perpendicular to the plane of the wall (F2), resulting in tension in the connection. This tension force prevents the wall from pulling away from the building and is designed for the greater of the wind or seismic force acting perpendicular to the wall element. Additionally, the 1991 Uniform Building Code (UBC) specifies minimum design forces for anchoring masonry and precast walls to diaphragms. In addition to the above lateral forces, the diaphragm to wall connection may also be required to resist the vertical gravity load (F3) being transferred to the wall.

6.3.3 Method

Typical diaphragm types used with masonry or precast walls include wood, cast-in-place concrete, precast concrete, and metal deck with or without concrete topping. In masonry and tilt-up construction, the diaphragm to wall connection is often achieved by embedding an anchor such as a bolt, rebar or strap into the masonry or precast wall and attaching it to the diaphragm structure. There are many ways to make this connection for each of the different diaphragm types.

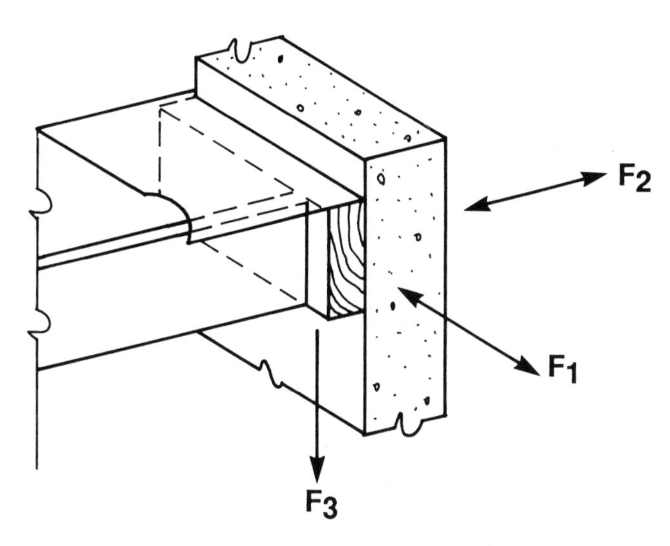

Figure 6.3.1 Schematic of forces involved in a floor to wall connection

When precast planks are used as the floor structure, the individual planks must be interconnected in order to provide a continuous path for diaphragm forces to reach the shear walls. A cast-in-place concrete topping can tie the floor diaphragm together. If this is done, in seismic zones 3 and 4 the UBC requires that the topping acting alone be proportioned and detailed to resist the design forces. If no topping is added, then the planks must be tied together mechanically or with a grout key between planks. Precast wall panels can either be connected at vertical joints or not connected in which case the panels act as individual units in resisting the lateral shear force.

6.3.4 Behavior

Regardless of the type of floor/roof diaphragm used, it is important that the connection to the masonry or precast wall be made in such a manner that it prevents brittle failure of the connection. This is particularly important for connections that may be subject to cyclic earthquake loads. For example, when bolts are embedded in a wall, adequate embedment must be provided to prevent the bolt from pulling out of the wall. In precast construction, welding is often used to connect two members. Welding can cause expansion and distortion in the steel which can degrade the steel to concrete bond. In addition, welded connections between precast elements may fail due to concrete shrinkage. Therefore, extra care is required in detailing welded connections as well as implementing the welding in the field. Using thick steel sections and low heat welding can reduce the detrimental effects of the welding process.

The Uniform Building Code contains a general requirement that concrete and masonry walls shall be anchored to floors and roofs that provide lateral support for the wall with positive direct connections. When connections are subjected to both shear and tension, the design of the connections should account for the simultaneous action of both loads. In seismic zones 3 and 4, there is a specific requirement which states that where wood diaphragms are used to laterally support concrete or masonry walls, the anchorage cannot rely on toe nails or nails subject to withdrawal and the connection cannot result in cross-grain bending or cross-grain tension. The following sections will present some of the issues pertaining to specific connections and provide examples. There are many methods of achieving the connections discussed below, and the illustrations shown should not be considered the only or necessarily the best method for a given situation.

6.3.5 Masonry Connections

6.3.5.1 Wall to Diaphragm Connection

When a wood diaphragm is used with masonry walls, there are many ways to make the diaphragm to wall connection. Often, a bolt is embedded into the masonry wall and attached to the wood diaphragm. The wall should be fully grouted around the bolt if this is done. The Uniform Building Code requires at least one inch of grout between embedded anchor bolts and the masonry.

If a wood ledger or blocking runs along the wall, a bolt can attach the ledger to the wall as shown in Figure 6.3.2 (left). The bolt then serves to carry the shear from the diaphragm into the wall as well as the vertical gravity load. In order to prevent cross-grain bending in the ledger due to out-of-plane wall loads, a hold down or joist anchor should be used to transfer the out-of-plane force directly into the joist as shown in Figure 6.3.2 (right). When joists run parallel to the wall, blocking can be added and the hold downs or joist anchors are attached to the blocking.

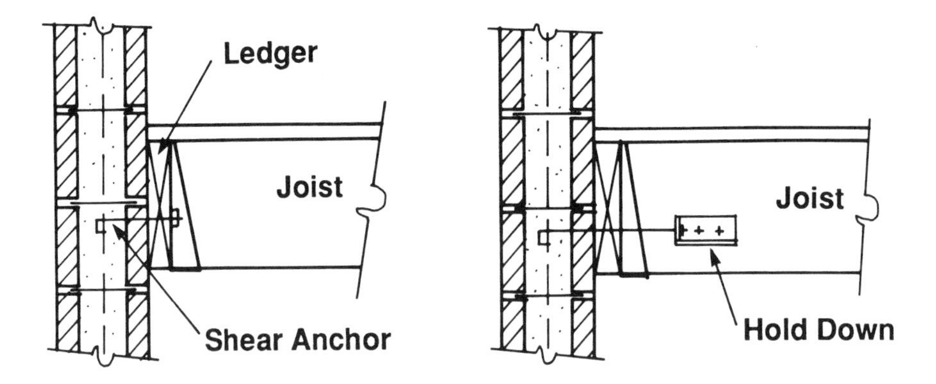

Figure 6.3.2 Floor to masonry wall connection

Design example 6.3.1

Assume the connection shown in Figure 6.3.2 is to be designed to resist an in-plane diaphragm shear of 300 #/ft (force F1 in Figure 6.3.1) and an out-of-plane force of 825 #/ft (force F2 in Figure 6.3.1). Dead and live vertical loads are 600 #/ft (force F3 in Figure 6.3.1).

Assumptions:
- 3/4" diameter anchor bolts into the wall
- 2 - 5/8" bolts from the hold down to the joist
- 4x ledger and 2x joists, joists are 16" on center
- f'm = 2000 psi, 6" bolt embedment into wall

(a) Lateral and gravity load shear transfer:
The resultant shear force = $(300+550)^{1/2}$ = 626#/ft.

(i) Allowable shear based on the 3/4" diameter bolt in the masonry, from UBC Table 24-E, V_{all}=1900#.

(ii) Allowable shear based on the bolt in the wood ledger, V_{all}=952# (using Hankinson's formula).
Minimum spacing = 1.33x952/626 = 2'-0"

(b) Out-of-plane Anchorage:

(i) Allowable tension based on the 3/4" bolt in the masonry, from UBC Tables 24-D-1 and 24-D-2, T_{all}=2520#.

(ii) Allowable tension based on the 2 5/8" diameter bolts from the hold down to the joist, T_{all}=2x996=1992#.
Minimum spacing = 1.33x1992/600=4'-5" say 4'-0" (every third joist)

If a concrete diaphragm is used, dowels placed in the diaphragm can be bent into the masonry wall as shown in Figure 6.3.3. If precast planks are used with a cast-in-place topping, the topping is usually assumed to carry the diaphragm shear and the rebar in the topping can then be turned up into the wall as shown in Figure 6.3.4. These dowels can serve to tie the wall and diaphragm together for loads both parallel and perpendicular to the wall. If precast planks are used without topping, the rebar within the planks can be turned up into the wall. Alternatively, weldments can be used to tie the wall and planks together, but special attention must be paid to these welded connections as discussed above and they may not be appropriate in highly seismic areas.

6.3.5.2 Wall to Foundation Connection

At the base of the masonry wall, a direct connection to the supporting substructure is necessary. The wall will usually be supported by a concrete footing. The connection is typically achieved by providing dowels which are cast-in-place in the concrete footing as shown in Figure 6.3.5. The dowels serve to resist both in plane and out of plane wall forces. The dowels need to be fully grouted in the masonry and extended above the footing a sufficient distance to develop the lap splice required with the vertical bars in the wall.

6.3.6 Tilt-Up Connections

The tilt-up details discussed in the following paragraphs deal with single panel construction where the individual wall panels are the full height of the structure. The panels can be either isolated units or coupled units depending on whether the panels are tied together along their vertical edges. The roof diaphragm for this type of construction will generally be of wood construction. If the structure is two stories, the second floor will often be constructed with precast floor units with a cast in place concrete topping.

6.3.6.1 Diaphragm to Wall Connection

The principles for diaphragm to wall connections are the same for tilt-up construction as they are for masonry wall construction. However, due to the fact that the wall has already been fully constructed when the diaphragm is attached, the details of the connection are slightly different. In the case of wood diaphragms, the connection to the walls is similar to those described above for masonry walls. In this case, the anchor straps and anchor bolts have to be cast in the wall during fabrication. If the building has a precast concrete floor with topping, the dowels can be bent from the wall into the cast in place topping to transfer forces both in the plane and perpendicular to the wall. The dowels should be extended an adequate distance into the topping slab to develop the bar in tension. The same type of detail would be applicable to a floor consisting of a steel deck with a concrete topping. Precast floor units without topping can also be used but untopped decks are not recommended for use in high seismic areas. The tilt-up wall and precast floor element are normally connected using embedded weld plates.

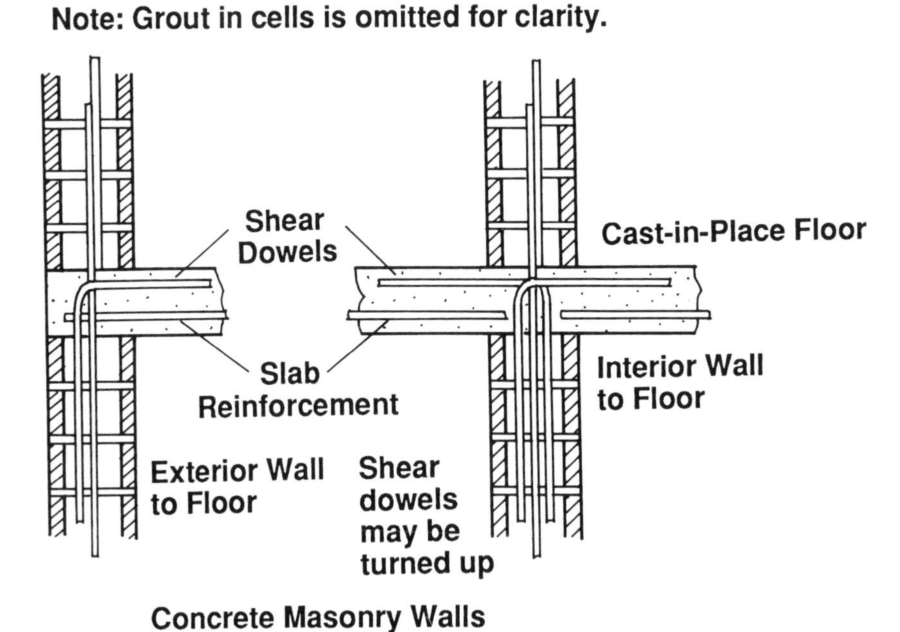

Figure 6.3.3 Connection between cast-in-place concrete floor and a concrete masonry wall

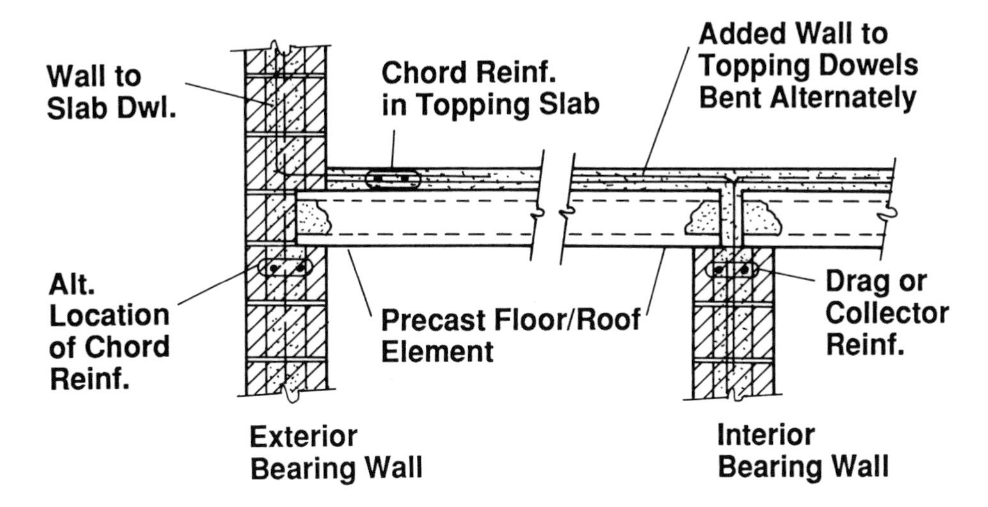

Figure 6.3.4 Typical detail for connecting precast concrete floor to concrete masonry walls

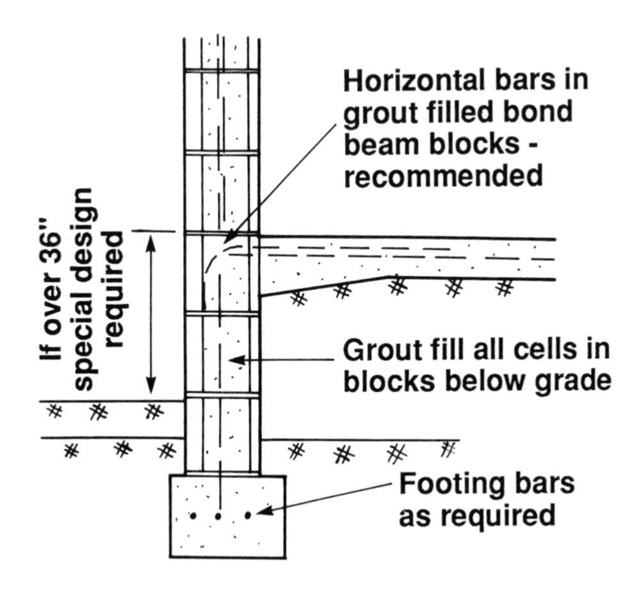

Figure 6.3.5 Concrete masonry foundation wall

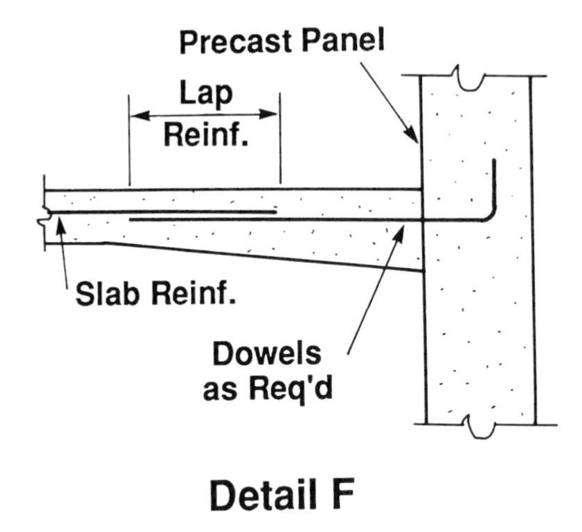

Figure 6.3.6 Precast concrete panel to concrete slab joint reinforcing

6.3.6.2 Wall to Foundation Connection

A tilt-up wall panel can be supported on either a continuous footing or on isolated footings near the end of the panel. A continuous footing will provide a more continuous distribution of forces to the supporting soil and is the preferred type of support. The wall to foundation connection must allow for transfer of shear forces parallel and perpendicular to the wall as well as overturning forces at the panel ends. The wall to foundation connection can be made by using weld plates at the base of the wall. When weld plates are used, because of the level of deformation capacity, the design force should be 2.5 times the UBC required force to ensure elastic performance of the connection (see Chapter 5). Alternatively, dowels can protrude from the bottom of the tilt-up panel and be encased in the foundation which is poured after the panels have been lifted into place. An effective method of transferring shear forces is to bend the rebar out of the wall and encase it in the poured in place floor slab closure strip as shown in Figure 6.3.6.

6.3.6.3 Floor Panel to Floor Panel Connections

Typical types of precast diaphragm units are hollow core planks, solid planks, and double tee sections. The precast units must be tied together in order to transfer shear forces across their edges. The most reliable method for achieving this is to provide a cast-in-place topping over the entire floor. The cast-in-place topping will typically be 2" to 4" thick and is usually assumed to carry the full lateral seismic force in the diaphragm. In low seismic areas it is possible to use precast diaphragm elements with no topping. The interconnecting weld plates must then be designed to transfer the entire diaphragm shear between elements. Figure 6.3.7 shows a typical floor panel to floor panel welded connection. Another possibility is to provide a grouted key between precast elements for shear transfer.

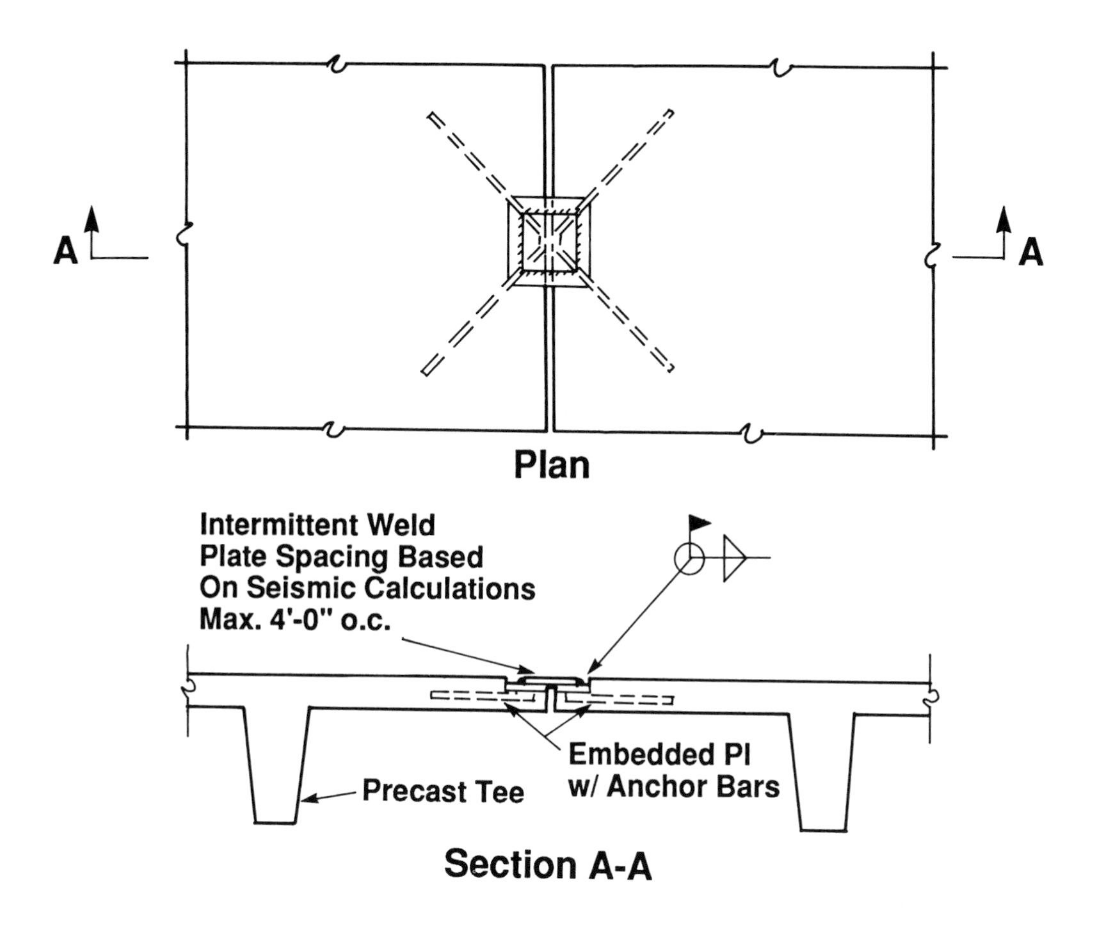

Figure 6.3.7 Connection between precast concrete elements

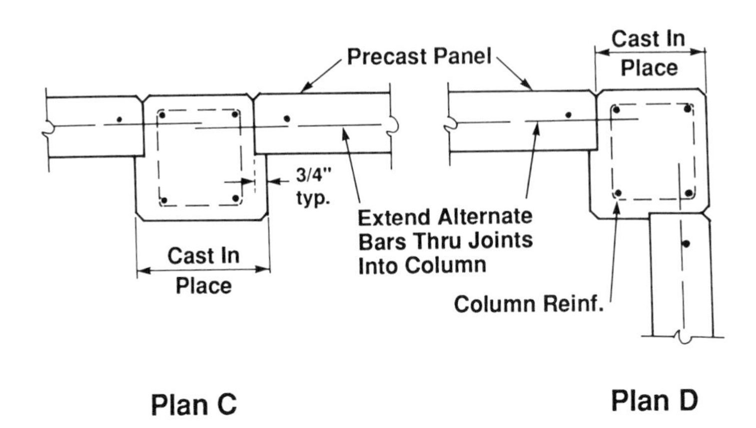

Figure 6.3.8 Connection between precast concrete element and cast-in-place concrete

6.3.6.4 Wall Panel to Wall Panel Connections

It may not be desirable to tie wall panels together. However, when tying panels together, consideration must be given to stresses at connection points due to shrinkage and temperature effects. If a rigid connection such as a weld plate is used, degradation of the connection can occur in the form of cracking around the weld plates.

One method of tying panels together is to have a cast-in-place pilaster located between the panels. Horizontal reinforcement from the panels can then be extended into the pilaster as shown in Figure 6.3.8. In order to prevent cracking in the panel, the amount of reinforcement extended into the pilaster can be reduced and bond can be prevented in the wall reinforcement. If weld plates are used, they can be slotted which will facilitate yielding of the connection and help prevent non-ductile behavior in the joint.

6.4 CONNECTION OF STEEL MEMBERS

6.4.1 Introduction

Many different circumstances arise in the connection of the elements of a steel building, and these give rise to a variety of solutions. In addition, low-rise buildings are often composites of different materials. The book by Hart, Henn and Sontag (1978) gives an excellent summary of a variety of connections with a discussion of the conditions for which these connections would be appropriate. Only the issues which distinguish the design of laterally-loaded low-rise buildings from gravity loaded ones will be treated in the following discussion. References to existing literature on steel connections subjected to lateral loading conditions are provided to supplement the overview presented here.

Seismic excitation places demands on the structure that require a ductile response. Generally, the forces on the connections for seismic excitation are severe but few in number. In most cases, the response of the structure as a whole will vary between the cyclic type and the incremental collapse type as shown in Figure 6.4.1. In both seismic and wind environments, low-cycle fatigue is likely to be the final

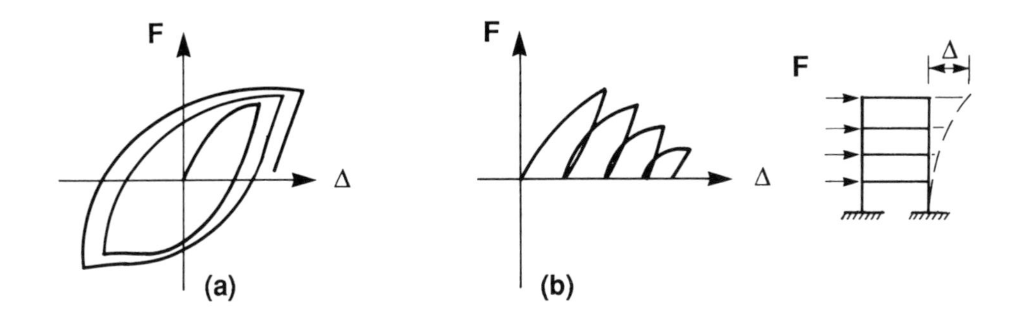

Figure 6.4.1 Cyclic versus incremental collapse

cause of failure of the connections (if the connection fails) regardless of the method of connection.

<u>The Method of Connection.</u>

There are two ways of connecting steel members together: welding and bolting. Often the most economic and practical steel connection has some welded and some bolted components. There are advantages and disadvantages to both methods. Bolted connections slip; they suffer from reduced sections due to the presence of holes; the bolts cause large stress and strain concentrations locally; and the connections are sometimes difficult to implement compactly. On the other hand, bolted connections are less prone to the development of fatigue cracks and the bolting process does not disturb the metallurgical properties of the metal. Welded connections can be executed compactly and can be easily designed to develop the full capacity of the member. However, the welding process alters the nature of the material in the heat affected zone and small imperfections in the quality of the weld can initiate fatigue cracks which could lead to failure.

<u>The Geometry of the Connection.</u>

There are a large number of different details which can be used to connect steel members. Because there are so many geometries, it is impossible to know how each one will perform in a severe wind or seismic event. Laboratory tests have been performed on certain types of steel connections, but practical applications often deviate from laboratory conditions. While the amount of laboratory data on the cyclic behavior of different geometries is scant, extrapolation of concepts from the known cases is often possible. In addition, design offices often have their own standard details which have been developed from years of successful practice.

The single most important consideration in selecting the geometry of a connection in steel (or any other material) is the provision of a smooth and continuous load path for the transfer of forces from one member to another. A tortuous path for the flow of stress invariably generates stress and strain concentrations which often cause the initiation of failure in the connections. A good load path can usually be assured if the form of the stress in the adjoining members is understood. Simple

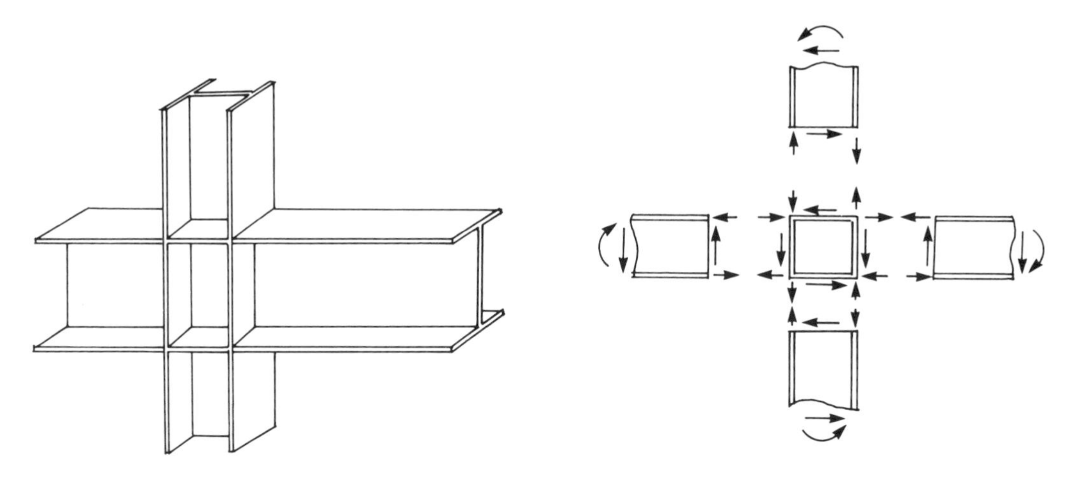

Figure 6.4.2 Moment-resisting connection

free-body diagrams are indispensable tools for assessing the connection for the transfer of loads.

As an example, consider the moment-resisting connection shown in Figure 6.4.2(a). The forces in the members being connected can be represented as shown in Figure 6.4.2(b). The forces are transferred from the beam flange through the weld joining them. The column web stiffeners help transfer the force into web shearing and the web shearing eventually generates normal stresses in the flange of the column remote from the joint.

An important geometric consideration in steel connections is the need for providing close tolerance fit-up for field fabrication. A good connection design will allow some latitude in field tolerances. Figure 6.4.3 shows two connections, one of which is difficult to execute in the field because of the need for close tolerance fit-up and the other of which suffers less from these problems. In most cases the economics of making the field connections is of primary importance in the selection of a detail; consequently the connections should be considered early in the design process.

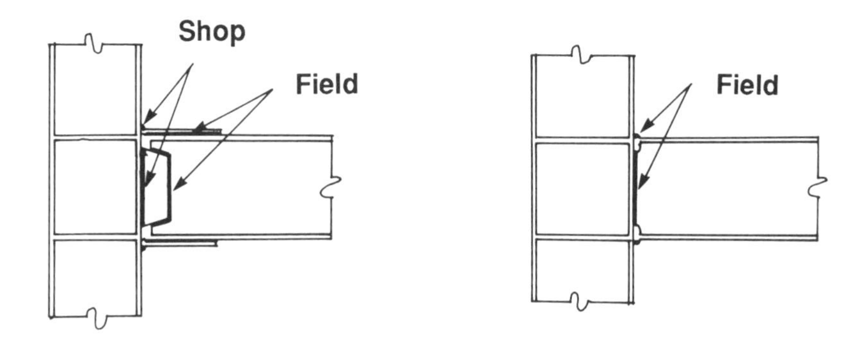

Figure 6.4.3 Easy versus difficult fit-up for field welding

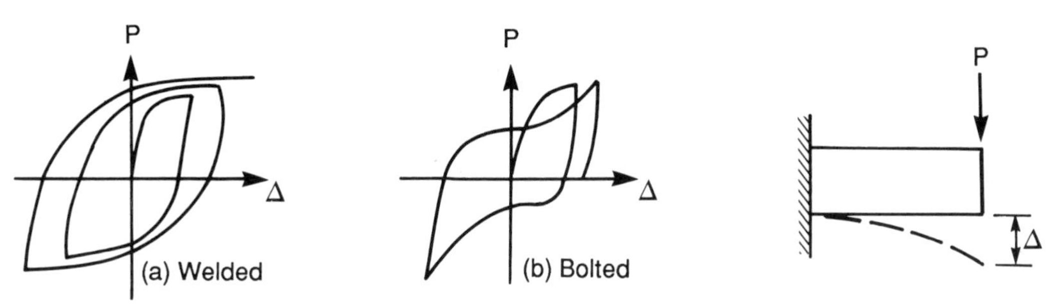

Figure 6.4.4 Typical hysteresis behavior

General Behavior of Steel Connections.

For cyclically loaded connections the additional requirements of ductility and low-cycle fatigue resistance are important. It is essential to the survival of the structure that the connection respond to repeated application of severe load levels without significant loss of strength. Sudden loss of capacity can generally be attributed to material tearing or fracture, a result of low-cycle fatigue. Tearing can occur in welded elements due to the propagation of fatigue cracks or by excessive straining. Bolted connections can fracture at regions of minimum net area.

The ability of a connection to sustain seismic loading is generally deduced from its hysteretic behavior under cyclic loads. The connection which is best able to dissipate energy is generally the best for seismic load environments, as the energy dissipation capacity is a direct measure of the toughness of the connection. Two generic force-deformation responses of connections are shown in Figure 6.4.4. The first demonstrates greater energy dissipation capacity over the second since the loop encloses a greater area. Bolted connections are more likely to exhibit the second type of behavior because of their tendency to slip. For this reason, welded connections are often preferred over bolted in seismic applications.

Many types of connections have degrading stiffness and strength properties. However, loss of strength by itself is not a valid reason to condemn a connection. Rather, the rate of loss of resistive properties should be considered when assessing the suitability of a certain connection configuration. Some laboratory data are available to aid in this assessment (e.g. Radziminski and Azizinamini, 1986, and Popov and Pinkney, 1969). The stiffness of a connection can be important to the stability of a structure, and hence it is important to know whether the stiffness characteristics will degrade under cyclic loading.

One must be aware of potential hazards which result from the discrepancy between the idealized behavior of a connection used in structural analysis and the actual behavior. For example, while it may appear that simple connection is conservative, it may not be. Flexible connections which are designed as simple connections may attract unanticipated forces due to their finite stiffness. Shear tabs will provide some bending resistance; the bending stress induced may lead to tearing at the extreme fibers of the tab and eventually impair its ability to carry shear. In most cases, the importance of these problems can be diminished by proper detailing.

6.4.2 Design of Steel Connections

Steel building structures can be classified into three main categories. The distinction among the categories arises largely from the way the structures resist lateral loads. In addition, each type has distinct issues to be addressed in the design of the connections. The three most common lateral load resisting frame configurations, shown in Figure 6.4.5, are the moment-resisting frame, the concentrically-braced frame, and the eccentrically-braced frame. The moment-resisting frame resists its loads through the development of bending moments which are transmitted among the beams and columns of the structure. The connections between the beams and columns are necessarily moment resisting or rigid. The concentrically-braced frame resists lateral force through trussing action of the triangulated frame and it is not essential that the beam-column connections be moment resisting. The bracing connections are the critical connection elements of the structure. The eccentrically--braced frame is a hybrid of the other two and hence has connection issues similar to both. It resists lateral loads through a combination of flexure and axial trussing action. The eccentric beam segment between the brace and the column must be given special consideration in detailing.

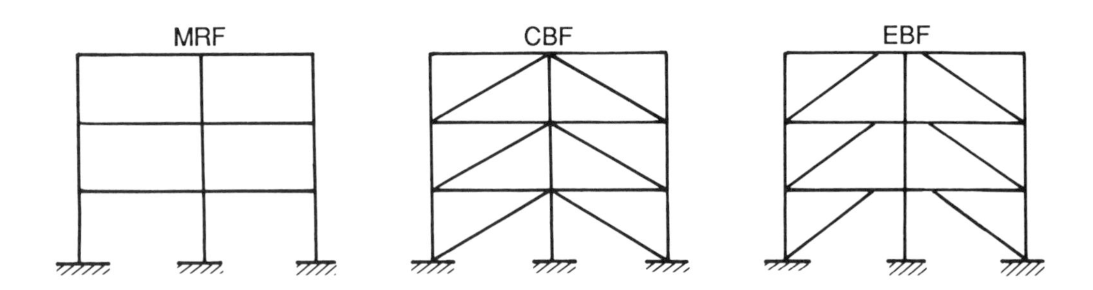

Figure 6.4.5 Types of steel framing systems

Beam-to-Column Connections.

The traditional approach to the categorization of beam-to-column connections in steel is based upon the stiffness of the connection. Accordingly, connections between beams and columns can be categorized into the two general types: Fully Restrained (FR) (AISC, Type I) and Partially Restrained (PR) (AISC Type II). The distinction between the two types is fuzzy but is usually described in terms of the percentage moment transmitted by the connection for a specific test case. For example, consider the partially restrained beam shown in Figure 6.4.6 which has a bending rigidity, EI, length, L, and connection stiffness, k. When two equal and opposite bending moments are applied to the beam ends they are resisted partly by the beam flexure and partly by the moment developed from the rotation of the connections. The ratio of the transmitted moment to the applied moment is termed the percentage of restraint, R. A fully restrained (FR) connection is one which has a restraint $R > 0.90$ while a partly restrained (PR) connection has a restraint $R < 0.90$.

Lateral-load bearing systems usually comprise fully restrained connections. They are essential for moment-resisting and eccentrically-braced frames. While it is not important to provide moment resisting connections in concentrically-braced frames, fully restrained connections are often used to provide a ductile system. Richard (1986) points out that the presence of gusset plates for the connection of the

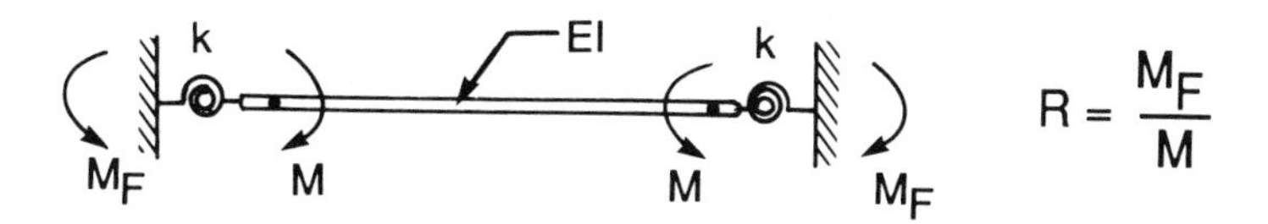

Figure 6.4.6 Definition of restraint

braces to the joint generally render the connection rigid, regardless of how the beam is attached to the column. Partially restrained connections are being used increasingly but they suffer from the need to know the moment-rotation characteristics for the connection in order to accurately assess the joint forces from an analysis of the structure. The moment-rotation relationship can generally be obtained experimentally or through a simplified analytical model of the connection.

Fully restrained beam-to-column connections can be all welded, all bolted, or a combined welded and bolted system. An example of a connection of a beam to the flange of a column is shown in Figure 6.4.7. Notice the presence of stiffener plates for the transfer of the beam flange forces to the column. Figure 6.4.8 shows that this type of connection can provide a large amount of ductility and has good cyclic stability in its force--deformation behavior.

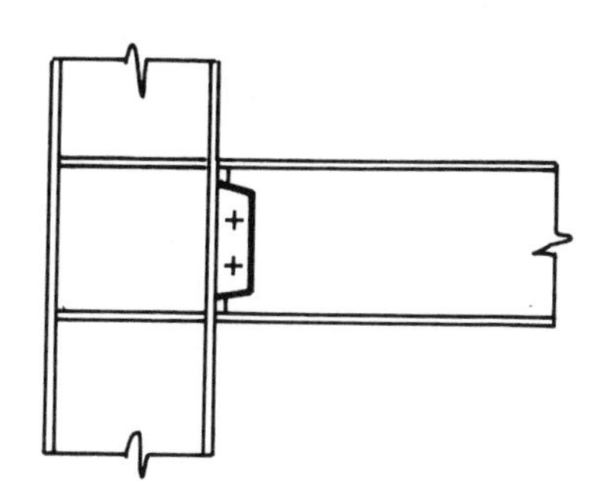

Figure 6.4.7 Beam-to-Column Flange Connection

Connections must often be made between a beam and the weak axis of a column. In such a case the beam is usually attached to the column web by means of a stiffener plate which is welded to the column web and both flanges as shown in Figure 6.4.9. The question of ductility of these weak axis joints is far from settled. Some experiments have suggested that this type of connection is much less ductile than its strong-axis counterpart. To avoid re-entrant corners, the stiffener is often tapered from the column flanges to the point of welding to the beam flange, as shown in the figure.

Other geometric irregularities, such as differing beam sizes on either side of the joint, can occur when attaching a beam to a column. It is a good idea to provide at least a partial width stiffener for each connection of a beam flange to the column to transmit the flange forces.

In many instances the performance of the adjoining members will directly affect the performance of the joint. Often there are steps that can be taken in the design of the connection to diminish these effects Some of the most important of these issues are discussed below.

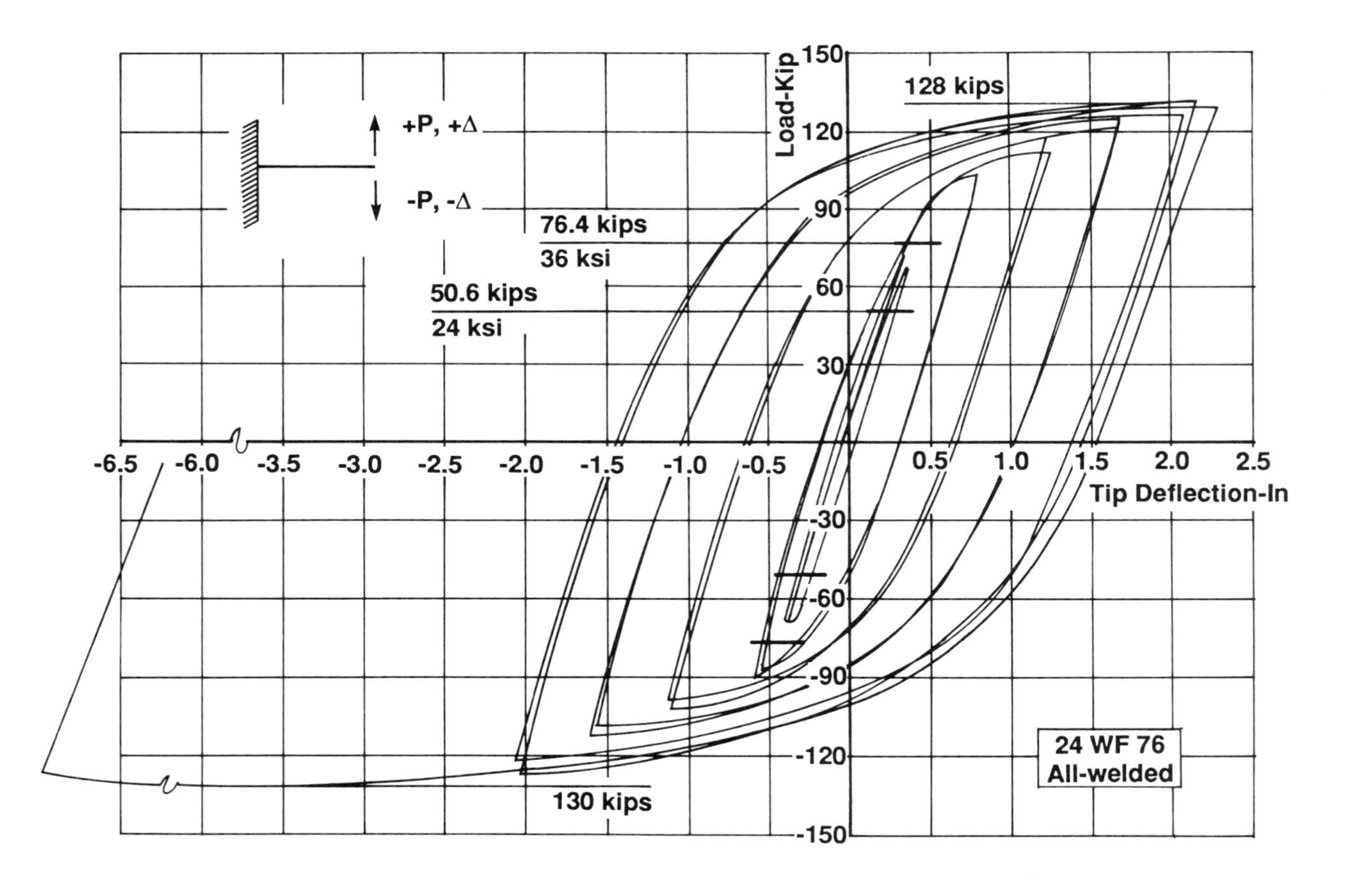

Figure 6.4.8 Experimentally derived hysteresis behavior

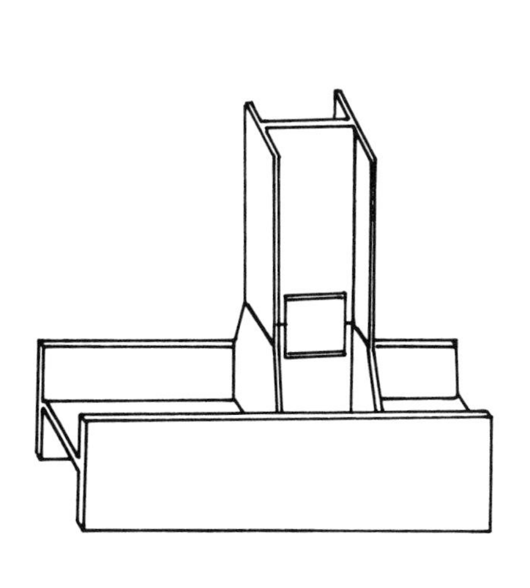

Figure 6.4.9 Weak Axis Beam-Column Connection

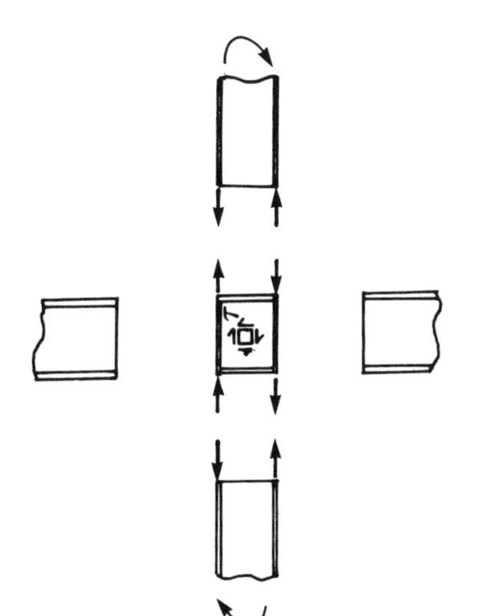

Figure 6.4.10 Column Panel Zone

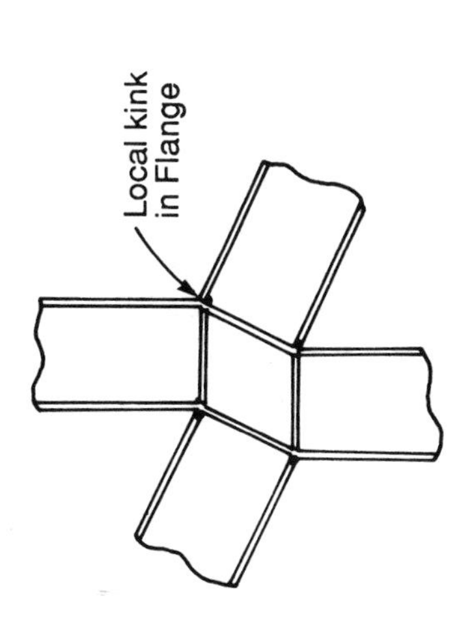

Figure 6.4.11 Deformation of Column Panel Zone

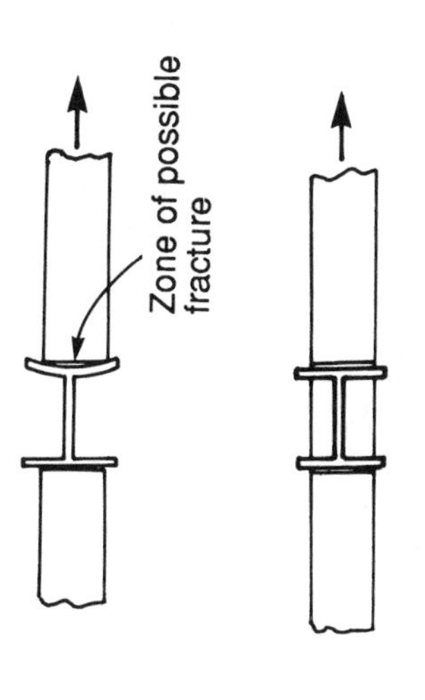

Figure 6.4.12 Bending of Unstiffened Column Flange

A large shearing deformation in the column panel zone can accrue in an interior beam-column joint because the imposed moments on either side of the joint are in the same sense for lateral loads. As shown in Figure 6.4.10 these moments are equivalent to a system of forces that cause high shear in the column panel zone. If the geometry of the connection is such that yielding of the panel zone in shear occurs prior to the development of a plastic hinge in the beam or column, then the ensuing deformation pattern is as shown in Figure 6.4.11. Large shearing deformations can cause undesirable drifts in the structure as a whole, but the moment resisting connection is most affected by the severe kink that develops at the weld joining the flange of the beam to the flange of the column. Cyclic repetitions of this condition can lead to early fatigue fracture of the weld. The addition of doubler plates to increase the thickness of the column web is suggested by the AISC for the column webs whose shear strength is less than the shear induced by the beams when their plastic moment capacity is reached, assuming that the entire moment in the beams is transmitted through the flange forces and is resisted by beam shear alone. While such an assumption appears to be quite conservative, particularly for columns with thick flanges, experimental evidence to justify relaxing the requirements is lacking.

If stiffeners are not provided for continuity of the flange forces through the column then the condition shown schematically in Figure 6.4.12 can occur. As the column flanges bend, the stress in the weld material changes, the material at the web flange junction taking most of the load. Again, this condition can lead to early fracture of the weldment. This problem has long been recognized for gravity loads but is even more important in a cyclic load situation.

Local buckling of adjacent elements can also cause distress in the elements of moment connections. Local buckling of the flange due to severe flexural strains almost always occurs near the connection. The local bending of the flange places severe demands on the weld material connecting the flange of the beam to the flange of the column. On the other hand, the local buckling of the member generally reduces its load carrying capacity and thereby diminishes the stress level experienced by the weldment.

<u>Eccentric Connections</u>.

The short beam segment in the eccentrically braced frame is often considered to be a connection rather than a member. The presence of very high shearing forces and deformations in this region of the structure gives rise to special connection considerations. Stiffeners should be provided in the link region to prevent web buckling [Hjelmstad and Popov, 1983]. The connection of the link to a column should be moment resisting. Generally, welded connections are preferred. A typical connection detail for a link to column flange is shown in Figure 6.4.13.

While experimental data are limited, it appears that ductile eccentric connections can be made between a beam and the weak axis of a column [Popov and Malley, 1983]. The same considerations apply to these weak axis connections as apply to standard beam-to-column connections. A typical weak axis connection is shown in Figure 6.4.14.

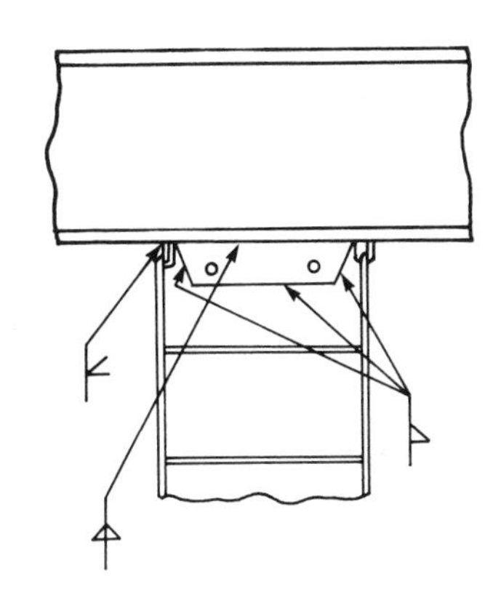

Figure 6.4.13 Typical Short Link-to-Column Flange Connection Detail

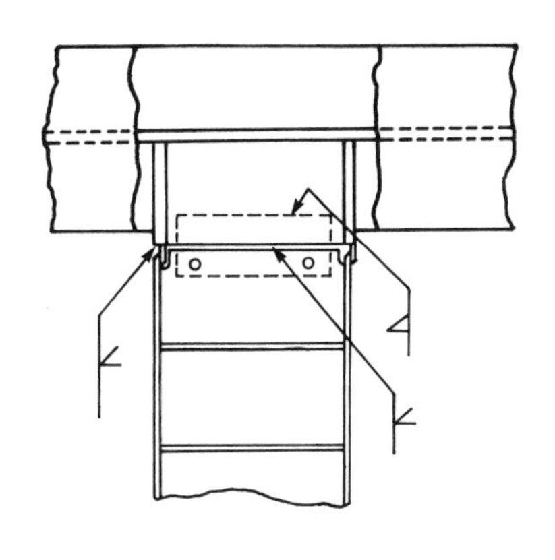

Figure 6.4.14 Link-to-Column Web Connection Detail Shown Has Limited Ductility Under Severe Cyclic Loads

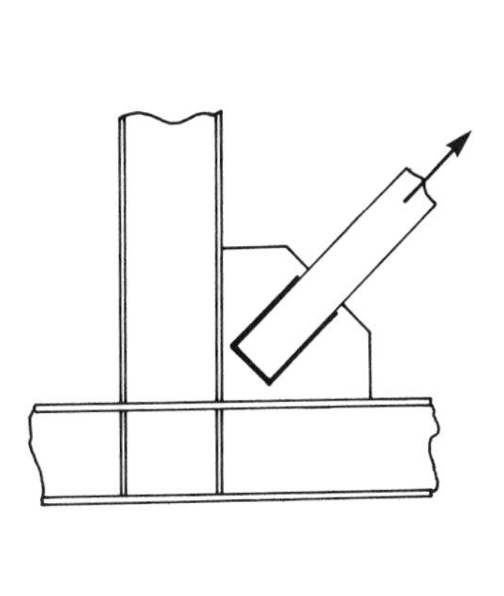

Figure 6.4.15 Brace Connection with Gusset Plate

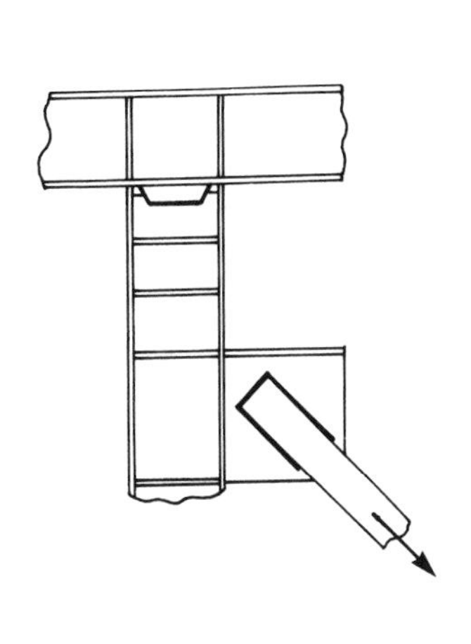

Figure 6.4.16 Structural Tee used for Gusset Plate in Eccentric Connection

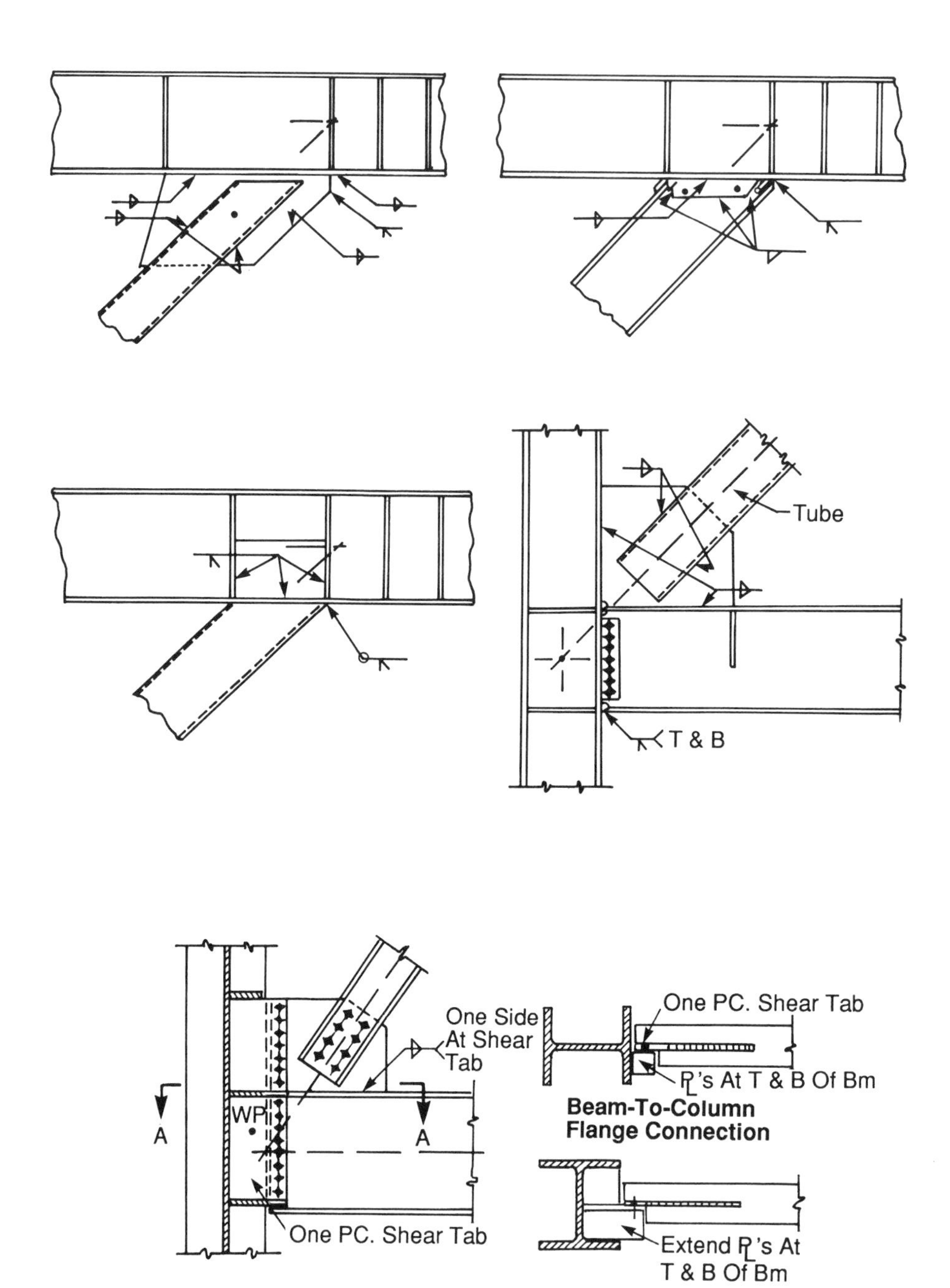

Figure 6.4.17 (a) (top left) Improved tube brace-gusset plate connection
(b) (top right) Detail of WF brace-to-link connection
(c) (left) Detail of tube brace-to-link connection
(d) (right) Detail for brace at beam-column moment connection
(e) (bottom) Bolted detail for beam-column-brace connection

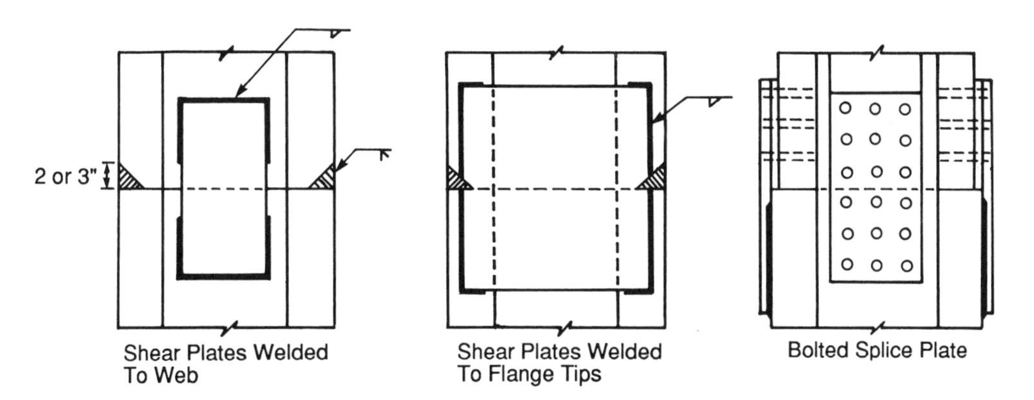

Figure 6.4.18 Column splices

Bracing Connections.

One of the most popular ways of connecting a diagonal brace to the beam--column joint is through a gusset plate as shown in Figure 6.4.15. The brace might be bolted or welded to the gusset plate and the gusset plate might be welded or bolted to the beam and column. In general, the presence of the gusset will make the joint rigid regardless of the method of connection of the beam to the column. The gusset must be examined for its ability to sustain both a tensile and a compressive load from the brace. When the brace is in tension, the system might suffer a block shear failure of the gusset, weld fracture, or bolt fracture. When the brace is in compression the gusset is at risk of buckling. A connection using a structural T section as shown in Figure 6.4.16 will improve the buckling resistance of the gusset. When two braces meet, in the middle of a beam, they can be attached directly to the flange of the beam without gussets. When the bracing joint occurs in the span of the beam, as it would in a K-braced frame or any eccentrically braced frame, the beam would be prone to lateral buckling if the brace force becomes misaligned either due to buckling or initial imperfections. Some typical details are shown in Figure 6.4.17. Figure 6.4.17(a) shows the connection of a tube brace to a beam in an eccentrically braced frame using a gusset plate; Figure 6.4.17(b) demonstrates the connection of a W section brace to a beam; Figure 6.4.17(c) shows a direct welded connection of a tube brace to a beam; Figure 6.4.17(d) shows a concentric brace connection at a beam-column intersection; and Figure 6.4.17(e) shows a bolted brace connection to a bolted weak axis connection between a beam and a column.

Beam and Column Splices.

Beams and columns are generally spliced at a point which is convenient for making the connection. Splices are generally designed on the basis of an elastic analysis of the structure which often leads to minimal splice requirements because of the proximity of the splice to a point of inflection. The location and design of column splices in moment frames that are to resist seismic forces are governed by the requirements of the UBC (1991). The designer should realize that as the structure yields, the moments in the frame are redistributed. The new moment distribution may be more critical than the initial elastic one for the design of the splice. It is generally not practical to design a splice to develop the full capacity of the section but some additional conservatism would be well placed in this important application.

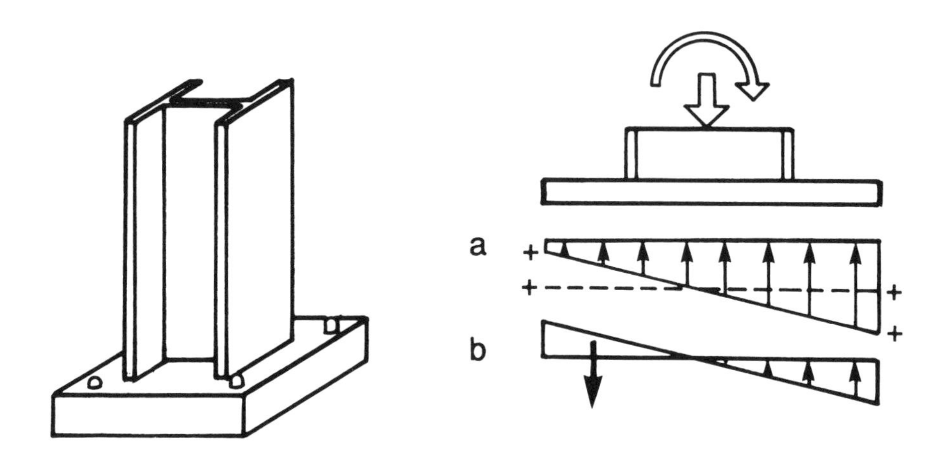

Figure 6.4.19 Column base plate connection

The need to splice columns may have practical implications on the selection of member sizes used in the structure. For example, it is much more convenient to splice two W sections having the same depth and differing weights than it is to splice two sections of different depths. In the latter case, it is necessary to provide an additional plate between the two elements to provide a connection surface and this may affect the seismic performance (ductility and strength degradation). Some examples of column splices are shown in Figure 6.4.18.

Column Base Plates.

Unlike the case of gravity loads, columns in buildings subjected to lateral loads can be subjected to uplift forces; hence the column base plates must be designed to accommodate this. Many failures in low-rise buildings during past earthquakes have been due to the fact that the structures were not properly tied to the foundations. An example of a column base-plate connection is shown in Figure 6.4.19.

Beam to Slab Connections.

The floor slab in a low-rise building might be either steel decking, concrete, or concrete covered steel decking. The connection of the steel beams to a concrete slab is accomplished through shear studs. The spacing and size if these studs must be sufficient to carry the shear flow induced by the composite action of the system. Steel decking is usually spot welded to the beam.

6.4.3 The Issue of Quality

One of the most important and most elusive factors affecting the performance of connections, especially welded ones, is quality control. Poor workmanship will render even the best connection detail ineffective. In general, it is easy to inspect bolted connections but relatively difficult to inspect welded connections. The designer should always consider the issue of quality when conceiving a connection detail. While good performance of a connection cannot be assured, poor performance can often be avoided by using details which are simple to fabricate.

6.5 CONNECTIONS FOR LATERAL FORCES IN CAST-IN-PLACE CONCRETE LOW-RISE BUILDINGS

6.5.1 Introduction

This section addresses the design of connections in cast-in-place concrete structures subjected to lateral loads, earthquake or wind. Seismic forces, which are proportional to a building's mass, are generally greater than wind forces in low-rise concrete buildings, and thus will govern the design in high seismic zones. Also, Section 5.1 stipulates that the detailing of lateral force resisting systems conform to earthquake requirements unless the wind loads exceed the seismic loads by 150% (a factor of 2.5). Therefore, the remainder of this section will be limited to a discussion of seismic detailing provisions.

The single most important feature in concrete structures is ductility. Ductility is a measure of a building's ability to undergo large inelastic displacements without degradation of strength below a level necessary to maintain structural integrity. Building Codes are based on the premise that details should be proportioned such that ductile failures occur first and forces available to cause brittle failures are limited [ATC 11, 1983]. Special provisions for seismic design of concrete structures may be found in Chapter 21 of the Building Code Requirements for Concrete, ACI 318-89 (1989), and Section 2625 of the Uniform Building Code [UBC, 1991]. For simplicity, reference will be made to ACI-318-89 only for the remainder of this section.

Ductile moment resisting frames and shear walls are two types of lateral load resisting systems common in concrete construction. This section will identify some of the critical elements in each system and describe detailing practices and code provisions which improve their dynamic performance. At the end of this section, a sample calculation will be given for a beam-column joint and a shear wall boundary element to demonstrate the application of code provisions.

6.5.2 Concrete Ductile Moment Resisting Frames

The response of reinforced concrete frame structures under seismic loading is complex. Past earthquakes have shown that one of the most critical regions in moment resisting concrete frames is the beam-column joint. A recent example that illustrates this point is the collapse of the Cypress Street Freeway during the Loma Prieta (San Francisco) Earthquake in October of 1989 as shown in Figures 6.5.1 and 6.5.2. One of the primary problems in this two-story rigid frame structure was joint detailing [Civil Engineering magazine, 1989]. The columns and beam-column joints had improperly spaced lateral ties that led to the loss of concrete confinement and the formation of plastic hinges in these undesirable regions. In addition, there were deficiencies in the continuity of the primary reinforcing steel. Modern codes and current practice include provisions to ensure that these types of failure mechanisms are suppressed.

Current code requirements for beam-column joints may be found in Section 21.6 of ACI-318-89. This section identifies three specific areas requiring special attention in the design of joints in frames resisting earthquake motions: (1) transverse reinforcement, (2) shear strength and (3) development length of bars in tension. The provisions of Section 21.6 are based on ACI-ASCE Committee 352 Report published in 1976 and reaffirmed in 1981. That report, which is titled

Figure 6.5.1 Joint Failure in Cypress Street Freeway
Photo courtesy of EERI

Figure 6.5.2 Collapse of Cypress Street Freeway
Photo courtesy of EERI

Recommendations for Design of Beam-Column Joints in Monolithic Concrete Structures, was updated in 1985 and reflects the latest advances in joint detailing.

Transverse hoop reinforcement within the joint provides concrete confinement which is necessary to preserve the integrity of the joint when it is subjected to inelastic strains. ACI-318-89 stipulates that the total cross-sectional area of the hoop reinforcement shall not be less than that given by the following equations:

$$A_{sh} \geq \begin{cases} 0.09\,s\,h_c \dfrac{f_c^{'}}{f_{yh}} \\ \text{or} \\ 0.3\,s\,h_c \left(\dfrac{A_g}{A_{ch}} - 1\right) \dfrac{f_c^{'}}{f_{fh}} \end{cases}$$

Additionally, the code imposes that the spacing of the reinforcement shall not exceed 4 inches or one-quarter of the minimum column dimension. The requirements for beam-column joints are identical to those for columns except that the code permits increasing the spacing of the transverse reinforcement to 6 inches and allows a 50% reduction in the amount of transverse reinforcing for joints with members and all four sides.

Secondly, to prevent brittle shear failure, the joint must have adequate capacity to resist the forces imposed on it by the connecting members. The code prescribes that the joint be capable of developing the ultimate strength capacities of these adjoining members. The forces acting on a typical joint are shown in Figure 6.5.3. The design shear, V_u, is computed on a horizontal plane at the mid-height of the joint by considering the shear forces in the columns and the normal tension and compression forces in the beams framing into the joint. ACI-318-89 prescribes the nominal joint shear as $\gamma\sqrt{f_c^{'}}A_j$, where γ varies with the degree of confinement provided by joining members. Thus, the joint shear should be proportioned such that $V_u < \phi V_n$, where $\phi = 0.85$.

Finally, longitudinal reinforcing bars must be properly developed at the joint to ensure that bond slip is kept within tolerable limits during stress reversals resulting from intense seismic loading. The code requires that the development length of a bar which terminates in a standard hook at a joint shall not be less than $8d_b$, 6 inches or $l_{dh} = f_y d_b / 65\sqrt{f_c^{'}}$. The critical section for development is taken at the face of the confined column core as shown in Figure 6.5.4.

6.5.3 Concrete Shear Walls

As in moment resisting frames, the primary objective in shear wall detailing is to develop a system that will have sufficient ductile energy dissipating characteristics when subjected to cyclic loading. One means of improving the overall shear deformation behavior of structural walls is the inclusion of boundary elements [Oesterle et al, 1985]. However, before discussing boundary members, a discussion of some of the key principles and detailing requirements in shear walls will be given.

The total shear resistance of a structural wall is a combination of two components, concrete shear strength and reinforcing strength. The formula given in the code for determining the nominal shear strength of a wall is $V_n = A_{cv}(2\sqrt{f_c^{'}} + \rho_n f_y)$.

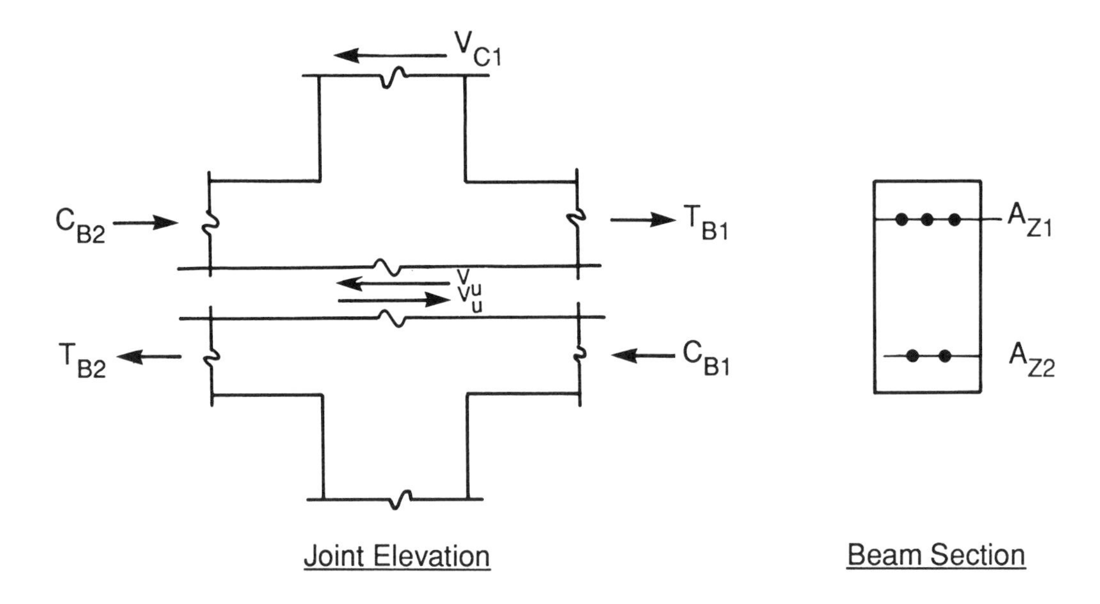

Figure 6.5.3 Forces acting on a Typical Joint

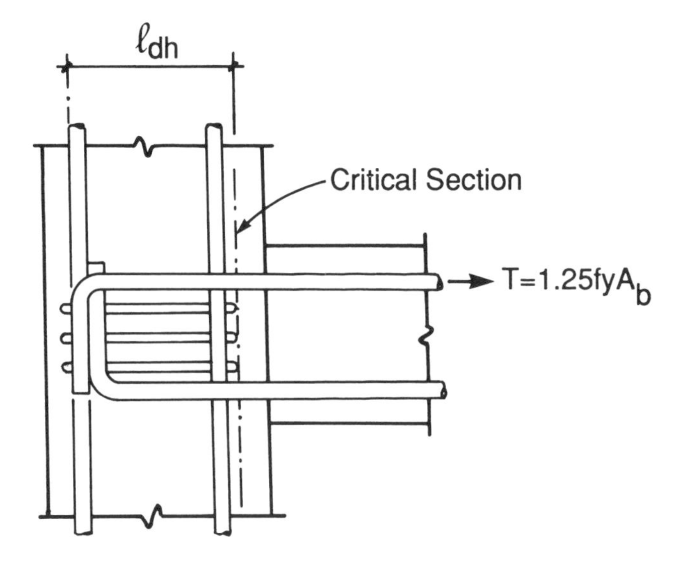

Figure 6.5.4 Critical Section for Development

At least two curtains of steel reinforcement are required by the code and the minimum area of reinforcement which is required in either direction is $0.0025A_{cv}$ [ACI 318, 1989]. Moreover, a common philosophy in shear wall design is to proportion a wall for the shear force associated with the maximum ultimate flexural moment, thereby suppressing brittle shear failure modes during an earthquake [Paulay, 1981].

By definition, boundary elements are specially detailed regions where vertical reinforcement is concentrated near the end of the wall as illustrated in Figure 6.5.5 [Oesterle et al, 1985]. They may be thought of as wall flanges which provide additional rotational capacity during an earthquake. Boundary members are required by ACI-318 (1989) when the maximum extreme fiber stress from the factored seismic loads exceeds $0.2f'_c$. The boundary member may be discontinued where the calculated compressive stress is less than $0.15f'_c$.

In order to perform effectively during severe earthquakes, vertical boundary members must be confined properly. As in frame columns, transverse reinforcement within the boundary element provides confinement and lateral restraint of the vertical reinforcing steel to inhibit buckling under compression. The minimum area and maximum spacing of transverse reinforcement is the same as frame columns. Boundary elements also provide a region to anchor the horizontal wall reinforcement. The code recommends that the horizontal web reinforcement be extended across the boundary element and terminated with a standard 90° bend as shown in Figure 6.5.6.

Design Example 6.5.1 Interior Beam-Column Connection

An interior beam-column joint is loaded as shown in Figure 6.5.7.

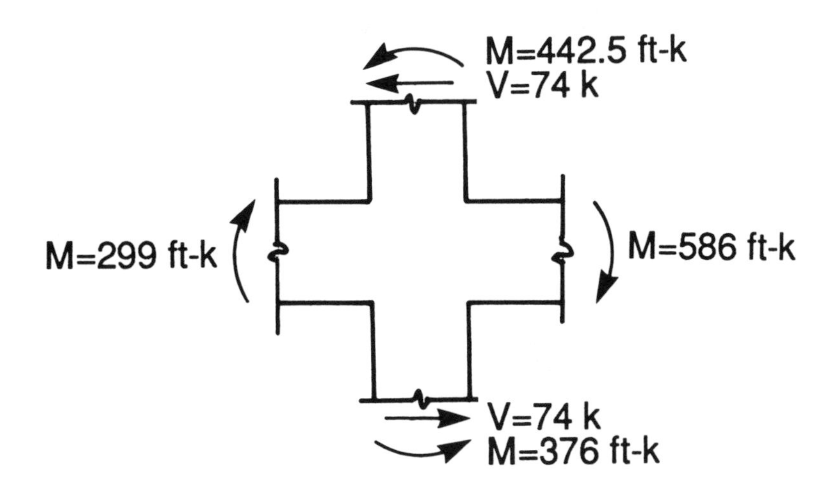

Figure 6.5.7 Loading on Interior Beam-Column Joint

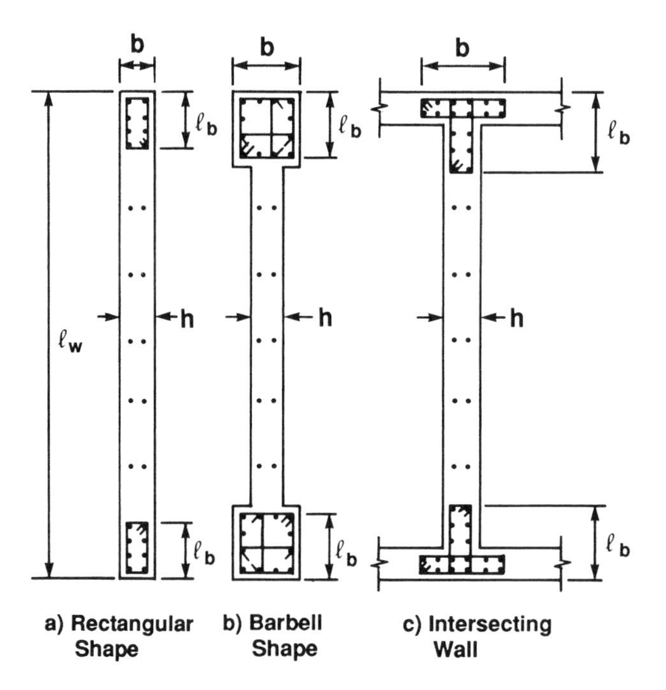

Figure 6.5.5 Horizontal Cross-sections of Structural Walls

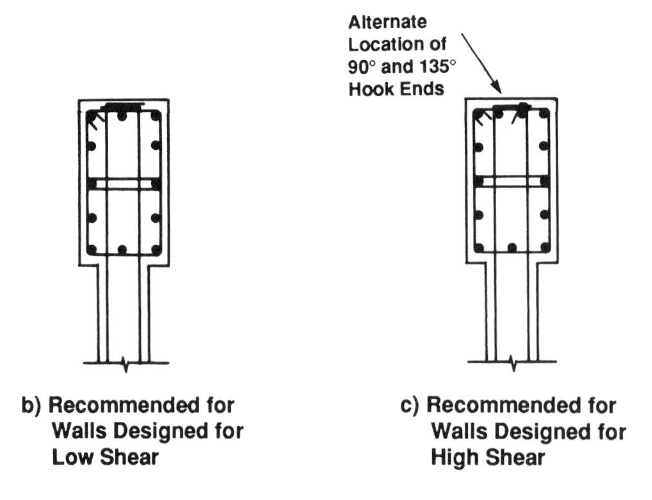

Figure 6.5.6 Examples of Details for Horizontal Reinforcement

Given: f'_c = 4 Ksi, f_y = 60 Ksi.
Column size: 26" x 26" with 8 No. 11 bars
Beam size : 20" x 24" with 5 No. 9 bars (top)
and 3 No. 8 bars (bottom)

a. Determine the transverse reinforcement requirements.

s_{max} = 4 in. (**governs**)
< 1/4 in. smallest dimension of column = 26/4 = 6.5 in.

Assuming No. 5 hoops with 1½" cover,
h_c = 24" - 1.5" - 1.5"-0.31" = 20.7"
A_{ch} = 21" x 21" = 441 in^2

Calculate A_{sh}:

$$A_{sh} = 0.09 s h_c \frac{f'_c}{f_{yh}} = (0.09)(4)(20.7)\frac{(4)}{(60)}$$
$$= 0.50 \text{ in.}^2$$

$$A_{sh} \geq 0.3 s h_c \left(\frac{A_g}{A_{ch}} - 1\right)\frac{f'_c}{f_{yh}}$$
$$= (0.3)(4)(20.7)\left(\frac{676}{441} - 1\right)\frac{4}{60}$$
$$= 0.89 \text{ in.}^2 \qquad (\textit{governs})$$

Since the joint has beams framing in on all four sides, with widths greater than ¾ column width, it is adequately confined and a 50% reduction in reinforcing steel is permissible.

∴ A_{sh} = 0.5 x 0.89 = 0.45 in^2
A_{sh} provided by No. 5 hoops = 2 x 0.31 = 0.62 in^2.

b. Check shear strength of joint.

Net shear across x-x: $T_1 + C_2 - V_{col}$ = 478 Kips (V_u)
Shear strength of joint confined on all four faces:
$\phi V_c = \phi\, 20 \sqrt{f'_c} A_j$
= $(0.85)(20)(\sqrt{4000})(26^2/1000)$ = 727 Kips

V_u = 478 K > 727 K **O.K.**

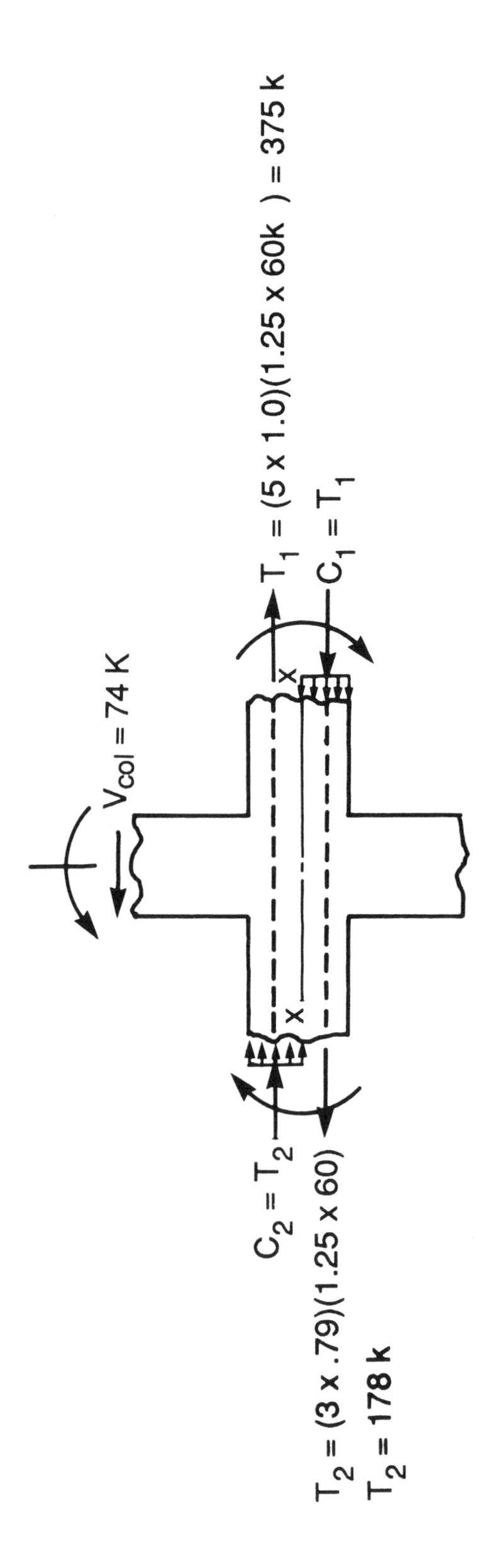

Figure 6.5.8 Forces on Joint

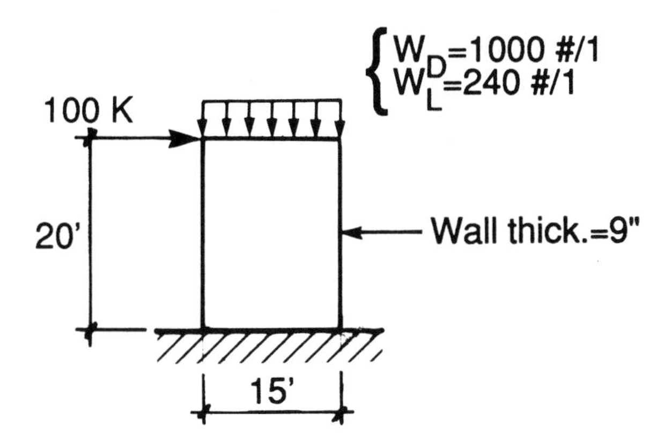

Figure 6.5.9 Details of concrete shearwall

Design Example 6.5.2 Concrete Shearwall

A single-story concrete shearwall is loaded as shown in Figure 6.5.9.

Given: f_y = 60 Ksi, f'_c = 3 Ksi

a. Check if boundary elements are required

Calculate the extreme fiber compressive stress.

Governing load combination: 1.4 (D + L + E)
P_u = 1.4 [(1000+240)15 + (9/12)(150)(20)(15)] = 73 Kip
M_u = 1.4 (100 x 20') = 2800 ft.Kip

Wall properties: Area, A = 180" x 9" = 1620 in^2
Section modulus, S = 180^3 x 9/12 = 48600 in^3
$\therefore$ f = P/A $\pm$ M/S = (73/1620) $\pm$ 92800 x 12/48600)
= 0.736 Ksi (compression)
0.646 Ksi (tension)
0.20 f'_c = (0.20)(3 Ksi) = 0.6 Ksi
$\therefore$ since 0.736 Ksi > 0.6 Ksi, boundary elements are required.

b. Boundary member design

<u>(i) compression</u>: max P_u = 1.4(D+L+E)
P_u = (73 K/2) + (2800 ft.Kip/15') = 223 K
Try boundary elements having dimensions of 15" x 18" (A_g = 270 in$_2$)
$\rho_{min} < \rho_{st} < 0.06$

Try ρ_{min} = 0.01

A_{st} = 0.01 x 15 x 15 = 2.25 in_2 : try 4 No. 8 bars (A = 3.16 in_2)

Axial load capacity of boundary element acting as a short column:

$\phi P_n = 0.80\,\phi\,[0.85\,f'_c\,(A_g - A_{st}) + f_y A_{st})]$

= 0.80 (0.70)[0.85 x 3 ksi (270 - 3.16) + (60 Ksi)(3.16 in_2)]

= 487 K

P_u = 223 K < 487 K **O.K.**

<u>(ii) Tension</u>: P_u = 1.4 E - 0.9 D

P_u = 2800/15 - 0.9 [1000 x 15/2 + (33.75 K)/2] = 165 K

A_s = 165 K / (0.9 x 60 Ksi) = 3 in_2 -- 4 No. 8 bars **O.K.**

c. Transverse reinforcement

- similar to Example 6.5.1.

7 Nonstructural Elements

7.1 CLASSIFICATION

7.1.1 Problem Definition

Nonstructural elements include a large number of diverse items built into buildings or added to them after construction by the occupants. Essentially, everything other than the structure itself which holds the building up and resists horizontal forces due to wind, earthquake, or other loads is nonstructural, and thus elevators, ceilings, partitions, cladding, equipment, and contents are included within this category.

The influence of nonstructural elements on the earthquake and wind performance of low-rise buildings is potentially more important than for high-rise construction because relatively less design attention is devoted to these smaller buildings. In addition, the economic consequences of damage to nonstructural elements in the case of low-rise construction is relatively more important than for high-rise buildings. Some of the factors which contribute to the need for improved design procedures for low-rise buildings are: the inattention of codes to the unique problems of low-rise buildings; and large building plan forms with large torsional eccentricities (between both center of pressure and center of mass with respect to center of rigidity) which increases the relative interstory displacement. Diaphragm response somewhat influences the design of the displacements and response at floor levels. A number of references of broad scope have treated this topic, including: Ayres et al. (1973), Meehan (1973), McGavin (1981, 1987), Eagling (no date), Rihal (1980), Stratta (1987), Reitherman (1983, 1984, 1988), EERI [Earthquake Engineering Research Institute, 1984].

7.1.2 Differences in Wind and Seismic Loads

Several obvious differences between the effects of wind and earthquake on nonstructural elements should be noted.

Wind loads only directly impinge on exterior nonstructural elements (unless major damage occurs and destruction of the envelope allows wind to enter the interior.) Wind-induced deflection of the structure can cause deformation of partitions or other nonstructural elements, though this is rare in low-rise buildings. Even less common is nonstructural damage caused by accelerations imparted to equipment or contents. Because earthquakes exert their forces more pervasively, attacking the generator mounted in a basement as well as equipment mounted on the roof, the discussion of interior elements, contents, and equipment in this chapter is generally limited to the earthquake topic.

The dynamic response of low-rise buildings to wind is rarely a major issue: The wind load is represented statically and the building's dynamic properties, such as the period of vibration, or damping, are rarely considered in design. Dynamic considerations are more often significant with regard to the seismic design of low-rise buildings and the seismic protection of individual nonstructural components.

Other distinctions pertain to the wind and seismic topics on a more detailed level. Glass breakage is an important problem in both fields, though more common with regard to the hazard of wind. Out-of-plane pressure or impact from wind-blown debris are more common causes of glass failure in high winds, especially for low-rise buildings, than failure caused by the drift of the structure. In the earthquake field, drift-induced failure (and thus, structural response) is the basic damage mechanism. Seismic loads are inertially originated, and thus heavy equipment is subject to large seismic loads. Mass is generally a stabilizing benefit in the field of wind design and a penalty in seismic design. As an example, an elevated water tank that is full when a hurricane strikes is less vulnerable than if it is empty; the opposite is true with earthquakes. While damaging earthquakes occur outside the Western U.S. and are increasingly receiving the design attention this hazard deserves in the South, Northeast, and Midwest, it is true that the historical development of seismic design in this country has been centered in California. Because of differences in regional construction practices, non-western areas of the country now dealing with the earthquake problem must sometimes devise solutions appropriate for different cladding systems, different architectural styles, and different habitual installation methods in the construction trades. Wind methods have not evolved with a similar regional specialization.

7.1.3 Categorization

One means of categorizing nonstructural items is to observe typical damage patterns. From this viewpoint, there are two basic types of nonstructural components: Those that are vulnerable to the inertial forces imparted by the motion of the building, and those that are vulnerable to the distortion of the building's geometry induced by the seismic stresses in the structure. In earthquakes, freestanding equipment overturns and is damaged because of the inertial mechanism, while windows and partitions are usually damaged because of the structural drift imposed upon these vertically oriented, rigidly built-in nonstructural items. Wind damage can also be caused by drift (though usually in taller buildings), and the basic wind analog for seismically caused inertia is wind pressure.

Another way to classify nonstructural components is in terms of the type of loss they may cause if they are damaged. An emergency power generator rarely presents a direct life safety threat (injury hazard) to occupants. The property damage that would be caused to a non-resistant generator in an earthquake is usually also a minor consideration, but the functional loss that would be occasioned by damage to the generator is almost always a major concern. In the case of most modern lightweight partitions, cracking of the partition does not generally pose a major life threat, but the property damage can add up if most of the partitions in a building are damaged. Damage to glass is an example where life safety is generally the paramount concern. Thus, nonstructural components can be classified according to their potential contribution to life safety, property loss, or functional loss risks.

Another categorization scheme, useful for the purpose of dividing up the scope of work in nonstructural earthquake vulnerability studies of hazard reduction programs, differentiates the permanent or built-in nonstructural components such as a ceiling or the mechanical (HVAC) system from the occupants' furnishings and equipment. This differentiation matches the division of ownership in cases where a building is owned by one party and leased and occupied by another. For buildings occupied by owners, retrofit projects for permanent nonstructural components typically require more engineering expertise, while at least some of the contents are usually suitable for self-help, simplified hazard reduction efforts.

While all of these classification schemes have their merits, for the purpose of this chapter, a three-part organization will be used

1. Built-in interior nonstructural components.
2. Built-in exterior nonstructural components.
3. Contents and equipment related to building occupancy.

1. Interior Components

Built-in interior components in low-rise buildings include such features as:

- Partitions

Based on observed behavior and performance, these may be further divided into the following categories:

Light Partitions, e.g.: Metal-stud framed or wood-stud framed partitions with various facing panels, such a gypsum wall-board, plywood, plaster, tile, etc.

Heavy Partitions, e.g.: Masonry partitions, which are likely to act in a structural manner even when called "nonstructural." Figure 7.1.1 illustrates the failure of a masonry block non-structural infill-wall/position during the Fruili, Italy, earthquake of 1976. The damage shown was probably caused by excessive torsional displacements and/or out-of-plane motions experienced by such components.

Based on observed behavior and performance of ceiling systems in low-rise buildings during previous earthquakes, the predominant ceiling type found to be prone to repeated wide-spread damage is suspended ceilings with tee-grid framing & lay-in acoustical tiles. Figure 7.1.2 shows representative ceiling damage observed during the Coyote Lake, California earthquake of 1979, in an older building where the more modern suspended ceiling system was not present. Provision of vertical support, with allowance for horizontal movement, was not provided.

Figure 7.1.3 shows representative damage of suspended ceiling at the perimeter, in the OSA building in downtown Oakland, during the Loma Prieta, CA, earthquake of October, 1989. This damage was probably caused by excessive inter-story drifts (especially at the corner) which could not be accommodated by the ceiling perimeter details and the suspension system.

- Light Fixtures

Long lines of pendant-hung light fixtures in low-rise buildings have been observed to have suffered wide-spread damage during every major previous earthquake (e.g., Anchorage, Alaska [1964]; San Fernando, California [1971]; Coalinga, California [1983], etc.). Figures 7.1.4 and 7.1.5 show the representative severe damage suffered by the long lines of pendant-hung light fixtures in the Dawson Elementary School during the Coalinga, CA, earthquake of 1983. The damage seems to result from

Figure 7.1.1 Failure of masonry Block Non-Structural Infill-Wall/Partition, Fruili Earthquake, Italy, 1976

Figure 7.1.2 Damage to Light Fixture and Ceiling in a One Story Commercial Building, Coyote Lake Earthquake 1979

Figure 7.1.3 Suspended Ceiling Damage - OSA Building, Oakland, CA; Damaged in the Loma Prieta Earthquake, 1989 (Photos: S. Rihal)

excessive stresses at connections, caused by excessive movements of this pendulum-type system. The failure appears to have been initiated at the connection between the suspension-stems and the light fixtures; as well as at the connections between the light-fixture units.

2. Exterior Components

- Cladding and Veneer
- Windows/Glazing
- Doors

3. Contents & Equipment

Except for exterior equipment (such as roof-mounted mechanical equipment), this section deals almost exclusively with earthquakes.

It should be recognized that any classification of contents and equipment must be related to building occupancy and building type (e.g., public schools, medical facilities, science laboratories, computer facilities, libraries, etc.), as well as to the specific nonstructural item.

Some items in this category, such as emergency power generators, are very essential; others, such as museum contents, are very valuable; some, such as tall shelving, can be life safety threats.

Figure 7.1.5 Failure of the Pendant-hung Light Fixtures, Dawson Elementary School, in the Coalinga, CA, Earthquake, 1983

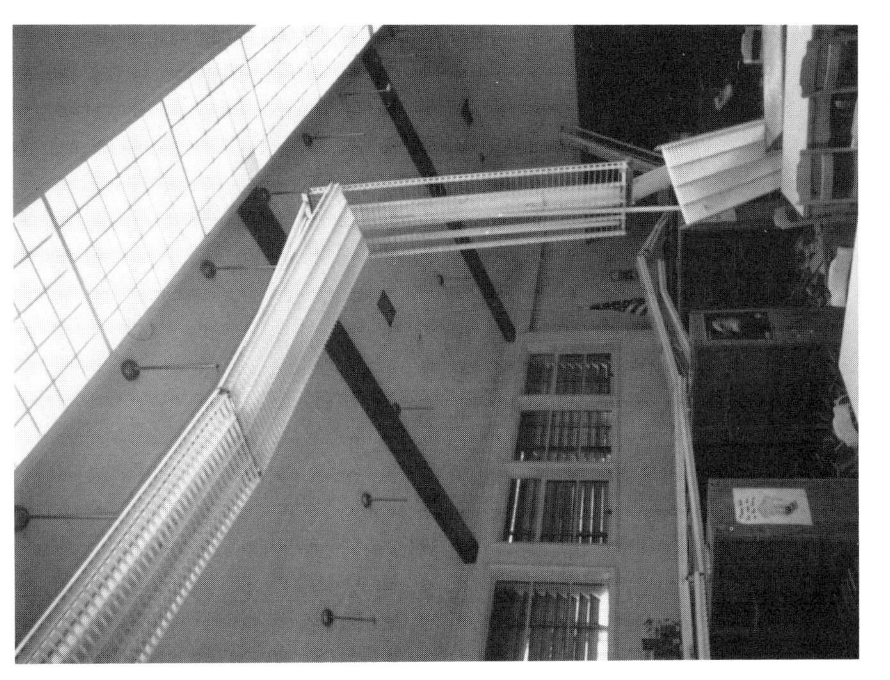

Figure 7.1.4 Failure of Pendant-hung Light Fixtures, Dawson Elementary School, in the Coalinga , CA, Earthquake, 1983

7.2 FUNDAMENTAL CONCEPTS

7.2.1 Acceleration Effects

Relatively heavy nonstructural components are prone to seismic inertial actions resulting form the acceleration levels experienced in the building. Included in this category are elements such as heavy partitions, suspended ceilings, exterior facades-/claddings, heavy exterior ornamentation and parapets. Freestanding slender contents, e.g., filing cabinets, shelving, or equipment, may overturn, causing injury and damage.

7.2.2 Drift Effects

Interior partitions, and exterior components, such as facades/claddings including windows and curtain-walls are subject to potential damage caused by distortions of the building's geometry during wind or earthquakes. These components are usually damaged because of effects of building drift imposed upon these vertically-oriented, rigidly built-in nonstructural elements. Nonstructural elements that cross separation joints in the structure, and which were not designed with flexibility in mind, are damaged by differential drifts. Piping that rigidly crosses a separation joint is one such case.

7.2.3 Interaction Between Nonstructural and Structural Systems

Detailed documentation of conceptual models of interaction between nonstructural and structural systems in buildings during earthquakes has been presented by McCue et al. (1978). Chen and Soong (1988) recently reviewed the methods available for analyzing the seismic response of secondary systems. Interaction between interior partitions and suspended ceilings with or without splayed-wire bracing has been investigated experimentally by Rihal et al (1984).

The important issue of cladding-structure interaction has been investigated by Goodno et al. (1983, 1984, 1986a, 1986b, 1988, 1989).

Rigid partitions as well as heavy cladding will interact with building structural systems and have significant influence on over-all building stiffness and ductility.

The increase in stiffness provided by nonstructural components such as cladding that is not thoroughly isolated from structural movements may be either desirable or undesirable in the case of earthquakes, because the shortening of period of vibration may either increase or decrease dynamic response, depending upon the frequency characteristics of the input motion. Not all low-rise buildings are stiff (short period), and so in some cases nonstructural stiffening may bring a building into the range of predominant energy of ground motion causing increased response of the building. For many relatively hard soil or rock sites, buildings with periods over one-half second respond less than stiffer structures. (To be accurate, one should keep in mind that it is the mass/stiffness ratio, not stiffness alone that determines a building's period.)

7.2.4 Wind Pressure Effects

Wind pressures primarily affect built-in exterior non-structural components, including cladding and veneer, windows and glazing, and doors. As described in Section 3.4.2, high local pressures are created by flow separation at sharp edges of the building. These high localized pressures often initiate a progressive failure of the anchors and connections that attach non-structural elements to the building structure.

Tributary areas associated with the anchors and fasteners of components and cladding are small compared to the tributary area of a main wind-force resisting sub-system such as a frame or a shear wall. The small tributary areas experience the full effect of small, but intense gusts that encompass the entire tributary area. In addition, a resonance condition may be achieved when the gust frequency matches the natural frequency of the component or cladding. Modern codes and standards recognize these conditions and account for their effects by giving special treatment to the design wind pressures on exterior components and cladding.

Windows, glazing and large overhead doors may fail by excessive wind pressure or by the impact of debris carried by the wind, resulting in a breach of the building enclosure. Breach of the building envelope causes a change in the interior building pressure. The pressure change can be either positive (increase in pressure) or negative (decrease in pressure), depending on the location of the opening relative to the wind direction. An opening in a windward wall results in an increase in internal pressure; openings in the other walls produce a decrease in pressure. The internal pressure combines with the external pressures caused by aerodynamic flow of wind over and around the building to create pressures larger than the external pressures on a totally enclosed building. Internal pressures are most adverse when the dominant opening is in only one wall. A dominant opening in a windward wall can result in an internal pressure equal to 75% of the stagnation pressure.

When openings occur simultaneously in two or more walls or the roof, air can flow through the building creating pressures on interior partitions, suspended ceilings, light fixtures and air handling ducts. When these non-structural components are not designed for some minimum interior wind pressure, significant wind damage to the building interior can occur. Channels of wind, where the air moves down narrow hallways, have reached speeds estimated as high as 50 mph. In addition to creating wind pressures, pieces of debris may be picked up and become missiles that are especially dangerous to occupants of the building.

7.3 INTERIOR COMPONENTS

7.3.1 Partitions

Current Practice and Problem Areas

Interior partitions and suspended ceiling assemblies in low-rise buildings are essentially designed on the basis of customary construction practices of interior systems sub-contractors. Fire-protection, acoustic, cost and appearance considerations are primary. Widespread use of metal-stud framed partitions for both full--height and partial-height partitions consists of 20 or 25-gauge metal studs spaced every 16 or 24 inches, positioned between 20-gauge metal channel runners fastened to the concrete floor with power-driven fasteners at 16 inches o.c. at the base and fastened similarly to the structural floor system above (for full-height partitions) or fastened with sheet metal screws to suspended ceiling framing (for partial-height partitions). The facing panels are usually gypsum wall-board, oriented either horizo-n-tally or vertically and fastened to metal studs and bottom channel runners. Connection details at the top and sides of full-height partitions are supposed to allow for movements due to gravity as well as lateral loads, though in practice installation details do not usually vary from one building to another where drift or other factors are different.

There is significant use of wood-framed partitions in low-rise buildings all over the country depending on local practices, economic conditions and the fire protection regulations pertinent to buildings of various sizes and occupancies. Wood-stud framed interior partitions consist of 2 x 3 or 2 x 4 wood studs spaced every 16 or 24 inches between bottom and top sill plates of the same size fastened to concrete floor systems at bottom and at top with shot-ins, or nailed to other members in wood frame construction. The facing panels on both sides of interior partitions, in the majority of cases, are gypsum wall-board, but exterior walls may include application of gypsum wall-board on one side and plywood on the opposite side fastened to wood-stud framing. A typical nailing schedule is given in Chapter 47 of the UBC.

By and large, light interior partition systems are likely to perform satisfactorily with minor damage only. Their damage will result primarily in an economic loss rather than severe life hazard. For the vast majority of relatively light building partitions, it is important to provide design details at top and sides that will accommodate movements resulting not only from lateral loads such as wind or earthquake, but also due to gravity loads and temperature effects. For heavy building components, e.g. masonry partitions and also partial-height partitions of all categories, design details and connections must ensure the stability of such components against motions normal to their own planes. Partitions to which significant masses are anchored (such as tall and heavy shelving) also require adequate bracing. Partial-height partitions that are braced by the suspended ceiling systems may rely on splayed-wire ceiling bracing for stability, or such partitions may be braced to the top by two-wire diagonal bracing or bracing using steel or wood struts. It should be noted that in-plane partition damage may also result due to the wire bracing system forcing the ceiling to have the same displacement as the diaphragm system.

Interior partitions and suspended ceilings may present some unique problems in existing low-rise buildings. Relatively heavy interior partitions, e.g. wood-stud framed partitions with stucco finish on one or both sides, as well as masonry partitions must be stabilized against effects of motion normal to their own planes. Such partition elements have significant in-plane lateral rigidity and therefore could possibly provide unintended lateral force resistance in existing low-rise buildings.

Analysis and Design (Examples)

The following analysis and design steps, as well as the similar summaries presented later for other nonstructural components, are intended to meet the needs of two basic design professional audiences or architecture/engineering applications: as a primer for those who are facing this topic for the first time, to summarize the procedures generally used in design; and as a checklist for use by designers who perhaps may already be well experienced. In the latter usage, the reader is encouraged to customize the checklist to include locally applicable code provisions or construction practices. Another useful type of customization is to include procedures typically used in a given design professional's office for checking with supervising designers in the office or with other design professionals. For example, in the preliminary design stages it is valuable to have a list of what the consulting engineer should bring to the architect's attention or vice versa. Only some of the relevant features of codes and practices are summarized here; for example, where the Uniform

Building Code is used, the exact text and its numerous footnotes or cross-referenced provisions should be consulted.

Criteria

First, what are the roles to be performed by the partitions? Where fire ratings or acoustic separation functions are important, it becomes more difficult to use sliding or other motion-tolerant joints as a solution for the imposed deformation problem. Where the partition is to be self-supporting only, rather than being relied upon to support the seismic loads of attached cabinetry, shelving, or anchored furnishings, bracing the tops of partitions that extend only to the height of a suspended ceiling becomes less difficult. Note that anchoring file cabinets and other very common furnishings and equipment to the wall may well be desirable when other components are considered, so the design of the partition relates to equipment and contents as well.

Should criteria more strict than the code's be used? Keep in mind that the underlying performance goal of most building codes is a reasonable level of protection of life and limb, and only secondarily damage limitation. Some clients may desire greater levels of safety or property protection.

Building Code Inertial Loading Criteria

If the Uniform Building Code is to be used, Table 23-P (1991 edition) (see Table 7.3.1) specifies a lateral force coefficient C_p of 0.75. The tabulated C_p values are for rigid components, which are defined as those having a fixed base period of 0.06 second or less, as specified in section 2336(b). This section also provides that the value of C_p for components supported at or below ground level may be two-thirds of the values tabulated in Table 23-P. This applies only for elements laterally self--supported at ground level. This value is combined with a zone (Z) factor and importance (I) factor, and multiplied with the weight or mass of the partitions, resulting in a design lateral seismic force for the component, as specified in UBC Section 2336(b):

$$F_p = Z\, I\, C_p\, W_p$$

The factor Z will be obtained from the UBC seismic zoning maps. The importance factor I is 1.0 for most occupancies and 1.25 for what are defined in Table 23-K as structures housing hazardous materials or essential emergency response and medical facilities. In the following example, the typical 1.0 value is used, along with a C_p value of 0.75 that applies to most equipment and piping.

$$F_p = Z\,(1)\,(0.75)\,(W_p)$$

or $F_p = (Z)\,(0.75)\,W_p$

This equation yields the design lateral force as a percentage of the weight of partition. In the highest ("worst") seismic zone, zone 4, the Z factor is 0.4.

If the building occupancy was in zone 4 and had an essential or hazardous category, an I factor of 1.25 would result in $F_p = 0.4\ (1.25)\ 0.75\ W_p$, or 3/8 W_p. For ordinary occupancies, 30% W_p is the typical result. The typical calculation product that pertains in other regions is: zone 3: 22.5% W_p; zone 2B: 15% W_p; zone 2A: 11-1/4% W_p. (Note that the building's I factor applies, except that if an individual non-structural item in an ordinary occupancy building has a critical life-safety or hazardous material containment role, then the I = 1.25 factor applies to that particular item.)

ELEMENTS OF STRUCTURES AND NONSTRUCTURAL COMPONENTS AND EQUIPMENT[1]	VALUE OF C_p	FOOTNOTE
I. **Part or Portion of Structure**		
1. Walls including the following:		
a. Unbraced (cantilevered) parapets	2.00	
b. Other exterior walls above the ground floor	0.75	2,3
c. All interior bearing and nonbearing walls and partitions	0.75	3
d. Masonry or concrete fences over 6 feet high	0.75	
2. Penthouse (except when framed by an extension of the structural frame)	0.75	
3. Connections for prefabricated structural elements other than walls, with force applied at center of gravity	0.75	4
4. Diaphragms	—	5
II. **Nonstructural Components**		
1. Exterior and interior ornamentations and appendages	2.00	
2. Chimneys, stacks, trussed towers and tanks on legs:		
a. Supported on or projecting as an unbraced cantilever above the roof more than one half their total height	2.00	
b. All others, including those supported below the roof with unbraced projection above the roof less than one half its height, or braced or guyed to the structural frame at or above their centers of mass	0.75	
3. Signs and billboards	2.00	
4. Storage racks (include contents)	0.75	10
5. Anchorage for permanent floor-supported cabinets and book stacks more than 5 feet in height (include contents)	0.75	
6. Anchorage for suspended ceilings and light fixtures	0.75	4,6,7
7. Access floor systems	0.75	4,9
III. **Equipment**		
1. Tanks and vessels (include contents), including support systems and anchorage	0.75	
2. Electrical, mechanical and plumbing equipment and associated conduit, ductwork and piping, and machinery	0.75	8

[1]See Section 2336 (b) for items supported at or below grade.
[2]See Section 2337 (b) 4 C and Section 2336 (b).
[3]Where flexible diaphragms, as defined in Section 2334 (f), provide lateral support for walls and partitions, the value of C_p for anchorage shall be increased 50 percent for the center one half of the diaphragm span.
[4]Applies to Seismic Zones Nos. 2, 3 and 4 only.
[5]See Section 2337 (b) 9.
[6]Ceiling weight shall include all light fixtures and other equipment or partitions which are laterally supported by the ceiling. For purposes of determining the seismic force, a ceiling weight of not less than four pounds per square foot shall be used.
[7]Ceilings constructed of lath and plaster or gypsum board screw or nail attached to suspended members that support a ceiling at one level extending from wall to wall need not be analyzed provided the walls are not over 50 feet apart.
[8]Machinery and equipment include, but are not limited to, boilers, chillers, heat exchangers, pumps, air-handling units, cooling towers, control panels, motors, switch gear, transformers and life-safety equipment. It shall include major conduit, ducting and piping serving such machinery and equipment and fire sprinkler systems. See Section 2336 (b) for additional requirements for determining C_p for nonrigid or flexibly mounted equipment.
[9]W_p for access floor systems shall be the dead load of the access floor system plus 25 percent of the floor live load plus a 10 psf partition load allowance.
[10]In lieu of the tabulated values, steel storage racks may be designed in accordance with U.B.C. Standard No. 27-11.

Table 7.3.1 UBC Force Factors for Non-structural Components [UBC, 1991]

An important task pertinent to all the non-structural items covered in this chapter is the appropriate modeling of a low-rise building and seismic analysis to determine amplification factors applicable to the non-structural components. The Tri-Services manual [Army, 1986 and 1982] contains more information on amplification factors and the various considerations that should be taken into account in determining whether the equivalent static force approach is adequate. Items located in taller or more flexible structures, especially at upper levels, or items that are internally flexible or are mounted flexibly, are cases where dynamic amplification should be raised as an important issue. While a dynamic analysis may indicate that higher values are needed, the 1991 UBC also places a limit on the lowering of nonstructural design loads where a dynamic analysis indicates that coefficients lower than the statically determined values are appropriate: At least 80% of the F_p calculated according to the static procedures outlined above must be used.

Aside from the specifically seismic regulations of the 1991 UBC, it is also necessary to check provisions covering partitions taller than six feet (Section 2309): a 5-psf load applied perpendicular to the partition must be resisted without a resulting deflection in exceedance of 1/240 of the span (brittle finishes) or 1/120 of the span (flexible finishes). For lightweight stud walls, this loading is similar in a high seismic zone to the above described F_p loading. In a low seismic zone, this general horizontal load criterion may govern for lightweight partitions, while with heavier partitions, the seismic criterion becomes more critical. The common code provision suggests that rational methods of engineering analysis can be used to justify designs using various materials and configurations. However, the UBC specifically bans the use of unreinforced masonry even as nonstructural partitions except in seismic zones 0 and 1.

In most cases, commonly accepted practices provide for sufficient strength to the partition for in-plane and out-of-plane forces. The usual problem is the partition that extends only up to the suspended ceiling, requiring diagonal braces at intervals along the top plate to provide out-of-plane bracing. Where shelving is to be attached to the wall—and keep in mind that this seemingly undesirable load on the partition is actually the easiest way to solve the vulnerability of these other nonstructural items—the studs or top and bottom connections may require strengthening above normal practice. An appendix to the 1976 Veterans Administration design handbook (1976, 1981) on nonstructural seismic design contains graphs that simplify the analysis of stud wall details.

Building Code Imposed Deformation Criteria

The UBC specifies that "only the elements of the designated seismic-force--resisting system shall be used to resist design forces," (Sec. 2312 h), but earthquakes are blind to what has been designated on drawings and in calculations: most partitions (with the exception of open plan freestanding partition dividers) are mounted with sufficient rigidity to cause them to act like shear walls, even if their low capacities cause them to quickly crack, and even if the designer did not rely on their capacities to resist design lateral loads. It is assumed here that the word "partition" implies "non-bearing;" any bearing walls would be considered structural rather than nonstructural elements and are not considered in this chapter. Cracking can be considered desirable by the structural engineer because it is energy absorbing and increases damping, but from the owner's, tenant's or earthquake insurer's viewpoints, a cracked partition is damage. Exit doors can also be jammed, which is a life safety concern.

In small metal or wood stud framed structures, non-bearing partitions with gypsum board sheathing may be designed to take a large portion of the over-all lateral loading on the structure. Although plywood sheathed walls are usually considered more reliable from an earthquake engineering standpoint, use of gypsum wallboard and plaster in a lateral load resisting system is permitted by building codes. Designing lightweight partition walls as structural components to take a building's lateral loads is outside the scope of this chapter.

Related to partition design is of course the over-all structure's drift: relatively stiff structures impose small deformations on their vertical nonstructural components such as partitions, glazing, cladding, or elevators. The term "structural-nonstructural interaction" usually refers to a modification of the structure's response by heavy or stiff nonstructural elements, but the topic is really a two-way street. The need to increase protection of non-structural components is one of the themes underlying the tightening of drift limitations in codes such as the Uniform Building Code in recent years. Along with purely structural concerns such as the P-delta effect, it is assumed that more stringent drift limitations result in reduced nonstructural damage. Further studies are needed to thoroughly investigate the effect of more stringent drift limitation on non-structural damage, because it is possible that as lateral stiffness is increased (stringent drift limitation) the responses may be increased rather than reduced (Ref. Chapter 4).

Arnold, Hopkins, and Elsesser (1987) describe in greater detail design approaches for partitions and other permanent nonstructural elements. Their work also provides a convenient summary of the differing approaches used in New Zealand, where isolation joints for partitions are more common than in the United States, although those special details are usually limited to important public buildings which must remain functional after earthquakes. Figure 7.3.1 from their report illustrates a partition isolation joint detail used on a 16-story government building in Wellington, New Zealand.

Partition Analysis and Design Checklist

1. Criteria

Other (non-seismic) roles: fire, acoustic

Will partition support attached mass?

Code: Code minimum? (General purpose lateral load, common to all codes)

Damage control, safety, or functional criteria in excess of code minimum?

Design Load: Selection of appropriate factors (UBC assumed here): Z, I, C_p, W_p.

Imposed deformation: Structure's stiffness will be relevant—nonstructural protection a criterion for selection of structural system?
Feasibility of isolation detail approach.

2. Analysis

Design Load: $F_p = Z\,I\,C_p\,W_p$; F_p = 30% of partition's weight, plus weight of attached cabinetry or other items, for ordinary occupancy, above ground level, zone 4 case; design force applied in most critical direction (typically out-of-plane).

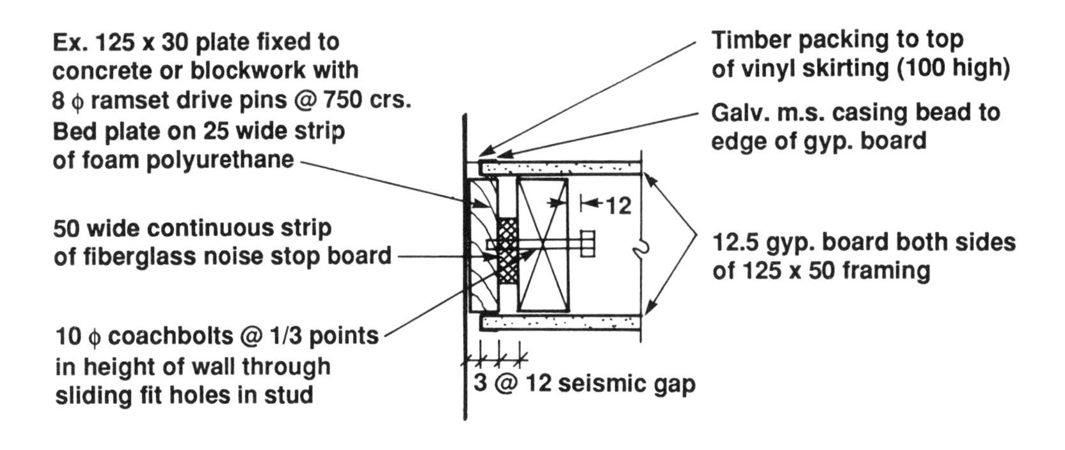

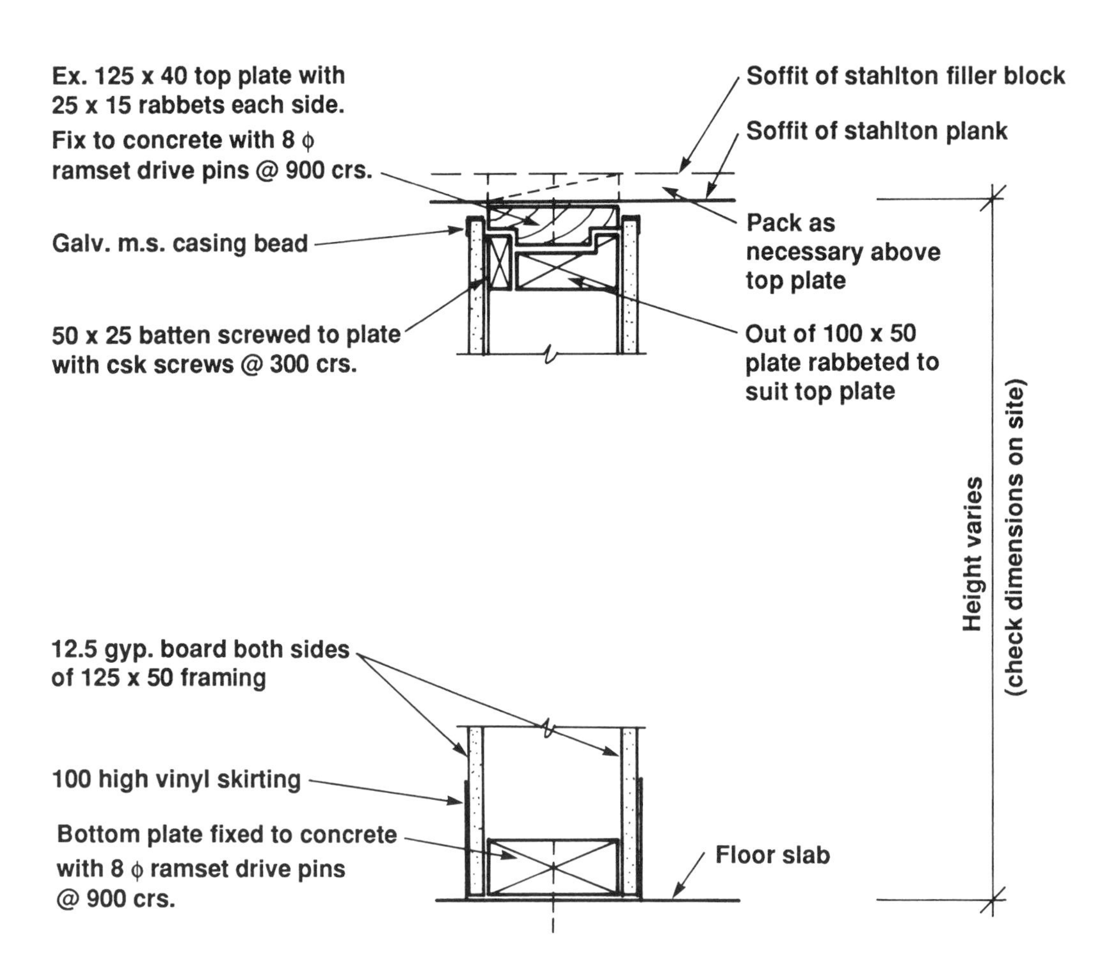

Figure 7.3.1 Seismic Isolation Joint Detail for Partitions [Arnold, Hopkins and Elsessor, 1987]

Other (such as 5 psf min.) loads.

Consider desirability of designing partitions strong enough to provide anchorage for future file cabinets or other typical nonstructural items that may be moved into the building, even if not shown on drawings.

Imposed deformation: Requires at least initial structural calculations to estimate drift.

Story drift used to define amount of isolation or sliding to accommodate, or to estimate damage if designed fixed top and bottom.

3. Design

Likely type for non-seismic reasons?

Fire resistance (1-hr. wall, 2-hr. wall, etc.); may influence stud wall/nonstructural partition versus masonry or concrete/structural wall decision; also may impose fire requirements on any movement isolation details.
Combustibility (wood versus metal studs).
Acoustic (mass of masonry or sealing of joints versus sliding joint details.)
Full-height versus ceiling-height
Ease of running ducts and other above-suspended ceiling components over partitions conflicts with aim of stabilizing partition with connection to underside of structure above; diagonal bracing of tops of partitions may be necessary.
Heavyweight partitions (concrete block, for example) require analysis of interaction effects with structure; reinforced masonry presumed for any seismic area.
Lightweight partitions (drywall-stud) may be designed as structural shear walls in small wood or metal frame buildings; isolation joint feasible but adds to cost.
Selection of stronger studs, top and bottom details, addition of blocking at convenient height to receive anchorage connections for shelving or cabinetry, should be considered for partitions which may be used as lateral support for file cabinets and other typical contents of significant mass.

7.3.2 Ceilings and Light Fixtures

Current Practice and Problem Areas

Suspended ceilings that need the most attention are tee-grid ceilings with lay-in acoustical tiles. Splayed-wire bracing (as specified in UBC Standard 47-18) is often not provided in suspended ceiling systems. Thus, typical suspended ceiling assemblies in most low-rise buildings are suspended from the structural floor system above with 12-gauge vertical suspension wires provided at intersections of all ceiling main runners and cross-tees. In areas where the seismic regulations of the UBC are enforced, bracing of the ceiling and adding two safety hanger wires to diagonally opposite corners of each lay-in fluorescent fixture is increasingly the rule.

Suspended tee-grid ceilings with lay-in tiles, covering large areas, are very prone to damage by motions likely to be experienced by these ceiling systems in low-rise buildings. Typically, suspended ceilings are likely to be damaged at the ceiling perimeter, as the entire ceiling including any fluorescent light-fixtures responds and swings like a pendulum under earthquake motions. Further damage results from hammering of ceiling framing into surrounding partitions, sprinkler heads, or other structural components.

Analysis and Design

While the designer is generally allowed by a code to design to meet performance or loading criteria, in practice most ceilings which have been installed with seismic protection features have been built according to standard, pre-approved details. In the case of the UBC, the UBC Standards (no. 47-18)[Int. Conf. of Building Officials, 1985] define the requirements for ceilings and ceiling light fixtures for both approaches, with Figure 7.3.2 showing an illustration of the basic features of the prescriptive approach. The performance and safety of these suspended ceilings can be greatly improved by addition of diagonal splayed-wire bracings every 12'-0" oc. as required by the UBC Standard 47-18, which now also requires vertical pipe-struts at the point of splayed-wire bracing, to control uplift of the ceiling plane, and also the addition of vertical suspension wires at 8 inches maximum from the unattached edges of the ceiling perimeter. Figure 7.3.2 shows details of suspended ceiling bracing specified by UBC Standard 47-18. The essence of the technique used to restrain the common two-by-four foot lay-in fluorescent light fixture is to provide redundant vertical support, but not to provide bracing per se. A 12-gauge hanger wire (which may be slack, thus serving only as a support wire in case the ceiling distorts and the fixture begins to fall) is attached to each of two diagonally opposite corners of the fixture and to the structure above. This is an example of a detail that is very inexpensive in new construction, but which is more difficult as a retrofit simply because accessing the above-ceiling space after the building is occupied is disruptive.

7.3.3 Stairs

Current Practice and Problem Areas

The imposed inter-story deformation, rather than inertial force developed within the nonstructural item, is the usual problem with stairs. Where the stairway is sufficiently stiff, and the building will drift significantly, and the stairway is rigidly attached to the building, a set of incompatible factors are present, unless the structural designer intends for the stairway to act as a strut in resisting lateral loads. Sliding "gangplank" details have been used where the aim is to prevent the stairway from resisting lateral loads.

The Olive View Hospital case in the 1971 San Fernando earthquake, where three of four exterior stair towers toppled over completely like felled trees and the fourth was left leaning as severely as the famous campanile in Pisa, is an example where the intent was to separate the stairways and their enclosing walls from the rest of the structure. The stair towers were not adequate to resist the earthquake forces as independent structures. Collision, or pounding between the main structure and its stair towers, perhaps also occurred. Such "seismic slap" is a problem to be overcome with sufficient clearances if the separation joint solution is selected. The basic analysis procedure is to estimate the drift of both the main structure and stair tower, assume both deflect toward each other at the same instant, and provide a greater separation distance. In calculating separation distances, note that a correction factor is specified in the code (section 2312 K in the 1988 UBC) to convert elastic-based code-level deflections to more realistic inelastic range values.

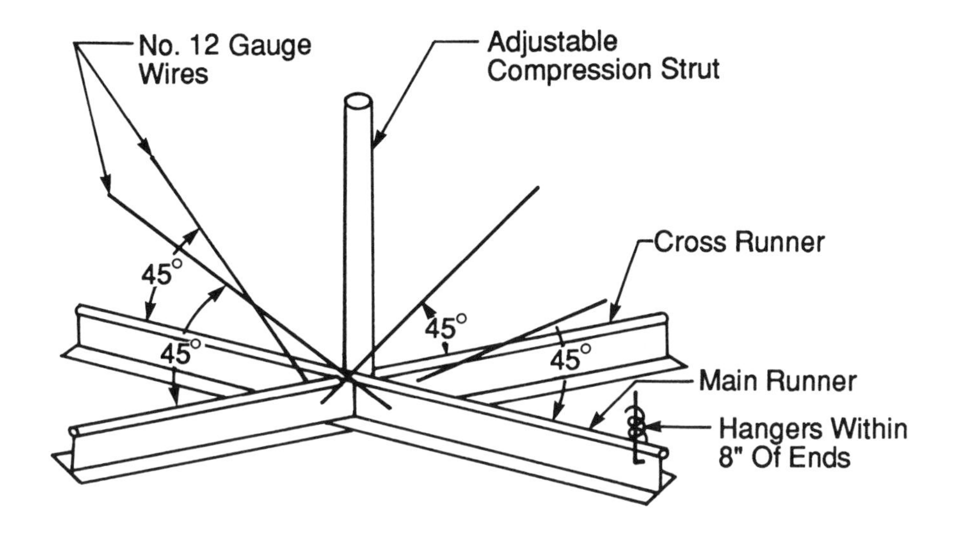

4-Way Diagonal Bracing Every 12 ft. On Main Runners, And Within 4 ft. Of Walls

Use 2-12GA. Fixture Support Wires At Diagonal Corners Of Each Fixture. For Fixture Weight Of 55 lbs. Wires May Be Slack. Use 2-12GA. Fixture Support Wires At All Corners Of Each Fixture Exceeding 55 lbs. In Weight. Wires To Be Taut. These Wires May Not Be Used To Support Any Other Item.

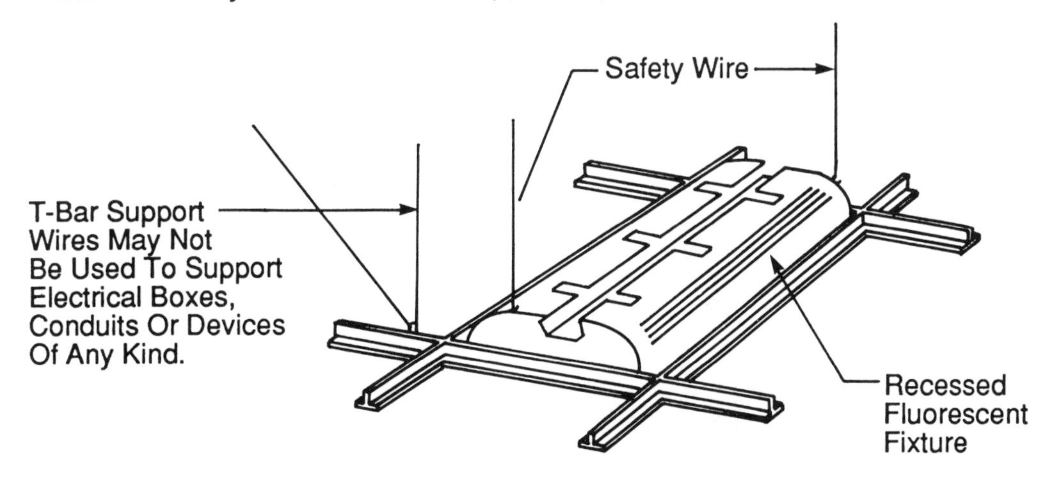

Figure 7.3.2 Ceiling and ceiling mounted light fixture restraint (a)(top) Basic features of suspended ceiling bracing [UBC 47-18, ICBO, 1985] (b)(bottom) Basic features of lay-in light fixture protection

Damage to stairways has not been as common in earthquakes as damage to ceilings or overturning of contents, but because of the importance of exit routes, the topic is of concern.

Analysis and Design

Figure 7.3.3 illustrates a design solution in use in New Zealand even in relatively stiff buildings, because of concern that strut-action in a rigidly connected stairway would cause damage and hinder egress.

For steel stairways, an easily accommodated detail is to connect one end of the stair to its support via slotted holes. Some designers have found it helpful to specify that both surfaces of the steel bearing plates be painted so that well-intentioned tightening of bolts (bolts which in original design documents were supposed to be only finger tight) will not defeat the purpose of the sliding joint.

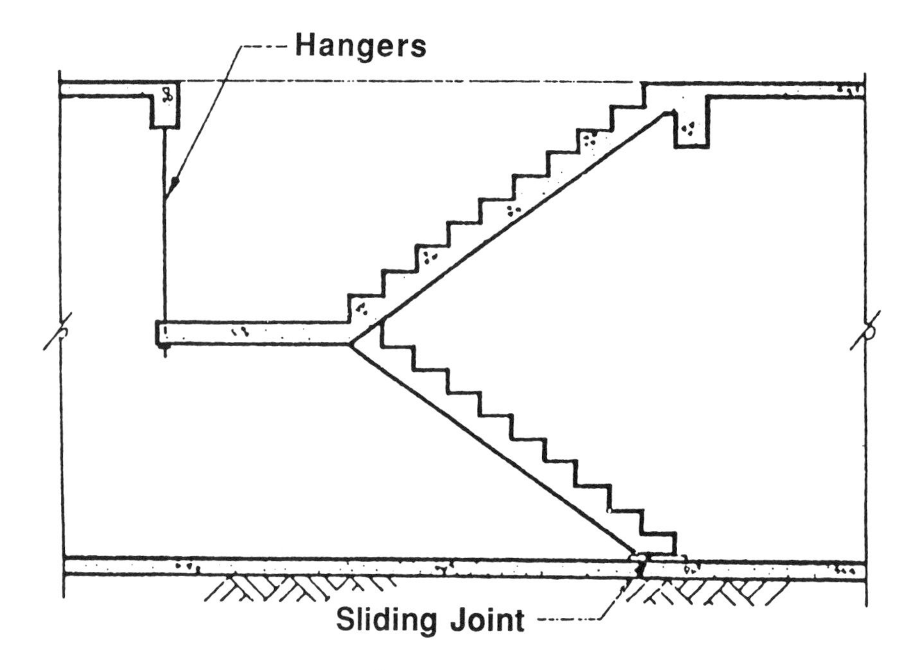

Figure 7.3.3 Typical concrete stair separation detail commonly used in a concrete walled building [Arnold, Hopkins and Elsessor, 1987]

7.3.4 Raised Floors

Current Practice and Problem Areas

As is common in the earthquake engineering field, the rapid development of new construction practices, including raised computer floors, has outpaced the evolution of appropriate seismic design practices and building code provisions. Of all the components in a building, the raised floor was formerly of least concern from a nonstructural earthquake protection viewpoint, while partitions, ceilings, windows, contents, etc., were problem areas. The raised floor, unless specifically designed to resist earthquake motions, is now a common high-vulnerability item and one which is difficult to retrofit if the original installation is not adequate.

While all raised floors are designed to carry large vertical loads because of the weight of mainframe computer equipment, associated air-conditioning units, or telecommunications equipment, the vertical capacity of this post and beam system does not inherently provide lateral restraint.

Because raised floor systems almost always support very essential equipment—the data processing equipment used to run key systems in a corporation or government agency, the communications equipment in a large fire or police dispatch or communications room, etc. — the vulnerability of such floors becomes more significant.

Analysis and Design

For new construction, the first approach is to consider selection of a pre-approved floor system which has been tested to meet the relevant earthquake requirements. In the case of the UBC, major brand access floor systems can now be specified with reference to manufacturer's details that have International Conference of Building Officials approval. In analyzing and designing customized solutions, note that the UBC's loading criteria for access floor systems call for the common 0.75 C_p value to be used in combination with a W_p that includes not only the dead load of the floor but also 25% of the live load and a 10 psf partition load.

A recent compendium of information on the seismic protection of computer facilities (FIMSC, 1987) contains more detailed information on this topic. Two means of bracing a raised floor are common, as shown in Figure 7.3.4. Cross bracing is a direct engineering solution to the problem, making a braced frame out of the post and beam access floor support system; however, the diagonal can get in the way of the extensive amount of cabling that runs under the floor.

Designing the stanchions or posts to act as base anchored vertical cantilevers is another possibility. In retrofitting raised floors, the feasibility of the vertical cantilever approach sometimes relates to whether the base plates for the stanchions have holes in them to make installation of drill-in-anchor bolts easy, to whether the floor is a one-foot high or two-foot high system and thus the magnitude of the bending moment, and to the strength of the steel pipe posts and their welded-on base plates.

In the 1989 Loma Prieta earthquake, some epoxy-glued pedestals performed well, though the reliability of adhesives as compared to anchor bolts is questionable.

Note that the restraint of computer equipment on the raised access floor is also a difficult problem. Two general approaches—anchor the equipment, or let it roll—are used, depending upon the particular situation and also according to the designer's preference and judgement since there is not yet a true consensus on this issue in the earthquake engineering field. Where high seismic loads are expected, at least some anchorage is generally found to be necessary for more slender equipment or items that cannot be allowed to move relative to the floor, for example tall telecommunications racks in the former case or heavy air conditioning units connected to water piping in the latter. Computer equipment protection is discussed later under Contents and Equipment.

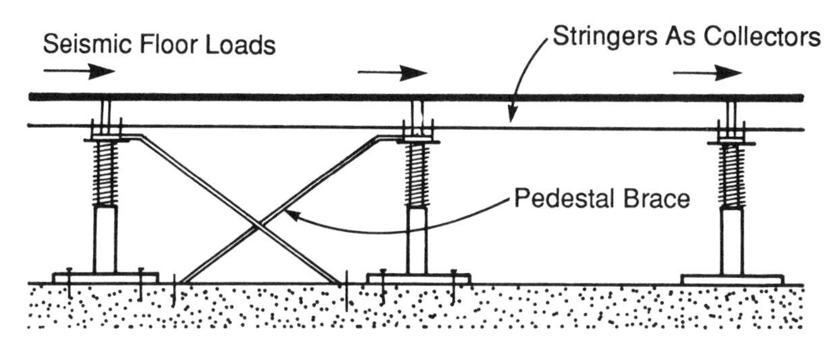

B. Braced Pedestal System

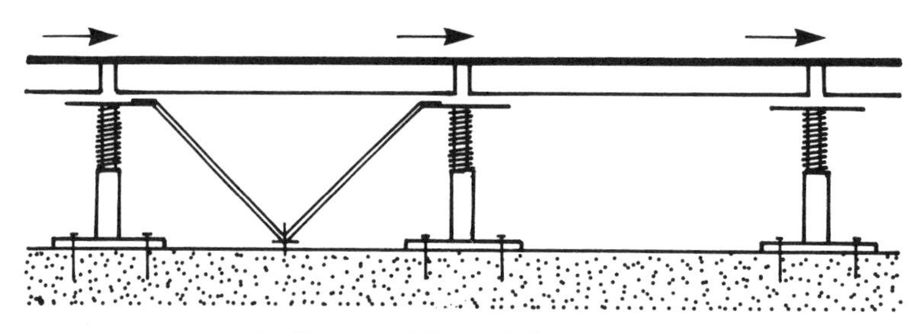

C. Braced Panel System

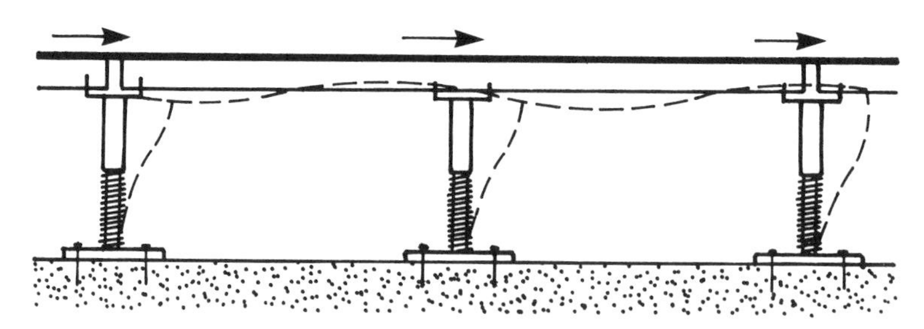

D. Pedestal-Stringer Frame

Figure 7.3.4 Basic raised floor bracing schemes [FIMSC, 1987]

7.4 EXTERIOR COMPONENTS

7.4.1 Cladding and Veneer

Current Practice and Problem Areas

There is considerable variety in the use of material and attachment systems for the exterior facade for low-rise buildings throughout the U.S. The selection of a particular system depends largely upon economy and architectural expression. Other factors include the availability of certain materials, regional construction expertise, and past experiences of designers with different cladding types. At present, for example, large precast concrete spandrel panels are popular in low-rise construction throughout the country, while GFRC (glass fiber reinforced concrete) is widely used on the West coast and brick is still common in much of the Eastern U.S. Attachment systems also vary depending on locale and the emphasis given to isolation of exterior panels in local building codes for the effects of interstory drift. In the West, ductile rod connections are widely used of late, while in the Eastern U.S. slotted connection angles are still used for the attachment of heavy facade components.

Exterior cladding systems for low-rise systems may be generally classified as either lightweight or heavyweight. Glass, metal, stucco and GFRC are examples of less massive systems for which relatively lightweight attachments are required. Masonry, stone, precast concrete, marble, and granite facades, however, are much more massive and require special handling, erection and attachments in order for these systems to work properly with the structure. In some cases, placement of heavy claddings must be given special consideration so that deflections of the underlying supporting structural framework can be accounted for during the erection sequence. At the present time, brick masonry, marble, granite and stone veneers are popular materials for heavy weight exterior cladding, although materials such as granite and marble are often regarded as too expensive for most low-rise building construction. Because of the penalty paid for mass in earthquakes, the heavy cladding components are the ones of most significance in terms of structural-nonstructural interaction. These heavy cladding systems can also interact with the structure by stiffening it. Heavy cladding is of less significance in wind design.

Exterior systems comprising the building enclosure/envelope, based primarily on architectural, energy, fire, weather-proofing and economic considerations, can be divided into two basic categories:

Light Exterior Facade/Cladding Systems

- Glass curtain-walls
- Steel-stud framed exterior enclosure with stucco finish
- GFRC panel systems

Light systems are generally more vulnerable to wind forces than earthquakes and must be adequately restrained for inward and outward pressures.

Heavy Exterior Facade/Cladding Systems

Heavy systems are typically more vulnerable to earthquakes and less vulnerable to wind.

Two basic configurations are found in low-rise buildings:

i. Spandrel Panel Systems

A significant percentage of construction of low-rise commercial office buildings seems to use this type of exterior system.

These can be further divided into the following categories:

- Precast concrete spandrel wall systems.
- Steel-stud framed systems with masonry/tile veneer or marble facing.

ii. Window-Wall Panel Systems

A significant percentage of low-rise commercial/office/institutional buildings are found to use this type of exterior system in a variety of configurations.

These can be further divided into the following categories:

- Story-high precast concrete panel system.
- Steel-stud framed systems with masonry/tile veneer or stone facing.

The slotted inserts which have been used for attachment of heavy precast cladding elements over the past 20 years or so have given way to the use of weld plate inserts combined with the use of slotted clip angles. In seismically inactive (and even moderately active) areas of the U.S., slotted systems are not used. Instead, the plate inserts in the cladding component and in the supporting floor slabs are directly joined by a plate welded to each insert. Panel sizes have become larger and larger in recent times, limited only by the ability of manufacture, transport, erect and economically attach these much larger components. Weld plate connections are more economical and contribute to faster erection times for heavyweight systems.

In low-rise buildings, specific attention must be given to the over-all behavior of the building. Much low-rise construction is comprised of an assemblage of diverse elements of unknown stiffness connected together by connections of unknown strength and ductility. The designer must ensure that the structure acts as an integral unit in resisting lateral forces, and that horizontal torsion effects are addressed in the design. If the structure platform is large or unusual in shape, the diaphragm flexibility may have to be considered in estimating the inter-story deformations.

Heavy facade components may also play a role in the lateral response of the structure, within the capacity of the cladding components and connections [Goodno et al, 1983, 1984, 1986a and b, 1987, 1988, 1989]. If this capacity is exceeded, the resulting damage and life safety hazard may be significant. This may also be the case for unreinforced masonry infill walls, which play a significant role in the lateral response of low-rise buildings. When the capacity of cladding or infill walls is exceeded, the dynamic characteristics of the structure can change abruptly, possibly increasing the effects of torsion and creating a more flexible structure which is then more sensitive to the ground motion. The consequences of failure of these so called nonstructural components should be considered in the original design unless they are isolated to prevent their interaction with the primary structure throughout its full range of motion. Code provisions for exterior cladding may be briefly outlined as follows:

UBC -	Provisions related to drift and design force levels for cladding & bodies of connectors.
ATC-3/NEHRP -	Provisions for ductility of connectors and specific requirements for nonstructural elements.

Analysis and Design

The situation in seismically active areas such as California has recently been described by Arnold, et. al. (1987). Ductile rod (push-pull) connections used for heavy claddings are economical and relatively inexpensive to install, and require a

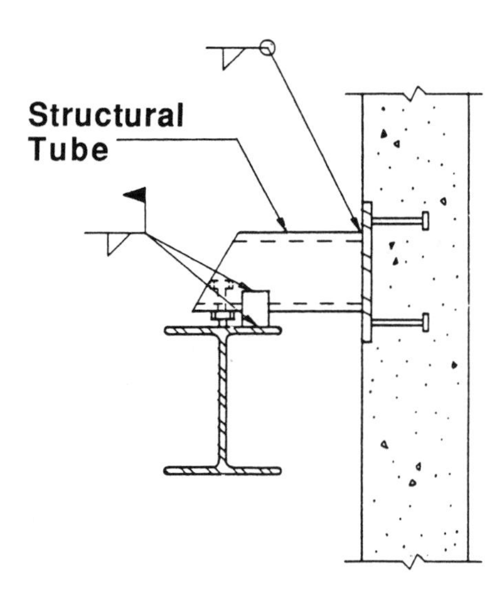

Figure 7.4.1 PCI recommended detail for eccentric bearing connection [PCI, 1987]

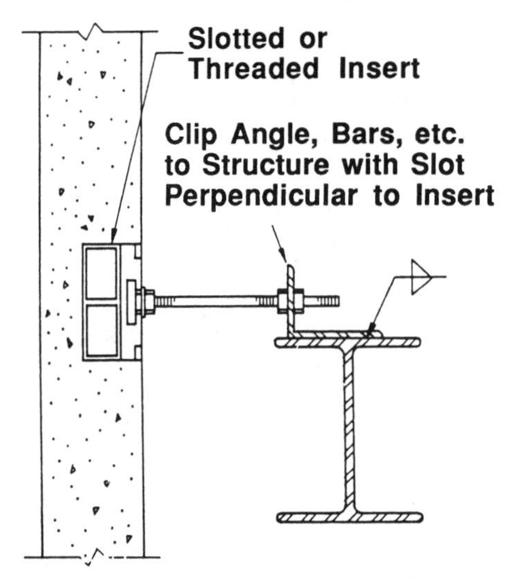

Figure 7.4.2 PCI recommended detail for bolted tie-back connection [PCI, 1987]

modest amount of construction inspection and maintenance. However, most low-rise building construction employs less massive facades comprising of a lightweight supporting framework to which metal panels or stucco is attached.

Typical design details of exterior components and connections examined by other researchers are presented elsewhere [Goodno and Palsson, 1986; Wang, 1987; and Rihal, 1988a and b].

Design details for cladding connections applicable to low-rise buildings have been recently proposed by the Architectural Precast Concrete Connection Details Committee of the Prestressed Concrete Institute (PCI). Two typical PCI recommended conceptual details for cladding connections are shown in Figures 7.4.1 and 7.4.2 [PCI,

1987]. An eccentric bearing type cladding connection is shown in Figure 7.4.1. A flexible (tieback) type cladding connection is shown in Figure 7.4.2. It should be noted that these PCI cladding details are conceptual details only and may have to be appropriately modified as needed in the seismic design process. These PCI connection designs illustrate common practice in attaching heavy cladding to buildings.

7.4.2 Windows and Glazing

Current Practice and Problem Areas

Glazing failures are related to the kind of putty or other edge material, amount of clearances, size of pane, and drift of structure.

Window panes are usually installed with glazing compounds or synthetic rubber gaskets. Recently advanced glazing systems, such as the silicone flush glazing system or suspended glazing system, are also utilized. Breakage of window panes occurs even in moderate earthquakes which causes no damage to the structural frame, endangering the building occupants and pedestrians around and below. As regards the safety of people, wired glass, tempered glass, and laminated glass are considered safe because the debris seldom scatters when breakage occurs. The damage survey for the Miyagi-Ken-Oki Earthquake in February 1978 reveals that most of the glass broken was mounted in hard-putty-glazed fixed window sashes as shown in Figure 7.4.3.

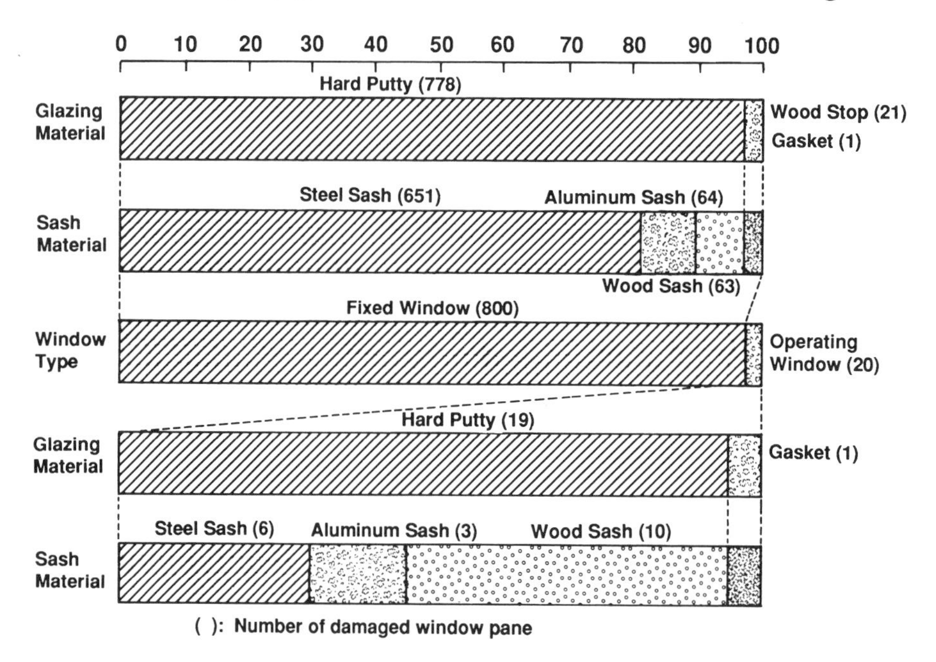

Figure 7.4.3 Damage to glass panes in the Miyagi-Ken-Oki earthquake, 1978 [Ito]

Based on previous studies, the hard-putty-glazed fixed windows are likely to be damaged at approximately a story drift index of 1/1000. The horizontal scatter distance of breakage debris was approximately half of the falling height (Figure 7.4.4). The hard putty was commonly applied till mid-1960's in Japan and still remains in many buildings. Notification No. 109, Ministry of Construction, which was revised in

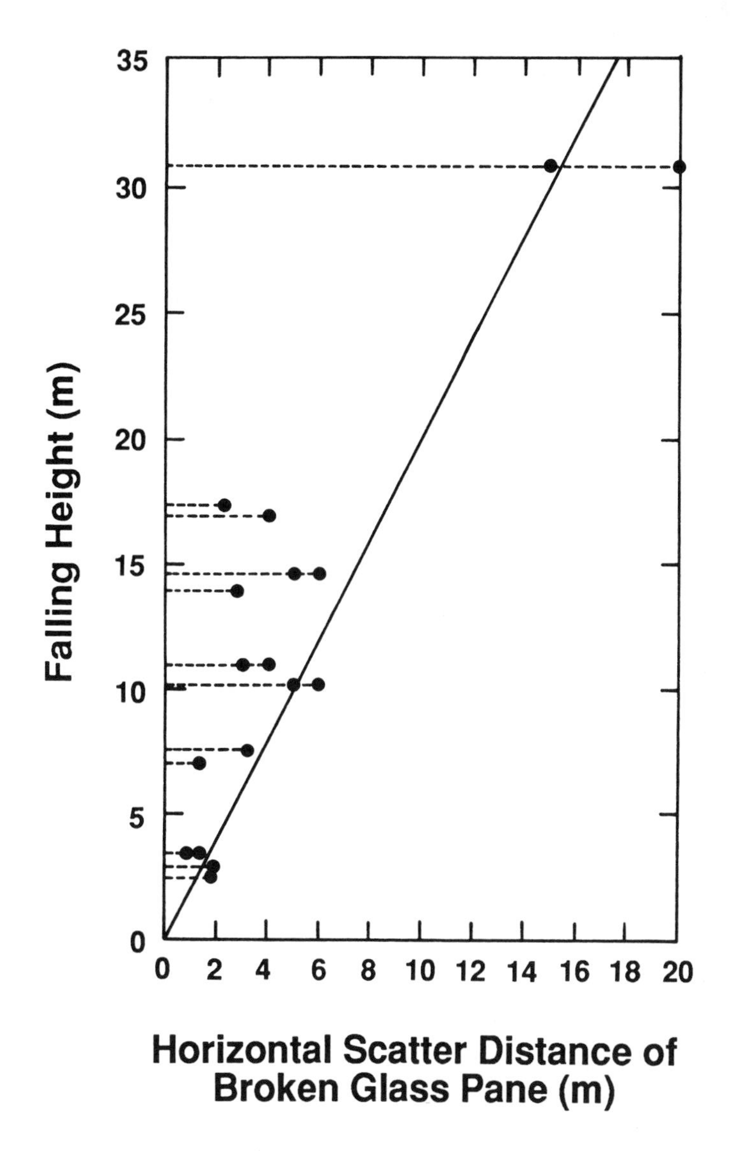

Figure 7.4.4 Relationship between horizontal scatter distance and falling height [Ito]

1978 on the basis of the lessons from the damage during two Miyagi-Ken-Oki earthquakes in 1978, prohibits the use of hard-putty-glazed fixed window in three-or-more story newly constructed buildings except for wired glass and laminated glass. Elastic glazing compounds and synthetic rubber gaskets are now generally used. Window panes surrounded by high stiffness components, such as precast concrete panels, have less problems because only slight deformation is transmitted to the window pane. A summary of reported glass damage in the Mexico earthquake of 1985 is found in Evans et al (1988). Two of the strongest indicators of damage were: vertical rather than horizontal pane orientation, and large rather than small panes.

Analysis and Design

An early study on the seismic safety of window pane was conducted by Bouwkamp (1961). Subsequent experimental studies in Japan have slightly modified the theoretical expression proposed by Bouwkamp, which provides the allowable in-plane deflection or drift of window sash. Figure 7.4.5 illustrating the movement of the glass pane within the sash frame shows that the final movement could be split into two stages:

a. horizontal movement of the glass until the glass pane touches vertical sash members
b. rotation of the glass until the glass pane hits the corners of the sash frame

As the glass itself has very high stiffness, slight deformation of the glass produces severe stress and failure. Assuming the failure occurs when the glass panel hits the diagonal corners of sash frame, the allowable in-plane drift, can be derived from the following expression:

$$\Delta = F \cdot 2\,c(1 + h/w\ d/c)$$

where c,d: clearance between glass edge and sash
w: nominal width of the glass pane
h: nominal height of the glass pane
F: reduction factor depending upon glazing compound ($0 < F < 1$).
for example, $F = 0$ for completely hardened putty
$F = 1$ for low modulus glazing compound

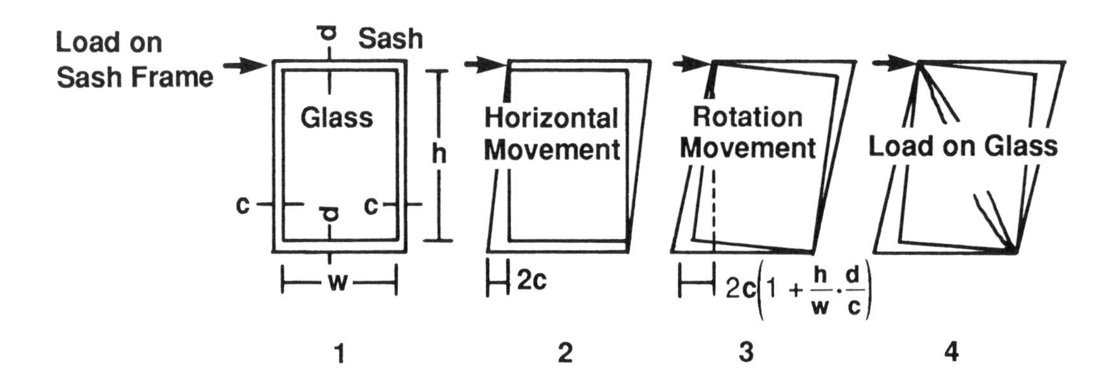

Figure 7.4.5 Movement of the glass panel within the sash frame for soft-putty-glazed window panel [Ito]

In practice, Δ/h for completely hardened putty is approximately 1/1000. The Architectural Institute of Japan [AIJ, 1985] recommends that the following formula omitting the rotation effect should be applied to 8 mm or thicker heavy glass preventing its rotation movement:

$$\Delta = F \cdot 2c$$

As for soft-putty-glazed fixed window, the formula can be applied directly. Operating windows allow additional deflection or drift.

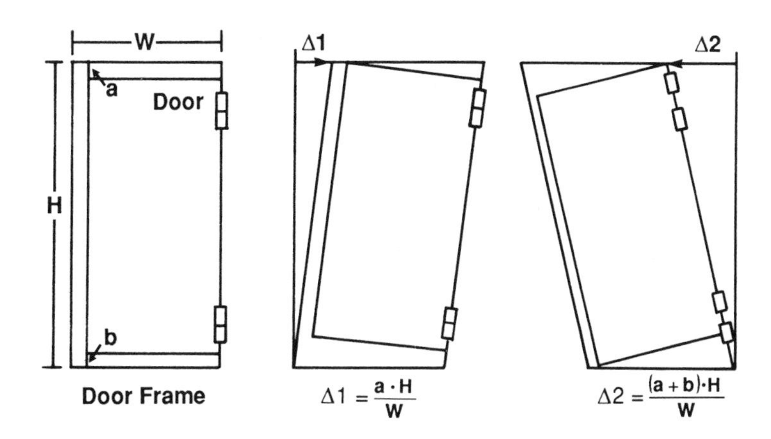

Figure 7.4.6 Door frame deflection and clearance [Ito]

7.4.3 Doors

Current Practice and Problem Areas

The problem related to doors is that jamming or damage to doors can make it impossible to evacuate a building. During the Miyagi-Ken-Oki Earthquake in June 1978, the damage of a nonstructural wall surrounding the entrance doors in a steel encased reinforced concrete apartment building caused the jamming of steel doors, and many residents could not escape from their apartments.

Analysis and Design (Examples)

The causes of door jamming are considered as follows:

a. Collision of the edge of door body against door frame.
b. Latch malfunction because of the collision of latch bolt or dead bolt against lip strike.
c. Enlargement of door swing resistance because of misalignment of the center of the hinges.

In case (a), the model shown in Figure 7.4.6 could be assumed; the door frame deflects like a parallelogram, the door body rotates, and finally the edge of the door body collides with the door frame. The following equation giving the relation between the allowable deflection angle of the door frame, Δ_1 , and the dimension of the door can be calculated from this geometrical relation.

$$\Delta_1 = a \sim H/W$$

a: upper door clearance
h: door height
W: door width

Test results have agreed with the above equation, making this a reliable prediction of the allowable deflection of the angle of door [Ito and Kazuo, 1986]. Recently, seismically protected doors are used in Japan whose features, compared with ordinary-type doors, are as follows;

a. Greater door clearance;
b. The hole of the lip strike is larger and allows more deformation;
c. Only two hinges, with an added tongue preventing thermal deformation for fire proofing.

7.5 CONTENTS AND EQUIPMENT

7.5.1 Current Practice and Problem Areas

Contents and Equipment

Contents and equipment in low-rise buildings do not present any significant unique considerations, as compared to the issues present in taller buildings. The height of the structure in which contents and equipment are located is not usually an important seismic variable in terms of design procedures. While the motion at upper floors of tall buildings may be greater, and with a lower frequency and greater displacement, and while many low-rise buildings employ shear walls rather than moment frames, there is no categorical difference between low-rise and high-rise buildings. While the taller building will generally show greater flexibility and have a longer period, even this generalization is subject to some qualification. Almost all tall buildings have relatively rigid concrete diaphragms, whereas many low-rise buildings have flexible diaphragms which may have relatively long periods of vibration. Thus it is possible for the roof-mounted equipment on a tilt-up structure of large plan dimensions and with flexible wood or light metal roof diaphragm to be subjected to a similar frequency of motion (say 1 Hz) as the same equipment located on top of a ten story building. Except for the fact that traction elevators are not generally used in low-rise buildings, most nonstructural equipment and contents are similar in their characteristics and their installation requirements in tall and short buildings. Hydraulic elevators are somewhat less susceptible to earthquake damage than cable-traction systems, so low-rise buildings have an advantage in that respect.

Figures 7.5.1-3 show representative damage to contents & equipment observed during previous earthquakes (e.g., Anchorage, Alaska (1964); San Fernando, California (1971); and Coalinga, California (1983)).

Contents

The predominant current practice, even in areas where seismic codes are used in the United States, can be summarized very simply: no attempt is made to protect contents from earthquake damage. While there are significant exceptions to this generalization—exceptions that show that seismic protection is technically feasible and relatively inexpensive—the general rule is that only some of the built-in nonstructural features such as ceilings or fire sprinkler systems are seismically designed. Most of the contents of buildings in areas where earthquakes are relatively common and seismic codes are in effect—contents such as file cabinets, laboratory chemicals, and computers—are identical in manufacture and in their installation details to those located in buildings in non-seismic areas. To a great degree, this is because the building code's scope is limited to the original construction number, or significant remodels of the permanent features of the building, and the building code enforcement process generally ends with the issuance of the certificate of occupancy. Following this building code process, the occupants move into the building and bring with them the contents related to their particular type of occupancy. To date, neither Federal OSHA nor state workplace health and safety regulations contain any significant provisions relating to earthquakes. This post-occupancy process of safety inspections and regulations using non-building code regulations could take up the process of providing nonstructural earthquake protection where the building code leaves off, but this is not done at present. In California, the state's five-year seismic

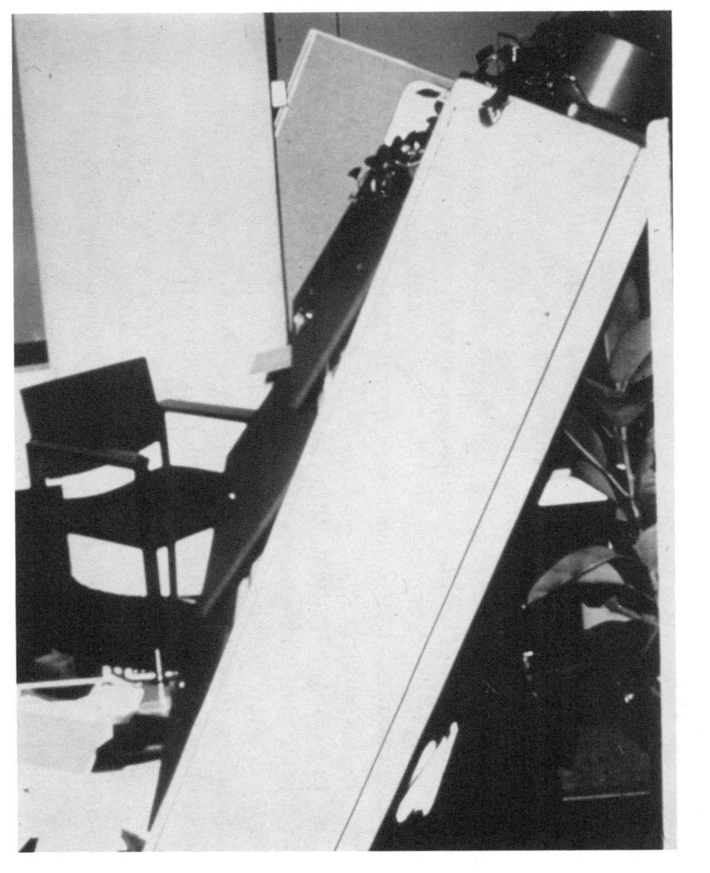

Figure 7.5.1 Overturned Laterally-opening File Cabinet, Morgan Hill earthquake, 1984 (Photo: W. Van Odsol)

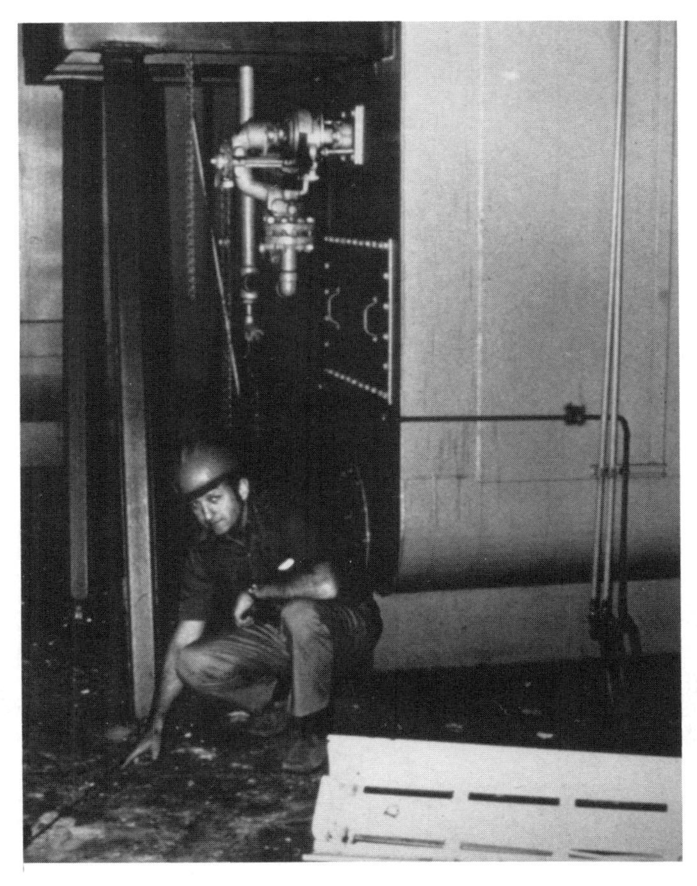

Figure 7.5.2 Boiler Slid Three Feet in the Olive View Hospital during the San Fernando Earthquake, CA, 1972 (photo: J.M. Ayres)

Figure 7.5.3 Failure of a Spring Mount for Mechanical Equipment in the Livermore Earthquake, CA, 1980 (Photo: W. Holmes)

safety plan identifies this post-occupancy nonstructural inspection process as a desirable objective [California Seismic Safety Commission, 1986].

In recent years, due to voluntary efforts beyond the scope of building codes, some effective means of protecting contents from earthquake damage have been devised.

As one example, museums have recently become aware of the great seismic vulnerability to which much of their collections are exposed. While paintings mounted in typical ways have generally performed well in earthquakes, the three-dimensional objects in museums are highly susceptible to damage from overturning or falling. These vulnerable items include statues, busts, figurines, vases and bowls, and antique furnishings. Recent conferences sponsored by the J. Paul Getty Museum [Getty, 1984], Cornell University [Jones, ed., 1984], the National Association of Museum Exhibition [NAME, 1987], and the Society of California Archivists [SCA, 1987] are signs of this trend. Contributions to the solution of typical museum contents problems written by those in the earthquake field include those of Reitherman (1984) and Schiff (1987), but the greatest source of innovation and development of technical solutions is the museums themselves, in particular a half dozen of the larger California museums whose curators, preparators, or registrars have taken an active interest in the subject and who freely exchange information with other museums. Only one portion of the solution involves earthquake engineering, or the adequate consideration of strength, frequency, deflection, and other seismic issues. Perhaps a larger portion of the solution involves many details of the aesthetics of restraining art objects, the extreme constraints presented by concerns over long-term chemical effects of various materials, and other issues with which museum staffs are already familiar.

Two new products or devices have been developed to deal with unique museum seismic problems. One is a commercial product by Tokico, a Japanese manufacturer of vehicular and seismic shock absorbers, which provides base isolation for individual statues (Tokico, n.d.). A custom-made device of similar purpose but of different mechanical design, has been produced by the Getty Museum (Stahl, 1987). One of the strongest arguments put forth in favor of base isolation of entire buildings is that nonstructural damage will be greatly reduced [Tarics, 1984]. Occupancies with sensitive nonstructural features are thus the most likely candidates for base isolation from this standpoint. This section needs a discussion related as to what seismic regions need an extensive study of possible damage to contents.

Equipment

In computer rooms, some organizations which operate computer installations have recognized the critical vulnerability computer functions would face in an earthquake. Not only are mainframe computers themselves typically installed without restraint, but the raised floors on which they sit are often designed only for vertical loads [Steinbrugge, 1982]. The Finance, Insurance, and Monetary Services Committee of the California Earthquake Task Force has been working to educate the operators of computer facilities as to the earthquake vulnerabilities faced by essential computer installations and to develop standardized approaches to seismic design approaches [FIMSC, 1983 and 1987]. This is a case where the critical function of the equipment has caused significant concern, even though building code regulations do not require retroactive floor bracing or equipment restraints being employed.

The example of computer room equipment, representing occupants' equipment in general, is similar to the example of museum objects mentioned above. In both cases, the unique nature of the objects to be protected must be recognized at the outset. In the case of museum objects, the standard solution of drilling holes in the bottom of the object to bolt it down obviously will not generally be appropriate, and this is also often true with computers. In some cases, with both art objects and computers, the objects are internally sensitive to vibrations, and even if the anchorages do not fail, damage can still result. In both cases, earthquake engineering expertise is not sufficient by itself to solve the problem, and the manufacturers or users of the contents or equipment must be involved in devising appropriate solutions.

7.5.2. Analysis and Design

Contents

Good and bad practice can be illustrated with regard to some common occupancies where nonstructural items are especially important.

In laboratories, compressed gas cylinders are ubiquitous. The most common means of restraint, found in non-seismic as well as seismic areas and intended simply to keep these tall, slender steel bottles from toppling when accidently bumped, is to put one loose chain around each cylinder or group of cylinders. The chain is often very weakly anchored to the wall, sometimes only with eyescrews inserted into plaster or gypsum board, and lightweight chain is often used. Even when the strength of the chain and its anchorage points is adequate, the chain is often loose enough for the cylinder to be able to slide and slip out of its grasp in an earthquake. Better practice is to use two chains, one near the top and one near the bottom, or two restraint bars. Also effective is a tightly cinched nylon belt, although a future edition of the Uniform Fire Code now being discussed may prohibit non-fire-resistive restraints for cylinders. This is an example where both the strength and the stiffness of the restraint is important. This is also an illustration of the common case with contents where occupants must properly attach and re-attach restraints, making nonstructural earthquake protection a training as well as an engineering task.

Chemicals in bottles on lab shelves are often unrestrained, even in seismic areas. Hazardous materials spills in earthquakes have occurred in the past [Reitherman, 1982] and in the next large earthquake to strike a heavily industrialized and technologically developed urban area, hazardous materials incidents could conceivably outweigh the other more direct casualties caused by the earthquake. Adjacent chemicals can produce toxic or flammable reactions if spilled and combined, so even small quantities can pose serious seismic hazards [Stratta, 1987. McGavin, 1981, 1987] and this has pointed out the many types of hospital equipment and containers which pose hazardous materials threats as well as functional loss problems. Some of the solutions suggested in a report published by the Veterans Administration on seismic protection of equipment and contents in VA hospitals over a decade ago [Veterans Administration, 1976] are still quite valid and workable, and merely require implementation. Since many typical kinds of contents in buildings are not seismically protected in any way, the examples of bad practice are quite apparent—they are the cases where the laboratory chemicals are completely unrestrained, the shelving is unattached to the wall, etc.

Increasingly common among larger companies in California are voluntary efforts to deal with these problems. A few fire departments have also begun to suggest or require specific restraints, such as plastic edges or lips along the fronts of chemical shelving or use of two chains rather than one chain on cylinders.

Equipment

Most of the solutions to the problem of equipment vulnerability involve anchorages. Most ordinary building equipment is not internally sensitive to vibration and if rigidly mounted to the structure of the building will perform adequately in earthquakes. The question of the amount of force for which anchorages should be designed has yet to be completely resolved, though current code requirements provide minimum legally required restraint forces. Further carefully documented observations from earthquakes as well as experimental work would be beneficial. The observations from earthquakes should do more than merely indicate that the "equipment was designed as per UBC" or for a certain force level. The actual details of the anchorages or bracing must be documented, with estimates of realistic capacities, to allow for an intelligent assessment of the observations of good performance or failure. When a piece of equipment was anchored to resist a lateral force of, say, $0.3W_p$ it may have been that ⅜-inch diameter expansion bolts were almost but not quite sufficient and so ½-inch diameter bolts were used instead. These ½-inch diameter bolts would typically have about twice the capacity of that of the next smaller typical size, the ⅜-inch diameter, which almost had enough capacity to meet the $0.3\ W_p$ criterion. The as-built design might have had anchorages close to twice as strong as required to meet the criterion, and thus if this anchorage fails in an earthquake, it would be misleading to term this a failure of a $0.3\ W_p$ lateral force design.

The UBC's seismic treatment of nonstructural elements is typical of most other seismic codes, specifying a lateral force factor to be used in the stability analysis of objects. Table 23-P is shown in Table I; its C_p values are one term in an equation which also includes an importance factor (I) and seismic zone factor (Z). For most equipment, located in zone 4 (where a significant percentage of California's population resides), an over-all lateral load factor of 30% of the weight of the object would be used as the design seismic force. Explicit consideration of vertical accelerations is not included in this formula but is a part of some other design procedures which are otherwise similar in many respects to those of the UBC, such as the VA design standards for hospitals. Because upward acceleration of an object in combination with horizontal pulses can be more significant in causing objects to overturn quickly, the vertical acceleration issue needs due consideration. In addition, the effects of velocity/acceleration ratios of the earthquake on instability of equipment also need due consideration.

The National Fire Protection Association has long published a standard governing the installation of fire sprinkler systems, including prescriptive or engineering rule of thumb details for seismic bracing [NFPA, 1983]. Standard details in conformance with California's Hospital Act have been published by trade organizations to facilitate the work of contractors in installing HVAC and piping systems [Hillman et al, 1982].

Checklist for Anchorage Analysis and Design

1. Decide criteria

Code: UBC see Chapter 23 ($Fp = Z\ I\ C_p\ W_p$)
Flexible or rigid?
Basis on which item is classified as flexible or rigid should be established. Is anchorage of item the only problem? Or will connecting pipe or other interface also require bracing or isolation to avoid relative movement problem?

Above-code criteria: life safety, functional loss, or property loss protection.

Non-seismic: Will object be moved frequently? Aesthetics? (have to hide connections?) OK to drill holes etc. in object?

2. Diagram component and loads:

Inertial load, through center of mass, most critical axis.
Inertial load, through c.g., most critical axis. Imposed deformation a problem? Interaction an issue?

3. Analyze overturning:

A matter of statics, once the dynamics of the situation are properly considered.

4. Analyze shear:

Friction is generally neglected (example assumes allowing item is not a stocky shape that can be safely allowed to slide and won't be anchored). Divide shear among number of anchors.

5. Design connections:

Check to see that connection detail meets rigid criterion if item was analyzed as rigid above.

Select bolt: combine tension due to overturning and shear due to translation; expansion bolt values should be carefully selected by using only code-approved values (manufacturer's brochures may contain other values); note assumptions made as to special or non-inspected installation; for wet locations, specify corrosion resistant model (stainless steel available for most sizes and models); for concrete slabs, note that installer should take precautions to avoid drilling into embedded electrical conduit.

For wall anchorages, verify type of wall; are studs full-height to structure above? accuracy in locating studs is important quality control issue in installation.

Design connection to object:

Ensure that object is strong enough to develop connection force; anchor to corner angles or other sturdy portion of cabinets.

Consider whether detail is installable (sufficient clearance for installer to reach behind or under item, need to use screws rather than bolts, etc.).

For sheet-metal cabinets, specify oversize washers to spread anchorage force over larger area and increase capacity of connection in cabinet wall as compared to capacity of anchor in wall or floor.

Check internal components to cabinet—will they be damaged by shaking even if cabinet remains stable?

Review criteria of number 1 above to see if all requirements met.

Review criteria of number 1 above to see if higher level of protection (higher force factor, inclusion of vertical effects) could be used without adding significantly to the expense of the detail devised in step 5.

See the example of Figure 7.5.4, which uses the Tri-Services (Army, 1982) criteria that are parallel to the 1975 version of the Structural Engineers Association of California "Blue Book." While the particular code criteria applicable to a particular case should be carefully checked, this example illustrates the basic steps involved.

7.6 CODE PROVISIONS

There are two types of codes or standards applicable to nonstructural elements. One type is the basic set regulations in a building code, such as the Uniform Building Code, which in one or more short seismic or wind sections deals with nonstructural items of various kinds. The other type of code or standard is the element-specific document that deals with only one type of element.

Chapter 23 of the 1991 Uniform Building Code contains procedures for calculating the lateral seismic forces for which selected nonstructural elements should be designed. An excerpt is shown in Table 7.3.1. Standard 47-18 in the cross-referenced

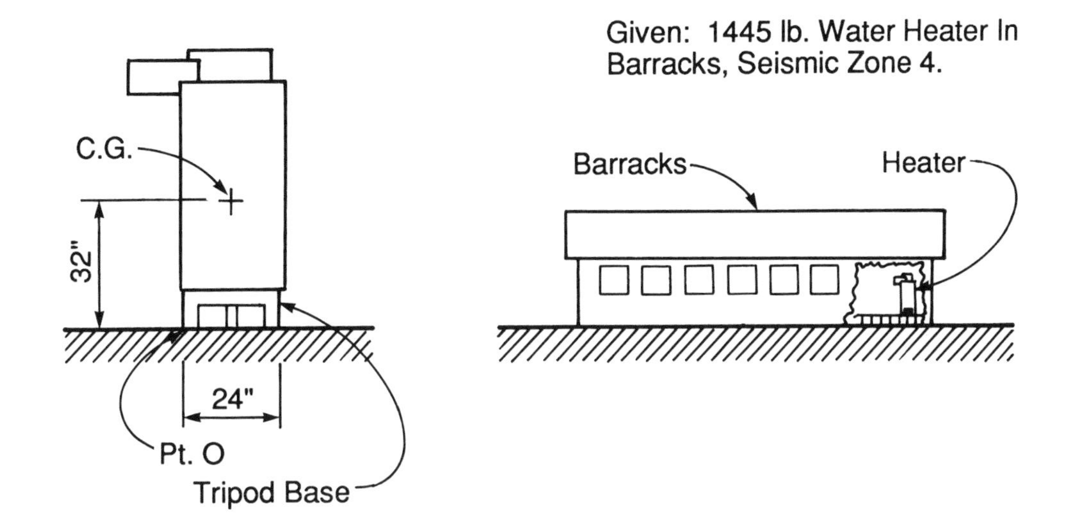

Required: Investigate The Water Heater For Seismic Loads.

Solution: Water Heater Will Be Classified As Being Equipment On The Ground And Will Be Considered To Be A Rigid Body. Since Friction Cannot Be Used To Resist Lateral Seismic Forces, The Water Heater Must Be Rigidly Attached To Its Foundation. Bolt Water Heater Legs To Floor. Refer To Para. 10-5a

$$F_p = ZI\,(2/3\,C_p)\,W_p \qquad (10\text{-}3)$$

$Z = 1.0$, $I = 1.0$, $C_p = 0.30$ (Table 3-4)

$$F_p = 1.0 \times 1.0 \times 2/3 \times 0.30 = 0.20W_p$$

Figure 7.5.4 Example of equipment anchorage [Army, 1982]

UBC Standards contains more detailed calculation procedures, as well as prescriptive details, for suspended ceilings and light fixtures.

An over-all survey of U.S. and New Zealand practices of design and detailing of exterior nonstructural elements has been presented recently by Arnold, et al. (1987). See Table 7.6.1. Seismic design provisions for nonstructural components are also highly developed in Japan. They have been summarized by the Architectural Institute of Japan [AIJ, 1985].

U.S. SEISMIC BUILDING CODES: NONSTRUCTURAL PROVISIONS

Issue or Components	UBC (1985)	NEHRP (1985)	Calif Title 24 (1979)	Tri-Services Manual (1982)	VA (1985)
Force Levels	Coefficient (Cp) varies 0.3-0.8 x importance factor 1-1.5	Coefficient (Cc) varies 0.6-3.0 x factor related to life safety 0.5-1.5	Coefficient (Cp) similar to UBC	Coefficient (Cp) = 2 CpWp, where Cp similar to UBC	Coefficient (Cp) larger than UBC, but no importance factor
Drift Limits	0.005 x story height	0.015 x story height (essential bldg.) 0.010 x story height (other bldgs.)	0.005 x story height 0.0025 head to sill of glazed openings	0.005 x story height (EQ 1) 0.010 x story height (EQ 2)	0.008 x story height
Separation of Elements	Accommodate 3.0/k x calc. elastic dis-placement, or 1/2" whichever greater	Accommodate elastic story drift x factor ranging from 1.25-6.5	Accommodate drifts computed by dynamic analysis or calc. elastic drifts x 2.0/k	Accommodate calculated elastic story drift x 4.0 or 1/2" which-ever greater	Accommodate calculated elastic story drift x factor ranging from 1-4 depending on structural type
Connections of	Multiply force determined by basic formula by: 1-1/3 for connector body 4 for fastener	Accommodate design story drift	As for UBC	As for UBC	As for UBC, or by modal analysis
Suspended Ceilings	Lateral force level with Cp = 0.3 + specific requirements	Lateral force value with Cc = 0.9 for fire related, 0.6 non-fire related	As for UBC, plus some prescriptive requirements	No specific requirements	Cp = 0.5 for 0.15g or over, Cp = 0.2 for 0.1 to 0.15g plus some pre-scriptive requirements
Partitions	Cp = 0.3 No specific separation requirements	Cc = 0.6 to 1.5 depending on importance of partitions	As for UBC	Cp = 0.2	Cp = 0.5 for 0.15g or Cp = 0.2 for 0.1 to 0.15g

Table 7.6.1 Summary of U.S. Non-structural Seismic Code Provisions [Arnold, Hopkins and Elsessor, 1987]

8 Summary

These guidelines have been written in response to a felt need for a document directed specifically to the design of low-rise buildings. They have been written as a resource document to supplement existing recommendations and codes, as well as extending the body of available information. Surveys of areas subjected to windstorms or earthquake motions indicate that low-rise buildings suffer the majority of damage loss; this is due both to the large number of low-rise buildings and to the complexity and lack of homogeneity when more than one material is used in their construction.

The guidelines:

- recognize that a typical low-rise building does not fit the analytical response model used in current design codes for earthquake loading;
- consider the effects of the structure rocking on the soil instead of assuming a fixed base;
- are directed towards the limitation of property damage when buildings are subjected to lateral loads. Life safety risks are considered to be reduced by design procedures that reduce property damage. However, special design considerations are given for building elements that have a higher than average probability of posing a life safety threat;
- recognize that the base shear approach for the design of all parts of the structural system for earthquake forces will, in general, provide an adequate level of strength for the main lateral force resisting system but will not be adequate for certain components of low-rise buildings;
- recognize that all buildings must be designed to be ductile (i.e. have energy dissipation capacity). This is a two-step process where structural systems with their load paths and structural detailing of connections and intersections are treated in separate steps.

The performance criteria for structures that are subjected to lateral forces are:

Earthquakes

- the structure should remain elastic, or nearly so, when subjected to small or moderate intensity earthquakes that have a high to moderate probability of occurrence; and
- the structure should respond in the inelastic range by local yielding, but retain its ability to sustain vertical load and not collapse when subjected a large intensity earthquake.

Windstorms

- the structure should suffer little or no damage when subjected to winds associated with a 50-year mean recurrence interval;
- damage from winds associated with a 100-year recurrence interval should not prevent the building from fulfilling its intended purpose, nor should any damage pose a threat to people assembled in the building; and
- the structure should not collapse if the building is subjected to wind speeds substantially greater than the design values (e.g. an intense thunderstorm) and there should be identifiable areas within the building for occupant protection.

Most loadings, especially lateral forces such as those due to wind and earthquake, fluctuate in time and space and are random in nature. Their effects on low rise buildings also have similar characteristics. Also, structural properties, such as stiffness, strength, energy dissipation capacity, etc, are important factors but are not normally known precisely. There can be significant deviation of the actual values from the nominal ones because of inherent variabilities in construction material properties as well as engineers' imperfect modelling of the structure and evaluation of these structural properties. In view of these uncertainties in the loading environment and the building systems, it is necessary to make use of probabalistic methods and concepts.

The selection of a design load is normally based on a given probability of the load level being in a prescribed period such as 5 minutes, 5 years or 50 years. Such design loads can be determined once the statistics and distribution of the maximum values of the loads over the period are known. Codes present this information in the form of maps of basic wind speed based on a 50-year return period and maps of seismic acceleration and velocity for 50 and 250-year return periods. The Guidelines give an outline of the characteristics of wind and seismic forces and the levels of risk implied by the current codes. This leads on to a discussion of the way in which multiple loads should be combined.

Wind forces are discussed in greater detail with an outline being given of the characteristics of extreme windstorms, covering straight winds such as cyclones, downslope winds, thunderstorms and downbursts, as well as rotating winds such as tornadoes and hurricanes. Several fundamental wind force concepts, such as frames of reference for wind speed, wind speed profiles, gust related effects and the variation of wind pressures are discussed. Wind-structure interaction is important and over-all wind effects, local wind pressures and internal pressures must be considered and the effect of these on actual buildings are illustrated.

Wind loading criteria are discussed in terms of the requirements of the various codes. The determination of wind pressures on low-rise buildings for design purposes is discussed with particular reference to ASCE 7-88. This covers the requirements for both the main wind force resisting system and components and cladding.

Earthquake forces and the development of building codes for earthquake-resistant design are treated in some detail. Those characteristics of earthquake motions that are important in considering the forces imparted to the structure during an earthquake are described. This covers ground motion parameters, earthquake intensity maps and seismic zoning, and the implications for building response. The earthquake response spectrum is discussed together with the use of a design spectrum and seismic coefficients for determining the lateral earthquake forces and their estimation as set out in the various codes. The distribution of this calculated force to the different levels of the structure, the possibility of horizontal torsional moment, and the limitations set by the codes on the building drift, are then outlined.

A number of factors that arise in modelling low-rise buildings are discussed with attention being drawn to the influence of differing materials, the effect of variations in actual material strengths, and the possibility of the roof or floor diaphragms being the main oscillator governing the vibrational behavior. Other factors that may need to be considered include the use of actual force-displacement relations to estimate the elastic periods (rather than code formulae), and the predicted state of stress should be checked to ensure that it correlates with the stiffness of the response model. Attention is drawn to the interconnection forces that may be developed and their importance to the over-all behavior of the structure.

Since design for wind forces is based upon conventional linear elastic behavior, whereas design for seismic forces assumes that some inelastic behavior will take place, this difference in expected performance means that energy absorption will be of great importance for earthquake resistant design. Seismic detailing should be provided even where the design wind forces exceed the design earthquake forces. The wind design forces should be compared with the earthquake design forces multiplied by a factor representing the reduction for inelastic action. This factor is about two; the 1991 UBC uses $3R_w/8$ to approximate this factor. The 1991 UBC also requires seismic detailing requirements to be followed even when wind forces govern the design.

Different types of structures will perform differently, and the codes take this into account in arriving at the level of design lateral seismic forces for a particular type of structure. Some of the advantages and disadvantages of the different types of structures are outlined and general requirements of the lateral force resisting systems discussed. The determination and distribution of the lateral force to the different parts of a structure are discussed in terms of the code requirements.

The particular requirements for steel and concrete frames with regard to detailing and drift control are outlined, along with the design and detailing of concrete and masonry shear walls, framed panel (wood) shear walls, concentric and eccentrically braced frames, the shears and deformations developed in diaphragms together with special considerations for the design and construction of diaphragms.

Six design examples are presented covering a one-story steel frame building, a one-story industrial building with mezzanine office, a three-story residential building, a five-story steel frame building, a five-story concrete frame/shear wall building, and a one-story wood frame building. These design examples cover all aspects of both wind and seismic analysis of the structures, including the possibility of torsional moment caused by the seismic forces.

The quality of the connections within the structural system is probably the most critical factor in the performance of a building subjected to high wind and seismic forces. This is addressed in the codes by their design requirements and discussion in their commentaries on the importance of tying the structure together as a unit. This takes on an even greater importance in low-rise construction since it commonly uses combinations of materials that have differing properties.

From the standpoint of connection design, two fundamental differences between wind and seismic forces should be noted: firstly, the high ductility demands on a system under seismic forces require special emphasis on the strength and ductility of connections in general, and, secondly, wind forces on lightweight building portions create special detailing requirements to address such issues as uplift effects on roof systems and negative pressures at discontinuities. Because the actual seismic forces on a building may greatly exceed the design forces, there must be strong emphasis on detailing the system to achieve maximum ductility. The inherent ductility demands require that the structural system as a whole, not just the individual members, should perform inelastically without collapse. To accomplish this, a continuous load path must be provided so that loads and forces can be transmitted through the structure down to the ground. This can be accomplished by satisfactory connection design.

The requirements for and the design of connections between building elements is discussed in relation to wood frame buildings, masonry and tilt-up buildings, steel members and cast-in-place concrete low-rise buildings. Several design examples are included covering a wood diaphragm-to-wall connection, a masonry wall-to-floor connection, and the design of concrete frame joint and wall. A number of illustrations of suitable connections are also provided. Attention is drawn to the need to ensure that the connections between the building elements will perform adequately under wind and earthquake forces to allow the structure as a whole to perform satisfactorily.

The performance of non-structural elements is discussed in some detail to cover how these elements are affected differently by wind and earthquake forces. The effect of accelerations, drift and interaction with structural elements is described for interior elements such as partitions, ceilings, light fixtures, stairs, and raised floors, together with exterior components such as claddings, veneers, windows, glazing and doors. The problems that can be experienced with building contents and equipment are also treated. Current practice as well as problem areas are discussed and, where possible, suitable steps in the analysis and design process are suggested.

References

Algan, B. (1982). "Drift and Damage Considerations in Design of Reinforced Concrete Buildings," dissertation submitted to the Graduate College, University of Illinois, Urbana, Illinois.

Algermissen, S.T. (1982). "Probabilistic Estimates of Maximum Acceleration and Velocity in Rock in the Contiguous United States," *Open file Report 82-1033*, U.S. Department of the Interior, Geological Survey.

Algermissen, S. T., Perkins, O. M., Thenhaus, P. C., Hanson, S. L., and Bender, B. L. (1987). "Probabilistic Estimates of Maximum Acceleration and Velocity in Rock in the Contiguous United States." *Open File Report 87-x*, United States Department of the Interior Geological Survey.

Allison, H. and Fisher, J. M. (1986). "Wind Drift Criteria for Steel-Framed Buildings." *Proceedings of the National Engineering Conference*, AISC, 4.1-4.19.

"Building Code Requirements for Reinforced Concrete." (1989). *ACI 318-89*, American Concrete Institute, Redford Station, Detroit, MI.

"Recommendations for Design of Beam-Column Joints in Monolithic Reinforced Concrete Structures". (1981). *ACI-ASCE 352R-76*, American Concrete Institute/American Society of Civil Engineers.

"Building Code Requirements for Masonry Structures." (1988). *ACI-ASCE 530-88*, American Concrete Institute/American Society of Civil Engineers.

Specification for the Design, Fabrication and Erection of Structural Steel for Buildings. (1978). American Institute of Steel Construction.

Load and Resistance Factor Design Specification for Structural Steel Buildings. (1986). American Institute of Steel Construction.

Engineering for Steel Construction. (No date) American Institute of Steel Construction.

Specification for the Design of Cold-formed Steel Structural Members. (1986 with addendum 1989), American Iron and Steel Institute.

American National Standard Minimum Design Loads for Buildings and Other Structures. (1982). ANSI A58.1-1982, American National Standards Institute, New York, NY.

"Performance Policies and Standards for Structural-Use Panels." (1988). *APA PRP-108*, American Plywood Association.

"Design Capacities of APA Performance-Rated Structural-Use Panels." (1988). *APA N375*, American Plywood Association.

Amrhein, J. E. (1978). *Reinforced Masonry Engineering Handbook*, Masonry Institute of America.

Anderson, J.C. and Naeim, F. (1987). "Design Criteria and Ground Motion effects on the Seismic Response of Multi-story Buildings." *Strong Ground Motion Seminars*, EERI.

Ang, A. H-S. and Tang, W. H. (1984). *Probability Concepts in Engineering Planning and Design, Vol. I, Basic Principles and Vol. II, Decision, Risk and Reliability*, John Wiley and Sons, Inc., New York, NY.

"Tentative Provisions for the Development of Seismic Regulations for Buildings". (1978). *Applied Technology Council, ATC3-06*, (NBS SP-510), also updates published under NEHRP in 1985 and 1988.

"Guidelines for the Design of Wood Sheathed Diaphragms." (1981). *ATC 7*, Applied Technology Council.

Recommendations for Aseismic Design and Construction of Nonstructural Elements. (1985). Architectural Institute of Japan.

"Seismic Design for Buildings." (1982). *TM 5-809-10/NAVFAC P-355/AFM 88-3 Chap. 13*, Departments of the Army, the Navy, and the Air Force.

"Seismic Design Guidelines for Essential Buildings." (1986). *Technical Manual TM5-809-10-1/NAVFAC P-355.1/AFM 88-3 Chap. 13*, Departments of the Army, the Navy and the Air Force, USA.

Arnold, Chris, Hopkins, D. and Elsesser, E. (1987). "Design and Detailing of Architectural Elements for Seismic Damage Control." *Architectural Detailing for Seismic Damage Control, Report*, submitted to the National Science Foundation, Building Systems Development, Inc., San Mateo, CA.

Atrek, Erdal and Nilson, Arthur H. (1980). "Non-linear Analysis of Cold-Formed Steel Shear Diaphragms." *Journal of the Structural Division*, ASCE, 106(ST3).

Ayres, J.M., Sun, T-Y and Brown, F.R. (1973). *Non-Structural Damage to Buildings, The Great Alaska Earthquake of 1964: Engineering*, National Academy of Sciences, Washington, D.C.

Basic/National Building Code. (1987). Building Officials and Code Administrators International, Inc., Homewood, IL.

Batts, M.E., Cordes, M.R., Russel L.R., Shaver, J.R. and Simiu, E. (1980). "Hurricane Wind Speeds in the United States." *NBS Building Science Series 124*, U.S. Dept. of Commerce, National Bureau of Standards, Washington, D.C.

Berg, G.V., (1983). *Seismic Design Codes and Procedures*, Engineering Monograph, EERI.

Bjorhovde, R. and Fisher, J. M. (1986). "Wind Drift Criteria for Steel-Framed Buildings: an Introduction." *Proceedings of the National Engineering Conference*, AISC, 7.1-7.13.

Blume, John A., Sharpe, Roland L. and Elsesser, Eric. (1961). "A Structural-Dynamic Investigation of Fifteen School Buildings Subjected to Simulated Earthquake Motion."

Blume, J. A., N. M. Newmark and Corning, L. H. (1961). *Design of Multistory Reinforced Concrete Buildings for Earthquake Motions.* Portland Cement Association, Chicago, 1961.

Bouwkamp, J.G. (1961). "Behavior of Window Panels Under In-Plane Forces." *Bulletin of the Seismological Society of America*, 51(1), 85-109.

"Code of Basic Data for the Design of Buildings, Chapter V. Loadings - Part 2: Wind Loads." (1984). *BSI CPE*, British Standards Institute, London, England.

National Building Code. (1990). Building Officials and Code Administrators, Country Club Hills, Illinois.

"Recommended Provisions for the Development of Seismic Regulations for New Buildings." (1988). *Part 1, Provisions*, Building Seismic Safety Council, Washington, D.C.

Butler, L. J. and Kulak, G. L. (1971). "Strength of Fillet Welds as a Function of Direction of Load", Welding Journal.

California At Risk: Reducing Earthquake Hazards, 1987 to 1992. (1986). California Seismic Safety Commission.

Chen, Yong qi and Soong, T.T. (1988). State-of-the-art Review: "Seismic Response of Secondary Systems", *Engineering Structures*, Vol. 10, October.

Chopra, A.K. (1981). "Dynamics of Structures -- A Primer." *Engineering Monograph*, EERI.

"The Great Quake: On-site Reports". (1989). *Civil Engineering, ASCE*, December.

Clough, R.W. and Penzien, J. (1975). *Dynamics of Structures*, McGraw-Hill, Inc.

Coalinga California Earthquake of May 1983, Reconnaissance Report, Earthquake Engineering Research Institute, El Cerrito, CA.

Cornell, C. A. (1986). "Engineering Seismic Risk Analysis." *Bulletin of the Seismological Society of America*, 58(5).

Corotis, R.B., Harris, M., and Bova, C. (1981). "Area-Dependent Processes for Structural Live Loads." *Journal of the Structural Division*, ASCE, 107(ST5).

Corotis, R. B. (1982). "Design of Timber Structures for Natural Hazards," *Structural Use of Wood in Adverse Environments*, R. W. Meyer and R. M. Kellogg, eds., Van Nostrand Reinhold Co., 327-352.

Tall Building Criteria and Loading. (1980). Council on Tall Buildings and Urban Habitat, Vol. CL, ASCE.

Craig, J.I., Goodno, B.J., Keister, M.J. and Fennel, C.J. (1986). "Hysteretic Behavior of Precast Cladding Connections." *Proceedings Third ASCE Engineering Mechanics Specialty Conference on Dynamic Response of Structures*, UCLA, CA, 817-826.

Davenport, A.G. (1960). "Wind Loads on Structures." *Technical Paper No. 88*, Division of Building Research, National Research Council of Canada, Ottawa, Canada.

Davenport, A.G., Surry, D., and Stathopoulos, T. (1977). "Wind Loads on Low-Rise Buildings." *Final Report of Phases I and II*, University of Western Ontario, BLWT-SS8, Ontario, Canada.

Davenport, A.G., Surry, D., and Stathopoulos, T.(1978). "Wind Loads on Low-Rise Buildings." *Final Report of Phase III*, University of Western Ontario, BLWT-SS4-1978, Ontario, Canada.

De Pineres, Oscar. (1987). "A Safer Earthquake Design Approach." *Civil Engineering Magazine*, ASCE.

Disaster Preparedness Workshop, 1987, annual convention of the Society of California Archivists, Asilomar, CA.

Durst, C.S. (1960). "Wind Speeds Over Short Periods of Time." *Meteorological Magazine*, London, England, 89, 181-186.

Eagling, Donald G., Editor. *Seismic Safety Guide*, Lawrence Berkeley Laboratory, University of California, Berkeley, CA.

Non-structural Issues of Seismic Design and Construction, (1984). Earthquake Engineering Research Institute, Workshop Proceedings, Publ. #84-04, June.

Ellingwood, B., Galambos, T.V., MacGregor, J.G., and Cornell, C.A. (1980). "Development of a Probability Based Load Criterion for American National Standard A58." *NBS Special Publication No. 577*, National Bureau of Standards, Washington, D.C.

Ellingwood, B., MacGregor, J.G., Galambos, T.V., and Cornell, C.A. (1982). "Probability Based Load Criteria - Load Factors and Load Combinations." *Journal of the Structural Division*, ASCE, 108(ST5), Proc. Paper 17068, 978-997.

Evans, Deane et al. (1988). *Glass Damage in the September 19, 1985 Mexico City Earthquake*, Steve Winters and Associates, New York.

Earthquake Damage Mitigation for Computer Systems. (1983). FIMSC, Inc.

Data Processing Facilities for Earthquake Hazard Mitigation. (1987). FIMSC, Inc., VSP, Inc., Sacramento, CA.

Federal Emergency Management Agency, "NEHRP Recommended Provisions for the Development of Seismic Regulations for New Buildings", 1988

Fisher, J. M. (1984). "Industrial Buildings - Guidelines and Criteria." *Engineering Journal*, AISC, 21(3), 3rd Quarter, 149-153.

Freeman, S. A. (1977). "Racking Tests of High-Rise Building Partitions." *Journal of the Structural Division*, ASCE, 103(ST 8), 1673-1685.

Freeman, S. A., Czarnecki, R. M. and Honda, K. K. (1980). "Significance of Stiffness Assumptions on Lateral Force Criteria." *Reinforced Concrete Structures Subjected to Wind and Earthquake*, ACI, SP-63, 437-457.

Freeman, Sigmund A. (1980). "Drift Limits - Are They Realistic." *Structural Moments*, Structural Engineers Association of Northern California, No. 4.

Fujita, T.T. (1971). "Proposed Characterization of Tornadoes and Hurricanes by Area and Intensity." *SMRP No. 91*, Satellite and Mesometeorology Research Project, The University of Chicago, Chicago, IL.

Galambos, T.V., Ellingwood, B., MacGregor, J.G., and Cornell, C.A. (1982). "Probability Based Load Criteria: Assessment of Current Design Practice." *Journal of the Structural Division*, ASCE, 108(ST5), Proc. Paper 17067, May, 959-977.

Getty. (1984). "Symposium on Protection of Art Objects From Damage By Earthquakes: What Can Be Done?" J. Paul Getty Center for the History of Art and the Humanities, Santa Monica, CA.

Goodno, B.J., Meyyappa, M. and Nagarajaiah, S. (1988). "A Refined Model for Precast Cladding and Connections," Proceedings of the 9th World Conference on Earthquake Engineering, Tokyo and Kyoto, Japan.

Goodno, B.J., Palsson, H.P. and Pless, D.G. (1984). "Localized Cladding Response and Implications for Seismic Design." *Proceedings of Eighth World Conference on Earthquake Engineering*, San Francisco, CA, Vol. V, 1143-1150.

Goodno, B.J., Craig, J.I. and Zeevaert Wolff, A. (1989). "Behavior of Architectural Nonstructural Components in the Mexico Earthquake." *Earthquake Spectra*, EERI, 5(1), 195-222.

Goodno, B.J. and Palsson, H. (1986). "Analytical Studies of Building Cladding." *Journal of Structural Engineering*, Paper 20498, ASCE, 112(4), 665-676.

Goodno, Barry, et. al. (1983). "Cladding-Structure Interaction in High-Rise Buildings." *Final Report to the National Science Foundation*, NTIS Report PB83-195891, Georgia Institute of Technology, School of Civil Engineering.

Goodno, B.J. and Streit, M.C. (1987). "Dynamic Analysis of Low Rise Buildings on Microcomputers." *Microcomputers in Civil Engineering*, Elsevier Science Pubishing Co., 2(1), 39-46.

Goodno, Barry and Pinelli, Jean-Paul. (1986). "The Role of Cladding in Seismic Response of Low Rise Buildings in the Southeastern U.S." *Proceedings, Third U.S. National Conference on Earthquake Engineering*, Charleston, SC, II, 883-894.

Gupta, A. K., editor. (1981). "Seismic Performance of Low-Rise Buildings," *Proceedings of the Workshop*, New York, NY, ASCE.

Ha, Kinh H., El-Hakim, Noor and Fazio, Paul P. (1979). "Simplified Design of Corrugated Shear Diaphragms." *Journal of the Structural Division*, ASCE, 105(ST7).

Hart, G.C. (1976). "Natural Hazards: Tornado, Hurricane, Severe Wind Loss Models." Prepared for the National Science Foundation (*NTIS No. PB294594/AS*), J.H. Wiggins, Company, Redondo Beach, CA.

Hart, F., Henn, W., and Sontag, H. (1978). *Multi-storey Buildings in Steel*. John Wiley and Sons, New York.

Heidebrechnt, A.C., Zhu, T.J., and Tso, W.K. (1988). "Effect of Peak Ground a/v Ratio on Structural Damage." *Journal of Structural Engineering*, ASCE, 114(5), 1019-1037.

Hjelmstad, K.D., and Popov, E.P. (1983) "Cyclic Design and Behavior of Link Beams". *Journal of the Structural Division, ASCE.*, Vol. 109, No. 10, 2387-2403.

Hillman, Biddison and Loevenguth. (1982). *Guidelines for Seismic Restraints of Mechanical Systems and Plumbing Piping Systems*, Sheet Metal Industry Fund of Los Angeles and the Plumbing and Piping Industry Council, Inc.

Hollister, S.C. (1979). "The Engineering Interpretation of Weather Bureau Records for Wind Loading on Structures." *Proceedings of Technical Meeting Concerning Wind Loads on Buildings and Structures, Building Science Series 30*, National Bureau of Standards, U.S. Government Printing Office, Washington, D.C.

Hudson, D.E. (1979). "Reading and Interpreting Strong Motion Accelerograms." *Engineering Monograph*, EERI.

Idriss, I.M. (1985). "Evaluating Seismic Risk in Engineering Practice." *Proceedings Eleventh International conference on soil Mechanics and Foundation Engineering*, San Francisco, CA.

Ingles, O. G. (1979). "Safety in Civil Engineering - Its Perception and Promotion, School of Civil Engineering." *Report R-188*, University of South Wales, Kenginston, Australia.

"Metal Suspension Systems for Acoustical Tile and for Lay-in Panel Ceilings." (1985). *Uniform Building Code Standards, number 47-18*, International Conference of Building Officials.

"Determination of the Characteristics Site Period T." (1985). *International Conference of Building Officials*, Uniform Building Code and Uniform Building Code Standard 23.1, Whittier, CA.

"Uniform Building Code." (1991). *International Conference of Building Officials - ICBO*, 1991 Edition, Whittier, California, May.

Ito, Hiroshi and Kazuo, Nishida. (1986). "Story Drift Accomodation of Steel Doors." *The 7th Japan Earthquake Engineering Symposium*, 1831-1836.

Jain, A., Goel, S. and Hanson, R. D. (1980). "Hysteretic Cycles of Axially Loaded Steel Members." *Journal of the Structural Journal*, ASCE, 106(ST8).

Jones, B. (1984). "Protecting Historic Architecture and Museum Collections From Natural Disasters." *Proceedings of a conference at Cornell University*, Stoneham, Massachusetts, Butterworths.

Kariotis, J.C. and El-Mustapha, A.M. (1989). "Relationship of Seismic zoning Parameters to Probable Earthquake Damage to Reinforced Masonry Buildings." *Proceedings 5th Canadian Masonry Symposium*, Vancouver, B.C., Canada.

Kasai, K. and Popov, E. P. (1986). "General Behavior of WF Shear Link Beams." *Journal of Structural Engineering*, ASCE, 112(2).

Kasai, K. and Popov, E. P. (1986). "Cyclic Web Buckling Control for Shear Link Beams." *Journal of Structural Engineering*, ASCE, 112(3).

Klinger, R. E. (1980). "Mathematical Modeling of Infilled Frames." *Reinforced Concrete Structures Subjected to Wind and Earthquake*, ACI, SP-63, 1-25.

Krawinkler, H.(1978). "Shear in Beam-Column Joints in Seismic Design of Steel Frames", Engineering Journal, AISC, Vol. 15, No 3.

"Live Loads Due to Wind." (1985). *Subsection 4.1.8, National Building Code of Canada, NRCC No. 23174*, National Research Council of Canada, Ottawa, Canada.

McCue, G.M., et. al. (1978). "Architectural Design of Building Components for Earthquakes." *Report submitted to the National Science Foundation*, MBT Associates, San Francisco, CA.

McDonald, A. J. (1975). *Wind Loading on Buildings*, Applied Science Publishers, London, England.

McDonald, J.R. (1985). "Extreme Winds and Tornados: An Overview." *A Minicourse presented at DOE Natural Phenomena Hazards Mitigation Conference*, Las Vegas, NV.

McGavin, G. (1981). *Earthquake Protection of Essential Building Equipment*, John Wiley & Sons, New York.

McGavin, G. (1987). *Earthquake Hazard Reductionfor Life Support Equipment in Hospitals*, Ruhnau McGavin Ruhnau Associates, Riverside, CA, for the National Science Foundation.

Malley, J. O. and Popov, E. P. (1984). "Shear Links in Eccentrically Braced Frames." *Journal of Structural Engineering*, ASCE, 110(9).

Mayyappa, M., Goodno, B.J. and Fennell, C.J. (1988). "Modeling and Performance of Precast Cladding Connections." *Proceedings of The Fifth ASCE Specialty Conference on Computing in Civil Engineering*, Alexandria, VA., 209-218.

Meehan, John F. (1973). "Public School Buildings." *The San Fernando, California Earthquake of February 9, 1971*, National Oceanic and Atmospheric Administration, Washington, D.C., I(B), 667-884.

Mehta, K.C., Editor. (1988). *Guide to the Use of the Wind Load Provisions of ANSI A58.1*. Institute for Disaster Research, Texas Tech University, Lubbock, TX.

Mehta, K.C., Minor, J.E. and Reinhold, T.A.. (1983). "Wind Speed-Damage Correlation in Hurricane Frederic." *Journal of Structural Engineering*, ASCE, 1099(2), 37-49.

Low-Rise Building Systems Manual. (1986). Metal Building Manufacturers Association (MBMA), Cleveland, OH.

"Survey Report on Damage to Window Glasses caused by 'Off-Miyagi Earthquake on Feb. 20, 1978'." (1978). *Kenchiku Kenkyu Shiryo or Building Research Institute Report*, Research Group on Earthquake Damage, Building Research Institute, Ministry of Construction, Japan (in Japanese).

The Building Standard Law of Japan. (1986). Building Center of Japan, edited by Building Guidance Division and Urban Building Division, Housing Bureau, Ministry of Construction.

Minor, J.E. (1974). "The Window Damaging Mechanism in Wind Storms", *Ph.D. Dissertation*, Texas Tech University, Lubbock, Texas.

Montogomery, James C. and Hall, William J. (1979). "Seismic Design of Low-Rise Steel Buildings", *Journal of the Structural Division*, ASCE, Vol. 105, No. ST10, Oct.

Naman, S.K. and Goodno, B.J. (1986). "Seismic Evaluation of a Low Rise Steel Building." *Engineering Structures*, 8(1), 9-16.

"Confronting Natural Disaster: An International Decade for Natural Hazard Reduction." (1987). National Academy Press, Washington, D.C.

Methods for Installing and Storing Objects to Mitigate Potential Damage From Earthquakes. (1987). National Association for Museum Exhibition, De Young Museum, San Francisco, California.

National Building Code of Canada. (1977 and Supplement 1985). NRCC Nos. 15555 and 23178 National Research Council of Canada, Ottawa, Canada.

National Building Code of Canada. (1980, and Supplement 1985). National Research Council of Canada.

"Standard for the Installation of Sprinkler Systems." (1983). *NFPA 13-1983*, National Fire Protection Association, Batterymarch Park, Quincy, MA.

"NEHRP Recommended Provisions for the Development of Seismic Regulations for New Buildings." (1988). *Part I: Provisions, Part II: Commentary*, Building Seismic Safety Council, Washington, D.C.

Newmark, N.M. and Hall, W.J. (1982). "Earthquake Spectra and Design." *Engineering Monograph*, EERI.

Oesterle, R.G., Fiorato, A.E. and Corely, W.G. (1985). "Reinforcement Details for Earthquake-Resistant Strucural Walls", *ACI Seminar on the Design of Concrete Buildings for Wind and Earthquake Forces*.

Palsson, H. and Goodno, B.J. (1988). "Influence of Interstory drift on Cladding Panels and Connections," *Proceedings of the 9th World Conference on Earthquake Engineering*, Tokyo and Kyoto, Japan.

Paulay, T. (1981). "The Design of Reinforced Concrete Ductile Shear Walls for Earthquake Resistance", *Research Report*, Department of Civil Engineering, University of Canterbury, Christchurch, New Zealand.

Peery, David J. (No date). *Aircraft Structures*. p. 207.

Petak, W.J., and Atkisson, A.A. (1982). *Natural Hazard Risk Assessment and Public Policy*, Springer-Verlag.

Popov, E. P. (1978). "Seismic Behavior of Structural Assemblages", *Journal of the Structural Division, ASCE*, Vol. 106, No. ST7, July.

Popov, E.P. and Black, R. G. (1981). "Steel Struts Under Severe Cyclic Loadings." *Journal of the Structural Division*, ASCE, 107(ST9).

Popov, E.P., Amin, N. R., Louie, J. C. and Stephan, R. M.(1986). "Cyclic Behavior of Large Beam-Column Assemblies", Engineering Journal,AISC Volume 23,No.1.

Popov, E.P., and Malley, J.O. (1983). "Design of Links and Beam-to-Column Connections for Eccentrically Braced Steel Frames", *Report No. UCB/EERC-83/02*, University of California, Berkeley.

Popov, E. P. and Pinkney, R. B. (1969). "Cyclic Yield Reversal in Steel Building Connections", *Journal of the Structural Division, ASCE*, Vol. 95, No. ST3, March.

Analysis and Design of Small Reinforced Concrete Buildings for Earthquake Forces. (No date). Portland Cement Association.

Proposed Connection Designs and Details. (1987). Architectural Precast Concrete Connection Details Committee, Prestresed Concrete Institute.

Architecture Precast Concrete Claddings, Its Contribution to Lateral Resistance of Buildings. (1989). Proceedings of International Symposium, Precast/Prestressed Concrete Institute, Chicago, Illinois, Nov. 8-9, 1989.

Reitherman, R. (1982). "Earthquake-Caused Hazardous Material Releases." *Proceedings of the 1982 Hazardous Material Spill Conference*, Milwaukee, WI.

Reitherman, R. (1983). "Reducing the Risks of Nonstructural Earthquake Damage: A Practical Guide," Prepared for California Seismic Safety Commission, Sacramento, CA.

Reitherman, R. (1984). "Protection of Museum Contents From Earthquakes." *J. Paul Getty Museum Symposium on Protection of Art Objects From Damage By Earthquakes: What Can Be Done?*.

Reinhold, T.A., and Ellingwood, B. (1982). "Tornado Damage Risk Assessment," *NUREG/CR--2944*, Brookhaven National Laboratory.

Richard, R.M. (1986). "Analysis of Large Bracing Connection Designs for Heavy Construction", *Proc. National Engineering Conference, AISC*, Nashville, Tennessee

Rihal, Satwant S. (1980). "Racking Tests of Non-Structural Building Partitions," *Final Technical Report Submitted to the National Science Foundation, Report ARCE R80-1*, Architectural Engineering Department, California Polytechnic State University, San Luis Obispo, California. (NSF Grant No. PFR-78-23085).

Rihal, Satwant. (1980). "Seismic Performance of Manufactured Components in Low-Rise Buildings." *Proceedings, Workshop on the Seismic Performance of Low-Rise Buildings*, American Society of Civil Engineers.

Rihal, Satwant S. (1986). "Earthquake Resistance and Behavior of Wood-Framed Building Partitions." *Proceedings, Third U.S. National Conference on Earthquake Engineering*, Charleston, SC, II, 1297.

Rihal, Satwant S. (1987). "Seismic Behavior and Design of Precast Concrete Facades/Claddings and Connections in Low/Medium-Rise Buildings." *Final Project Report submitted to the National Science Foundation, Architectural Engineering Department, Research Report, ARCE R88-01*, Cal Poly State University, San Luis Obispo, California.

Rihal, Satwant S. (1988). "Earthquake Resistance and Behavior of Heavy Facades/Claddings and Connections in Medium-Rise Steel-Framed Buildings." *Proceedings, Ninth World Conference on Earthquake Engineering*, Tokyo, Japan.

Rihal, Satwant S., et. al. (1984). "Experimental Investigation of the Dynamic Behavior of Building Partitions and Suspended Ceilings During Earthquakes." *Proceedings, Eight World Conference on Earthquake Engineering*, San Francisco, California.

Roeder, C. W., and Popov, E. P. (1978). "Cyclic Shear Yielding of Wide Flange Beams", *Journal of the Engineering Mechanics Division, ASCE*, Vol. 104, No. EM4, Aug. 1978.

Rosenblueth, E., editor. (1980). *Design of Earthquake Resistance Structures*, Chapters 2 and 6, Halsted Press, New York.

Rouse, H. (1938). *Fluid Mechanics for Hydraulic Engineers*. McGraw-Hill, New York, NY.

Rubin, C.B., Yezer, A.M., Hussian, Q. and Webb, A. (1985). *Summary of Major National Disaster Incidents in the U.S.: 1965 to 1985*, Federal Energy Management Agency, Washington, D.C.

Saffir, H. (1974). "The Hurricane Disaster-Potential Scale." *Weatherwise*, p. 196.

Schiff, A. (1987-in progress). *Guide For Protecting Museum Objects From Earthquake Damage*, research funded by the Naitonal Science Foundation and pending publication by the J. Paul Getty Museum, Malibu, CA.

Schneider, R. R. and Dickey, W. L. (1980). *Reinforced Masonry Design*, Prentice- Hall Inc.

Seed, H.B. and Idriss, I.M. (1982). "Ground Motions and Soil Liquefaction During Earthquakes." *Engineering Monograph*, EERI.

Simiu, E., Changery, M., and Filliben, J. J. (1977). "Extreme Wind Speeds at 129 Stations in the Contiguous United States." *Building Science Series Report 118*, NBS, Washington, D.C.

Simiu, E., Changery, M.J. and Filliben, J.J. (1979). "Extreme Wind Speeds at 129 Stations in the Contiguous United States." *Building Science Series Report 118*, National Bureau of Standards, Washington, D.C.

South Florida Building Code. (1979). Board of County Commissioners, Metropolitan Dade County, Florida.

Standard Building Code. (1986). Southern Building Code Congress International, Inc., Birmingham, AL.

Stahl, J. (1987). "Earthquake Mitigating Display Mounts for the Antiquities Collection at the J. Paul Getty Museum." NAME conference on Methods for Installing and Storing Objects to Mitigate Potential Damage From Earthquakes, De Young Museum, San Francisco.

Rules for Minimum Design Loads on Structures -- SAA Loading Code (Metric Units). (1989). "Part 2 - Wind Forces, AS1170." Standards Association of Australia, Standards House, North Sydney, Australia.

Steinbrugge, K.V. (1982). "Scenarios for Earthquake Related Problems at Computer Installations Used By Financial Institutions." Finance, Insurance, and Monetary Services Advisory Committee of the California Seismic Safety Commission Earthquake Preparedness Task Force.

Stratta, J. (1987). "Building Contents, Nonstructural Items." Chapter 8, *Manual of Seismic Design*, Prentice-Hall Publishing Co., Englewood Cliffs, NJ.

Recommended Lateral Force Requirements and Commentary. (1988). ("Blue Book") Seismology Committee, Structural Engineers Association of California, San Francisco, California.

Recommended Lateral Force Requirements and Commentary. (1990). Structural Engineers Association of California, San Francisco, California.

Taric, A. (1984). "The Acceptance of Base Isolation for Earthquake Protection of Buildings." *Proceedings of the 8th World Conference on Soil Dynamics and Earthquake Engineering.*

Tarpy, T. S. (1984). "Shear Resistance of Steel Stud Wall Panels", *Proceedings of the Seventh International Specialty Conference on Cold Formed Steel Structures*, University of Missouri-Rolla, 203-248.

Tecson, J.J., Fujita, T.T and Abbey, R.F. Jnr (1979). "Statistics of U.S. Tornadoes based on the DAPPLE Tornado Tape", *Preprints 11th Conference on Severe Local Storms*, American Meteorological Society, Boston, Mass..

Thiel, C. C. and Zsutty, T. C. (1987). "Earthquake Characteristics and Damage Statistics." *Earthquake Spectra*, 3(4), 747-792.

Thom, H.C.S. (1960). "Distribution of Extreme Winds in the United States." *Journal of the Structural Division*, ASCE, 86(ST4).

Tokico, product brochure (in Japenese) on the Shinkula motion isolator, catalog number U0073-01, Kawasaki, Japan.

Turkstra, C.J., and Madsen, H.O. (1980). "Load Combinations in Codified Design." *Journal of the Structural Division*, ASCE, 106(ST12).

Turkstra, C.J. (1985). "Design Load Combination Factors," *Proceedings, Symposium on Structural Safety Studies*, J.T.P. Yao, R. Corotis, C.B. Brown, and F. Moses, eds., ASCE, New York.

Veletsos, A.S. and Newmark, N.M. (1960). "Effect of Inelastic Behavior on the Response of Simple Systems to Earthquake Motions." *Proceedings Second World Conference on Earthquake Engineering*, II.

Vellozzi, J.W. and Cohen, E. (1968). "Gust Response Factors." *Journal of the Structural Division*, ASCE, 94(ST6), 1295-1313.

"Study to Establish Seismic Protection Provisions for Furniture, Equipment and Supplies for VA Hospitals." (1976). Veterans Administration.

Seismic Restraint Handbook. (1981). Veterans Administration.

Wang, Marcy Li. (1987). "Cladding Performance on a Large Scale Test Frame." *Earthquake Spectra*, 3(1), Earthquake Engineering Research Institute, El Cerrito, CA.

Wen, Y.K. (1990). *Structural Load Modeling and Combination for Performance and Safety Evaluation*, Elsevier Science Publisher, B.V.

Wu, J. and Hanson, R. D. (1987). "Investigation of Design Spectra for Possible Use in 1988 NEHRP Recommended Provisions," *Research Report UMCE 87-6*, Department of Civil Engineering, The University of Michigan, Ann Arbor.

Bibliography - Chapter 3

Abbey, R.F., Jr. (1976). "Risk Probabilities Associated with Tornado Wind Speeds." *Proceedings Symposium on Tornadoes: Assessment of Knowledge and Implications for Man*, Texas Tech University, Lubbock, TX, 177-236.

Allen, D.E. and Dalgliesh, W.A. (1973). "Dynamic Wind Loads and Cladding Design." *Research Paper No. 611*, Division of Building Research, National Research Council of Canada, Ottawa, Canada.

Anthes, R.A. (1982). "Tropical Cyclones: Their Evolutions, Structure and Effects." *Meteorological Monograph*, American Meteorological Society, 19 (41).

"Wind Forces on Structures." (1961). Task Committee on Loads and Stresses of the Structural Division, J.M. Biggs, Chairman, *Transactions, Paper No. 3269*, Part II, ASCE, 126, 1124-1198.

Wind Loading and Wind-Induced Structural Response. (1987). Committee on Wind Effects, American Society of Civil Engineers, New York, NY.

Brinkman, W.A.R. (1974). "Strong Downslope Winds at Boulder, Colorado." *Monthly Weather Review*, 102, 592-602.

Cermak, J.E. (1977). "Wind Tunnel Testing of Structures." *Journal of the Engineering Mechanics Division*, ASCE, Proc. Paper 134445, 103(EM6), 1124-1140.

Davenport, A.G. (1961). "The Application of Statistical Concepts to the Wind Load of Structures." *Proceedings Paper 6480*, Institution of Civil Engineers, Vol. 19.

Davenport, A.G. and Isyumov, N. (1967). "The Application of the Boundary Layer Wind Tunnel to the Prediction of Wind Loading." *International Research Seminar on Wind Effects on Buildings and Structures*, Ottawa, Canada.

"Interim Guidelines for Occupant Protection from Extreme Winds." (1975). *TR83A*, Defense Civil Preparedness Agency, Washington, D.C.

Dunn, G.E. and Miller, B.I. (1960). *Atlantic Hurricanes*, Louisiana State University Press, Baton Rouge, Louisianna.

Eaton, J.K. and Mayne, J.R. (1975). "The Measurement of Wind Pressures on Two-Story House at Aylesbury." *Journal of Industrial Aerodynamics*, 1(1), 67-109.

Ensminger, R.R. and Breen, J.E. (1971). "Hurricane Protection of Gulf Coast Buildings." *Civil Engineer*, ASCE, 4(12), 27.

Hoerner, S.F. (1965). *Fluid Dynamics Drag*, published by the author, Midland Park, NJ.

Hollister, S.C. (1979). "The Engineering Interpretation of Weather Bureau Records for Wind Loading on Structures." *Proceedings of Technical Meeting Concerning Wind Loads on Buildings and Structures*, Building Science Series 30, National Bureau of Standards, U.S. Government Printing Office, Washington, D.C.

Huschke, R.E., Editor. (1959). *Glossary of Meteorology*, American Meteorological Society, Boston, MA.

Kessler, E., Editor. (1982). *Thunderstorms: A Social, Scientific and Technological Documentary*, 3 Vol., U.S. Dept. of Commerce, NOAA-ERL, U.S. Government Printing Office, Washington, D.C.

Marshall, R.D. (1977). "The Measurement of Wind Loads on Full-Scale Mobile Homes." *NBSIR 77-1289*, National Bureau of Standards, Washington, D.C.

Minor, J.E., McDonald, J.R. and Mehta, K.C. (1977). "The Tornado: An Engineering-Oriented Perspective." *NOAA Technical Memorandum ERL NSSL-82*, National Oceanic and Atmospheric Administration, NSSL, Norman, OK (NTIS Accession No. PB 281 860).

Newberry, C.W. and Eaton, K.J. (1974). "Wind Loading Handbook." *Building Research Establishment Report K4f*, Her Majesty's Stationery Office, London, England.

"Commentary B: Wind Loads; Commentaries on Part 4 in the Supplement to the National Building Code of Canada." (1985). *NRCC No. 23178*, National Research Council of Canada, Ottawa, Canada, 153-186.

Pautz, M.E., Editor. (1969). "Severe Local Storm Occurrences 1955-1967." *ESSA Technical Memorandum, WBTM FCST 12*, Office of Meteorological Operations, Weather Analysis and Prediction Division, Silver Spring, MD.

Peterson, R.E., Editor. (1976). *Proceedings Symposium on Tornadoes: Assessment of Knowledge and Implications for Man*, Texas Tech University, Lubbock, TX.

Sherlock, R.H. (1982). "Variation of Wind Velocity of Gusts with Height." *Proceedings of the American Society of Civil Engineers*, ASCE, 18, 126.

Simiu, E. and Scanlan, R.H. (1978). *Wind Effects on Structures*, John Wiley and Sons, New York, NY.

Simpson, R.H. and Riehl, H. (1981). *The Hurricane and Its Impact*, The LSU Press, Baton Rouge, LA.

Smart, H.R., Stevens, L.K. and Joubert, P.N. (1967). "Dynamic Structural Response to Natural Wind." *Proceedings Wind Effects on Buildings and Other Structures*, University of Toronto Press, 1, 595-630.

Thom, H.C.S. (1960). "Distribution of Extreme Winds in the United States." *Proceedings of the American Society of Civil Engineers*, Structural Division, ASCE, 86(ST4).

Thom, H.C.S. (1968). "New Distribution of Extreme Winds in the United States." *Proceedings of the American Society of Civil Engineers*, Structural Division, ASCE, 94(ST7).